SYSTEMS DEVELOPMENT
Analysis, Design, and Implementation

SYSTEMS DEVELOPMENT
Analysis, Design, and Implementation

ALAN L. ELIASON
University of Oregon

LITTLE, BROWN AND COMPANY Boston Toronto

Library of Congress Cataloging-in-Publication Data

Eliason, Alan L.
  Systems development.

  Includes bibliographies and index.
  1. System design. 2. System analysis. I. Title.
QA76.9.S88E45 1987   003   86–20034
ISBN 0–316–23256–4

Copyright © 1987 by Alan L. Eliason

All rights reserved. No part of this book may be reproduced in any form or by any electronic or mechanical means including information storage and retrieval systems without permission in writing from the publisher, except by a reviewer who may quote brief passages in a review.

Library of Congress Catalog Card Number 86–20034

ISBN 0-316-23256-4

9 8 7 6 5 4 3 2 1

Don

Published simultaneously in Canada
by Little, Brown and Company (Canada) Limited

Printed in the United States of America

Acknowledgments

  Figure 6–5, p. 178. From Online Business Computer Applications, SRA, 1983, p. 363.
  Figure for exercise 10–4, p. 359. From Online Business Computer Applications, SRA, 1983, p. 439.
  Figure 16–5, p. 531. From Business Information Processing, SRA, 1980, p. 416.
All three figures reproduced by permission of Science Research Associates.

  Cover art: Jack Tworkov, Indian Red Series #2, oil on canvas, 72" × 72", 1979. Courtesy of Nancy Hoffman Gallery, New York.

Preface

THIS book was written to fill the need for a systems book that provides a life cycle approach to the process of systems development and considers, at length, how a structured methodology might clarify and simplify this approach. To this end, the book should be of interest to students who need to know more about what a system is, how a system can be studied in a systematic way, and which tools and techniques are essential to such a study.

The text was written for an audience familiar with beginning concepts of data processing. Most of the readers will probably have had some exposure to structured programming and design. Even so, this text should not be considered appropriate only for advanced students. Rather it is an introductory text, designed in large part to provide a survey of systems development methodologies.

A striking feature of the book is that it is not limited to a discussion of systems analysis at the expense of either systems design or systems implementation. A brief study of the Table of Contents shows that all three subjects are described in considerable detail.

Another important feature of the book is the art. The illustrations were designed to clarify written descriptions. Through the use of a large number of figures, it was possible to document, step-by-step, how one begins to analyze complex systems logically, break them down into clear working technical designs, and translate these designs into usable software.

The organization of each chapter should be of special interest. Each chapter begins with an overview, followed by a discussion of concepts and tools important to the chapter topic. Numerous examples and case studies are essential to this discussion. The Mansfield, Inc., case study, discussed within the body of each chapter, considers the design of a payroll and a customer billing system from inception to finished computer software. Each chapter ends with the World Interiors, Inc., case study. This case study provides student exercises, which can be answered, in part, by understanding how Mansfield, Inc., analyzed and resolved its systems problems. One important reason for using case studies is that they address real problems. Another reason is to provide the reader with a better appreciation of how structured systems development tools and techniques are applied. Still another reason is to help explain how people affect the development process.

In addition to case studies, each chapter contains review questions and student exercises. Why all these materials? An understanding of systems development comes from doing—problem solving, designing, testing, and installing. This text is committed to the principle of learning by doing.

All structured techniques described in the text are language dependent, though most of the examples refer to COBOL or Pascal. Likewise, the software designs were not prepared with a particular computer in mind. All designs can be translated into software using any of several languages or types of computers.

Many people were important in helping me to write and polish this manuscript. At this time, I would like to extend my appreciation. First, I wish to thank the undergraduates in computer science at the University of Oregon who have followed so faithfully the design methodology presented in this book. Their help was invaluable in defining and refining concepts and techniques important to the study of systems development. Many examples and illustrations contained in the book draw on this student work. Second, I wish to thank the reviewers of the manuscript. These include: William Fuerst, Price Waterhouse; Joan E. Hoopes, State University of New York at Binghamton; Paul Maxwell, Bentley College; George Miller, North Seattle Community College; Dennis Geller, Babson College; Robert Keim, Arizona State University; Larry Sanders, University of Washington; Tim Sylvester, College of Du Page; and Murray Berkowitz. Special appreciation must be given to the development team at Little, Brown. In particular, Denise Clinton, CIS editor, has been an enthusiastic backer of the project since its start. She has insisted on making it a quality product. Special thanks is given to Scott Huler, who helped with the work of making the manuscript very readable and consistent. Finally, I thank my family: my wife Jane, our two boys, Edward and Brian, and my mom and dad. During the writing of the book, my dad did all he could, but could not overcome a serious illness. In appreciation, I dedicate this book to him.

Alan L. Eliason

Brief Contents

1. Systems Analysis and Design 1
2. Business Systems 33
3. Tools of Structured Analysis and Design 64
4. System Requirements 101
5. Data Collection and Analysis 132
6. Data Organization and Documentation 168
7. Feasibility Analysis 204
8. Logical Design Specification 245
9. System Organization 284
10. Input/Output Design 322
11. Data File and Database Design 364
12. Computer Program Design 396
13. Processing Control Design and the Technical Design Specification 429
14. Programming and Program Testing 457
15. System Testing and Conversion 492
16. Systems Maintenance 522

Index 553

Contents

1 SYSTEMS ANALYSIS AND DESIGN 1

Introduction 1

Systems Analysis, Design, and Implementation 2
Systems Analysis 2
Systems Design 5
Systems Implementation 6
Systems Development Versus Systems Planning 9
Systems Development Versus Systems Maintenance 10

Project Life Cycle 10
Project Activities 11
Project Planning 12
Project Organization 16

Managing the Systems Effort 19
Systems Administration 20
Resource Management 22

The Mansfield, Inc., Case Study 25

Summary 28

Review Questions 29

Exercises 29

World Interiors, Inc.— Case Study 1 30

References 32

2 BUSINESS SYSTEMS 33

Introduction 33

What Is a System? 33
General Model of a System 34
General Model of a Business System 37

What Is a Business Computer System? 43
The Structure of a Business Computer System 46
Interrelations of a Business Computer System 47
The System Trinity 48

What Is a Computer System? 49
Computer Hardware 49
System Software 54
Application Software 54
Procedures 54
Data Processing Personnel 55

The Mansfield, Inc., Case Study 56

Summary 57

Review Questions 58

Exercises 59

World Interiors, Inc.— Case Study 2 62

References 63

ix

3 TOOLS OF STRUCTURED ANALYSIS AND DESIGN 64

Introduction 64

Tools of Structured Analysis 65
Data Flow Diagram 65
Data Dictionary 73
Data Structure Diagrams 73
Structured English 76

Tools of Structured Design 77
Structure Chart 77
Pseudocode 82
Input and Output Layouts 83
File and Database Layouts 84
Processing Controls 85

Traditional Tools of Systems Analysis and Design 86
System Flowcharts 86
Program Processing Menus 89
Program Flowcharts 91

The Mansfield, Inc., Case Study 93

Summary 96

Review Questions 97

Exercises 97

World Interiors, Inc.—Case Study 3 99

References 100

4 SYSTEM REQUIREMENTS 101

Introduction 101

Problem Classification and Definition 102
Defining the Problem 102
Evaluating the Problem 106

Problem Identification and Evaluation Tools 108
Causal Analysis 108
Marginal Analysis 110
Functional Analysis 113

Finalizing Project Requirements 118
Determining Project Feasibility 119
Project Specification 120

The Mansfield, Inc., Case Study 124

Summary 126

Review Questions 127

Exercises 128

World Interiors, Inc.—Case Study 4 129

5 DATA COLLECTION AND ANALYSIS 132

Introduction 132

Data Collection and Data Flow Diagrams 133
Data Collection and DFD Decomposition 135

Personal Interviews and Questionnaires 142
The Personal Interview 143
The Questionnaire 145
Advantages and Disadvantages 148

Forms and Procedures Collection 149
Forms Collection 150
Procedures Collection 151
Flowchart Collection 155

The Mansfield, Inc., Case Study 157

Summary 161

Review Questions 162

Exercises 162

World Interiors, Inc.—Case Study 5 163

References 167

6 DATA ORGANIZATION AND DOCUMENTATION 168

Introduction 168

Data Dictionary 169
Contents 169
Conventions 171
Organization 174
Reasons for Creating the Data Dictionary 176

Data Store Descriptions 177
Data Structure Diagrams 180

Transform Descriptions 188
Rules of Structured English 188
Structured English Conventions 189

The Mansfield, Inc., Case Study 193

Summary 197

Review Questions 198

Exercises 199

World Interiors, Inc.—
Case Study 6 202

References 203

7 FEASIBILITY ANALYSIS 204

Introduction 204

User's Performance Definition 205
Statement of New System Goals 206
*Determination of Appropriate Design
   Strategy 208*
*Identification of Specific Design
   Objectives 214*
*Ranking of Specific Design Objectives
   215*
Identification of Design Constraints 216

Designing the New System 216
Determining New System Outputs 217
*Developing Design Alternatives to
   Achieve Those Outputs 217*
Evaluating Each Alternative 221
Ranking Each Design Alternative 225
Selecting the Best Design 226

Social, Technical, and Organizational
Impact Analysis 226
Task Analysis 228
Technical Analysis 229
Organizational Analysis 232

The Mansfield, Inc., Case Study 234

Summary 237

Review Questions 238

Exercises 239

World Interiors, Inc.—
Case Study 7 241

8 LOGICAL DESIGN SPECIFICATION 245

Introduction 245

Preliminary System Design
Requirements 246
System Organization Requirements 246
Input/Output Requirements 250
*Data File and Database Requirements
   254*
Computer Program Requirements 258
Processing Control Requirements 259
Design Assumptions 260

Preparing the Structured
Specification 261
Executive Summary 262
Current Logical System 262
Feasibility Analysis 262
Proposed Logical System 263
System Design Requirements 263
Schedule and Budget 263
Hardware and Software 264
User's Guide 265

Design Schedule and Budget 265
*Preparing the Detailed Time Schedule
   266*
*Preparing the Detailed Project Budget
   272*

The Mansfield, Inc., Case Study 274

Summary 277

Review Questions 278

Exercises 279

World Interiors, Inc.—
Case Study 8 282

9 SYSTEM ORGANIZATION 284

Introduction 284

The Design Process 285
Preliminary Functional Design 287
Systems Design: The Process 294

Constructing HIPO Charts 296
Visual Table of Contents 297
Overview HIPO Diagrams 298
Detail HIPO Diagrams 299
Advantages and Disadvantages 300

Structured Walkthroughs 302
Types of Structured Walkthroughs 303
Conducting a Structured Walkthrough 303
Why Conduct a Walkthrough? 310

The Mansfield, Inc., Case Study 310

Summary 314

Review Questions 315

Exercises 316

World Interiors, Inc.—
Case Study 9 319

References 321

10 INPUT/OUTPUT DESIGN 322

Introduction 322
Where to Begin? 322

Design of System Output 323
Step 1: Defining Output Requirements 323
Step 2: Documenting the Logical Data Structure 324
Step 3: Defining the Physical Data Structure 327
Step 4: Representing System Outputs Visually 329
Step 5: Selecting the Most Appropriate Output Medium 331

Design of System Input 334
Step 1: Defining Data Capture Requirements 334
Step 2: Documenting the Logical Data Structure 334
Step 3: Defining the Physical Data Structure 337
Step 4: Representing System Inputs Visually 338
Step 5: Selecting the Most Appropriate Input Medium 341

Design of Interactive Dialogue 345
Four Types of Formal Messages 345
Constructing a Dialogue Tree 347
Picture-Frame Analysis 349
Screen Design 354

The Mansfield, Inc., Case Study 355

Summary 357

Review Questions 357

Exercises 358

World Interiors, Inc.—
Case Study 10 360

11 DATA FILE AND DATABASE DESIGN 364

Introduction 364

The Design of Physical Files 365
Physical Versus Logical Records 365
Record Structure 366
File Structure 369

Design of Relational Files 371
Entity-Relationship Diagrams 373
Relational Design Rules 376

Design of System Files 380
Determining File Processing Requirements 380
Database Considerations 382
Database Types 383
Preparing the Physical File Design 385

The Mansfield, Inc., Case Study 387

Summary 390

Review Questions 391

Exercises 391

World Interiors, Inc.—
Case Study 11 394

References 395

12 COMPUTER PROGRAM DESIGN 396

Introduction 396
Where to Begin? 396

Program Structure Charts 397
Principle of Partitioning 398
Principle of Coupling 399
Principle of Cohesion 402
Principle of Clear Labeling 403
Principle of Span of Control 404
Principle of Reasonable Size 404
Principle of Shared Use 405

Module Specification 408
Describing the Intent of Each Module 409
Describing the Implementation of Each Module 411
Program Flowcharts 412
Nassi-Shneiderman Charts 415

Evolving the Top-Down Program Design 416

The Mansfield, Inc., Case Study 418

Summary 421

Review Questions 422

Exercises 423

World Interiors, Inc.—Case Study 12 425

References 428

13 PROCESSING CONTROL DESIGN AND THE TECHNICAL DESIGN SPECIFICATION 429

Introduction 429

Types of Processing Controls 430
Source-Document Controls 430
Input (Transmission) Controls 433
Output Controls 435
Computer Program Controls 435

Data Validation 436
Edit Tests 436
Check-Digit Tests 438

Audit Considerations 440
Types of Audits 440

The Technical Design Specification 446
Executive Summary 447
System Organization 447
System Requirements 447
Coding and Test Plan 447
Test Schedule and Budget 448
Hardware and Software Specification 448
User Support Requirements 449

The Mansfield, Inc., Case Study 450

Summary 453

Review Questions 454

Exercises 454

World Interiors, Inc.—Case Study 13 455

References 456

14 PROGRAMMING AND PROGRAM TESTING 457

Introduction 457
Moving from Systems Design to Systems Implementation 458
Team Programming 458

Programming 459
Making Programs Easy to Read 460
Designing Programs Using Standard Constructs 462
Designing Programs Following a Top-Down Design 466
Documenting Computer Code 470
Testing Computer Code 472

Program Specifications 473

Program Testing 476
The Traditional Approach 477
The Incremental Approach 478
The Critical Path Approach 481

The Mansfield, Inc., Case Study 483

Summary 485

Review Questions 486

Exercises 486

World Interiors, Inc.—Case Study 14 489

References 490

15 SYSTEM TESTING AND CONVERSION 492

Introduction 492

System Testing 493
Types of System Tests 494
Equivalence Partitioning 497
Test Plan 498

Conversion 501

Database Creation 501
System Changeover 503
New Work Procedures 505
Completion of System Documentation 507
User Training 510

Acceptance Testing 513
Software Acceptance Criteria 513
Manual Procedures Acceptance Criteria 513

The Mansfield, Inc., Case Study 514

Summary 516

Review Questions 517

Exercises 518

World Interiors, Inc.—
Case Study 15 520

References 521

16 SYSTEMS MAINTENANCE 522

Introduction 522
What Is Systems Maintenance? 523
Types of Systems Maintenance 524

Adaptive Maintenance 524
User Requests for Service 524
Mini–Systems Development 526
Housekeeping Tasks 528

Corrective Maintenance 530
Critical Incident Reporting 531
Module Repairing 533

Perfective Maintenance 534
Modifying Program Data Structures 534
Modifying Programmed Procedures 538

The Mansfield, Inc., Case Study 542

Summary 545

Review Questions 546

Exercises 546

World Interiors, Inc.—
Case Study 16 549

References 551

INDEX 553

SYSTEMS DEVELOPMENT
Analysis, Design, and Implementation

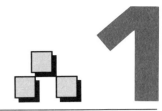

Systems Analysis and Design

INTRODUCTION

IMAGINE that you have been hired by a medium-size business firm to study the firm's computing problems. Before you can respond, various questions come to mind: What kind of computer? What kind of computing problems? Why me? What is the purpose of the study? Has a similar study been done in the past? Whom should I talk to? What can I read? Where do I begin?

Questions such as these indicate that you are searching for a systematic approach. They suggest that you are looking for not only a place to begin but a reason for beginning. They indicate that you are seeking a method of organizing your work. More than likely, you will want your results to be complete, be understandable to others, and feature a series of logical steps that explain how your study was carried out.

This chapter serves as an introduction to systems analysis and systems design. It begins with a discussion of the various stages involved in the study of business problems and in the design of solutions to these problems. It concludes with an examination of how systems activities typically are organized. When you complete this chapter, you should be able to

- describe the types of activities important to systems analysis;
- describe the types of work performed in systems analysis;
- define the role of a systems analyst, a systems designer, and a computer programmer;
- identify the stages of a project life cycle; and
- prepare an organization chart and discuss different ways of organizing the activities important to systems analysis and design.

SYSTEMS ANALYSIS, DESIGN, AND IMPLEMENTATION

In business environments, systems analysts and designers are employed to study current methods of operation and to develop and implement new, improved methods of operation. This work is done to improve the way in which a business operates—it leads to greater efficiency and effectiveness of business activities. Since the advent of the computer, there has been a rapid rise in the need for trained systems analysts and designers. *Systems analysts* specify the types of information that should be processed by computer and why that information needs computer processing. The work performed by analysts is called *systems analysis*. *Systems designers* determine how the computer will process this information. This work is known as *systems design*. Besides analysts and designers, *users* (individuals such as managers or office personnel) describe in detail their information processing requirements. Users must work with analysts and designers in preparing design specifications and in learning how to use new computer-based methods effectively.

Systems analysts and designers are vital in achieving the potential offered by electronic data processing. Many business firms have discovered that while the computer offers great hope for improving business efficiency, realizing that goal is often quite difficult. Before improvements are possible, a business must

1. understand how it currently processes data;
2. identify opportunities for using the computer;
3. develop plans and specifications to describe current and future uses of computing;
4. design computer software and test it to ensure that it works properly;
5. undergo the process of changing over to computer-based methods of processing; and
6. learn how to effectively use these new computer-based methods.

These six steps clearly suggest that the computer cannot accomplish it all alone. Rather, three parties at the very least are involved: systems analysts, systems designers, and users.

Systems Analysis

Let's examine the process of systems analysis and systems design in somewhat more formal terms. Systems analysis is a logical, conceptual activity. It typically is larger in scope than systems design and requires a thorough understanding of business systems and procedures. The purpose of systems analysis is to define the work to be done and the end product of this type of analysis is a logical design specification.

The process of systems analysis can be divided into five parts (see **Table 1–1**):

1. *system requirements analysis*, which determines the nature of the system problems and the initial actions to take;

2. *data collection and analysis*, which studies current methods of processing to understand what is being done;
3. *data organization and documentation*, which arranges collected data in a systematic way;
4. *feasibility analysis*, which determines whether current methods of processing need to be changed, and how those changes will be made; and
5. *logical design specification*, which defines an improved method of processing.

The following sections examine each of these five parts in greater detail.

System requirements analysis consists of the systematic review of users' and management's real and perceived needs for information and leads to the "definition" of the system to be analyzed. Typically, a study team comprising one or more systems analysts and one or more programmer/analysts (subordinate to the systems analyst) performs the system requirements analysis. When conducting this study, the team must remember that certain types of information are essential to users and managers, while other types are nice to have but are not critical. Thus, as part of the study, the team must pay close attention to how information is used in the organization and how it influences decision making. The team must also determine the root cause of system problems and decide whether those problems require further study.

Data collection and analysis is also known as the study of systems and procedures. It involves the detailed study of administrative and operational systems in an organization to determine whether information processing procedures are reliable, timely, economic, and complete, or

TABLE 1–1 Life cycle project activities

Phases in the Development Cycle	Project Activities
Systems Analysis	System requirements analysis Data collection and analysis Data organization and documentation Feasibility analysis Logical design specification
Systems Design	System organization Design of system input and output Design of system files and databases Design of system programs Design of system controls Technical design specification
Systems Implementation	Program coding and debugging Program and system testing System implementation and evaluation

whether they are haphazard, consistently late, costly, and vague. Data collection and analysis must pay close attention to how information flows in the organization. By carefully studying this flow, the systems analyst can determine which procedures are used in processing different types of routine transactions (e.g., payroll, monthly statement, and payables transactions) and different types of summary information (e.g., payroll summary information).

Data organization and documentation concerns the arranging of collected data in a systematic manner (that is, it must be properly organized and documented). Although the documentation of collected information is exacting and often time-consuming, the rewards of a job well done are significant. Through proper documentation, a study team can demonstrate how a system functions and indicate which alternative processing designs might be possible.

Feasibility analysis involves identifying, first, alternative methods of processing and, second, the optimal or most feasible method of processing. This dual purpose of feasibility analysis suggests that it is not enough to conclude that current methods are terrible, presupposing that there is only one feasible solution. Careful systems analysis calls for the identification of several feasible solutions, including, perhaps, the solution to keep an inadequate system up and running.

Logical design specification deals with the written description of the new method of processing. The specification details the major functions and the important subfunctions of the system. It explains the specific types of information required by the system and how information flows from one part of the system to another. Most systems analysts employ some type of *structured approach* in preparing a logical design specification. A structured approach generally features a top-down methodology, which begins with an overview of a system, followed by the breakdown of the system into functions and corresponding subfunctions. The primary benefits of a structured approach are twofold: it helps individuals understand the different parts of a system and it demonstrates how these parts are interrelated.

Figure 1-1 shows that the process of systems analysis consists of transforming user requirements into a logical design specification. System requirements analysis, data collection and analysis, data organization and documentation, and feasibility analysis are all vital parts of this translation process. These activities help to answer the questions of:

- Why is there a problem?
- What is being done?
- What methods are available to improve current methods of processing?
- Should a new method be implemented?
- How should a new method be designed?

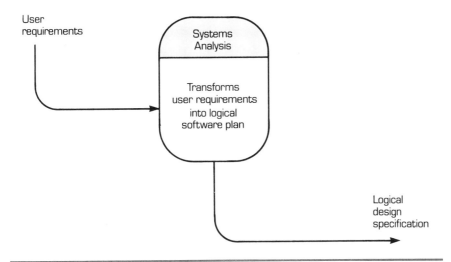

FIGURE 1–1 Systems analysis: the process

If the decision is made to design a new system, the systems analyst then decides which new methods are feasible, followed by which design alternative is most appropriate. The logical requirements of the best design must then be specified. This logical description represents a logical software plan. The plan explains the details of the system to be developed, and it describes how user requirements can be met following system implementation.

Systems Design

In preparing the logical design specification, the systems analyst specifies in logical terms system input and output requirements, data file or database requirements, data processing requirements, and processing control requirements. Systems design involves transforming these logical requirements into a technical software plan (see **Figure 1–2**). This second plan is called the *technical design specification*—a document that shows how computer programs are organized and how they are to be written; it specifies how all input, output, data file, and processing control requirements are to be designed.

Table 1–2 summarizes the five main types of requirements that must be logically described by the systems analysis team and later translated into a technical design specification by the systems design team. *Output requirements* identify what is to be produced from processing, whereas *input requirements* define the types of data that must be entered into processing. Compare these input/output requirements to the other three: *data file* and *database requirements* identify which types of data are to be stored on computer files, *computer program requirements* explain the

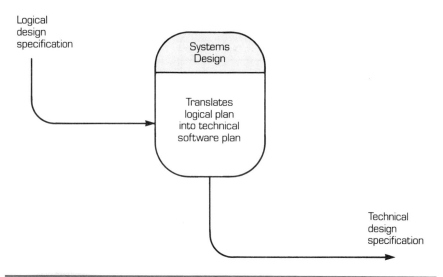

FIGURE 1-2 Systems design: the process

functions of each computer program in a system, and *processing control requirements* demonstrate how the software design is to be monitored once it becomes operational.

Table 1-2 suggests that systems design is more detailed than systems analysis. In systems design, a variety of questions related to hardware, software, and business matters must be settled:

- How will output be organized and printed or displayed by the computer?
- How will input be organized, entered into processing, and checked for accuracy by the computer?
- How will computer files be organized, linked together, and called into processing by computer programs?
- How will computer programs be organized and written?
- How will processing controls be utilized?

Systems Implementation

Besides preparing the technical design specification, the systems designer is usually responsible for directing the implementation of the design. Systems implementation consists of three components: programming, testing, and conversion. *Programming* involves the writing of instructions to be read by a computer. A collection of these instructions is better known as a *computer program*, or *computer software*. Most programmed instructions are written in a high-level language—that is, a language that bears some syntactic similarity to English. However, it is

TABLE 1–2 Five main types of system requirements

System Requirements	Systems Analysis	Systems Design
Output requirements explain what is to be produced from processing.	Describes output requirements and supplies logical drawings.	Determines the actual content of printed or displayed results; decides how output will look once it is produced.
Input requirements explain what types of data must be entered into processing.	Defines the types of data to be entered into processing.	Develops input formats and input documents; determines how data input and verification procedures are to take place.
Data file and *database requirements* explain what types of data are to be stored on computer files.	Determines file content and how stored data are to be used in processing.	Defines the methods of file storage, the format and layout of each file, and the links necessary to tie together different physical files.
Program processing requirements explain the function of each computer program in a system.	Specifies how each function and subfunction of the system might be performed.	Develops the design solution by indicating how inputs and outputs, including inputs and outputs from computer files, are handled, how each computer program is to be organized, and how the different computer programs are to be interrelated.
Processing control requirements show how the software design is protected against the entry of invalid data and how each stage in processing is to be verified for accuracy and completeness.	Defines the logical *audit trail*—the logical points in a process where computer output is to be checked for correctness.	Defines the programmed instructions needed to ensure the correctness of computer processing.

also possible to write instructions in a language that resembles binary expressions consisting of 0s and 1s.

Programming is performed by *computer programmers* rather than systems analysts or designers. Tasks essential to programming include writing the coded instructions, testing the code, and finally implementing the code. The completed program directs the computer to execute an *algorithm* (i.e., a complete statement of a procedure for computing a solution to a problem). Besides this, the program serves a second important function: it conveys the programmer's interpretation of the designer's technical design specification.

Testing consists of putting together the various coded pieces of a design, testing those pieces, and correcting the parts of the code (or the design) that are incorrect or inappropriate. Part of testing is to intentionally introduce errors into processing to determine whether they will be spotted by the program as errors. This addition of error-checking routines is a vital part of computer program design.

Conversion is the process of making the coded and tested computer programs operational. The conversion team must manage the changeover from the old system to the new system. Three types of documentation support the release of a new system:

1. Systems documentation—the historical record that describes the development of a project. It includes the logical design specification, the technical design specification, and computer program source code.

2. Operations documentation—instructions for setting up, operating, and maintaining the software product. Setup instructions show how the computer should be made ready before running a program, and operating instructions describe what should be done during program execution. Maintenance instructions are somewhat easier to visualize. These instructions document the required steps for keeping the software up and running; they describe suggested ways to "fix" the software should it fail to run properly.

3. User documentation—instructions for the user. Most user documentation begins with a section entitled "Getting Started," followed by "How to Enter Data into Processing," or "How to Create a Data Record." These materials are supposed to be "user friendly"; that is, they are easy to read and understand, and they provide an easy-to-follow set of illustrations and examples.

Figure 1-3 illustrates the process of systems implementation. Unlike systems analysis or design, areas that yield written specifications, systems implementation leads to the development of usable software. Be-

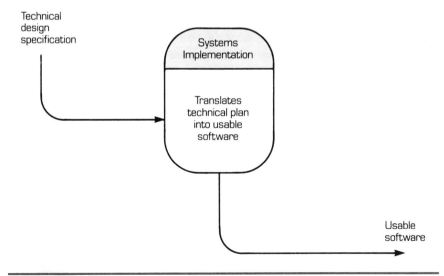

FIGURE 1-3 Systems implementation: the process

cause this software evolves from a carefully designed set of logical and technical specifications, it is often referred to as being engineered. *Software engineering* is the process of using a set of techniques to carefully document each step leading to the production of usable software.

Systems Development Versus Systems Planning

The overall process of systems analysis, systems design, and systems implementation is called *systems development*. As shown in **Figure 1–4,** systems analysis may be preceded by still another activity, known as *systems planning*. The purpose of systems planning is to specify the user's processing requirements from the user's point of view. Systems analysis, in turn, begins with a more detailed investigation of user requirements. The purpose of systems analysis is to determine whether a problem exists and, if so, to recommend a logical, improved method.

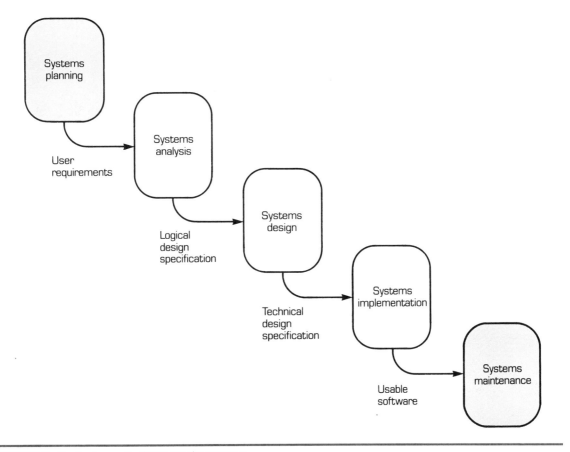

FIGURE 1–4 Systems development: the process

The specific features of the new method are then documented in the logical design specification. Systems design, in contrast, transforms the logical specification into a technical software plan called the technical design specification. The technical specification shows the design of system input and output, files and databases, programs, and controls. Finally, systems implementation, the final stage of development, produces usable and well-documented computer software. Computer programming, testing, and conversion all occur during this phase of development.

Systems Development Versus Systems Maintenance

A new system does not remain new forever. Eventually, it must be modified to handle new processing requirements. *Systems maintenance* deals with keeping the software operational. Central to this function are three types of maintenance activities:

1. undertaking preventive measures to keep computer programs current,

2. monitoring and fixing problems with computer programs, and

3. modifying programs in response to new user requirements.

One way to look at systems maintenance is to see it as a miniature systems development effort. Maintenance programmers must specify both logical and technical design changes. Once this is done, they must undertake programming, testing, and conversion to produce usable software.

PROJECT LIFE CYCLE

The realization that every software application must undergo the same process of analysis, design, and implementation leads to the concept of the *project life cycle*. The life cycle concept advances several important principles about how systems development should be organized and managed, including:

1. *Systems development is a planned activity.* The objectives to be accomplished for each phase of development should be determined, and the time required to complete each phase should be estimated.

2. *Systems development is a birth-death process.* The project begins (the birth) with the decision to analyze user processing requirements; the project ends (the death), provided all goes well, once a usable software product is implemented.

3. *Systems development is self-documenting.* Each major stage in development leads to a set of written documents. The first of these documents is the logical design specification; the second is the technical design specification. The third, computer program documentation, follows the programming, testing, and conversion of computer programs.

4. *Systems development is a managed activity.* Because each stage of development is unique and has a known objective, a predetermined point exists at which actual results can be evaluated (this point is called a *milestone*).

Of vital importance to the systems development process will be the ability to improve its management. In the past, a major criticism of software development was that it was largely left on its own, with some highly objectionable consequences. Computer programmers, for example, have been discovered to produce an average of only ten to fifteen lines of code per day;[1] and most data processing employees have been found to spend 75 percent or more of their time maintaining completed systems. In the eyes of most managers, these practices are not acceptable.

The keys to achieving higher productivity in the systems group are twofold. First, a structured approach is advisable—from the beginning of a project to its end. The average productivity of most computer programmers tends to jump dramatically when a structured methodology is used. Programmers averaging only ten to fifteen lines of code per day jump to forty or more lines of code per day.[2] Second, a project management approach is required—from the beginning of a project to its end. With project management, resources (people, money, computers, and so forth) are often used far more efficiently.

Project Activities

Breaking down a large complex system into smaller, well-defined project activities represents an initial step in understanding how a structured and a better managed approach work in combination. The process of systems development can be broken down into a set of smaller activities; each activity, in turn, can be viewed as a small project. And just as important, each activity can be assigned its own objective (i.e., its mission) and milestone (i.e., the point at which results can be evaluated).

Figure 1-5 shows the rough percentage of time required for the main activities of systems development. It may come as some surprise that program and system testing requires 30 percent of the total development effort, while program coding and debugging requires only 15 percent. Note, too, that 50 percent of the development process is required for systems analysis and design. Why the high cost for analysis and design? Consider the following: with careful attention given to systems analysis and design, the resources invested in testing can be well supported, since the purpose of testing is to prove that the design works as planned and is easy to maintain. However, suppose that inadequate attention is given to analysis and design. Programmers will then begin to write computer programs without having a clear idea of what each program must do and how the computer programs are to be tied together. The results of this practice are all too obvious. Programmers find themselves caught in an endless loop of writing a section of a program, running the pro-

12 Chapter 1: Systems Analysis and Design

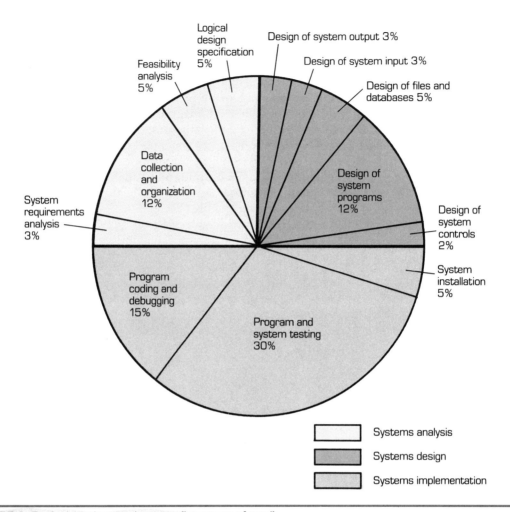

FIGURE 1–5 Activity time requirements (in percent of total)

gram to test the section, finding new errors, debugging (removing errors or "bugs" from a program), writing another new section, and so on. When this occurs, computer programming and testing can require 80 to 90 percent of the total development effort, three to four times more than anyone would have expected.

Project Planning

Project planning is required in most organizations in order to ensure that systems analysis and design receive adequate attention. The purpose of planning is to clarify what is to be done, how much it will cost, and how long it will take to complete. This process may seem difficult, and in practice, it is. It is especially difficult to estimate in advance which resources will be required, when little is known about the project to be studied.

Various tools and techniques are used in a project management environment to help systems analysts estimate systems development requirements. One of these tools is the *project planning and control chart*. As **Figure 1–6** shows, such a chart projects the number of labor hours and labor dollars required for systems development over the life of the project. For this project, 1,000 hours are projected for systems analysis, 1,000 hours are projected for systems design, and 2,000 hours are pro-

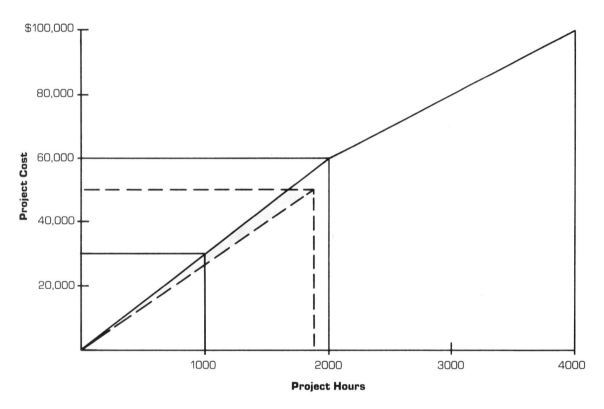

FIGURE 1–6 Project planning and control chart

jected for systems implementation. Likewise, a project cost of $100,000 is estimated: $30,000 for analysis, $30,000 for design, and $40,000 for implementation.

Project control takes on meaning when actual hours and costs are known. As shown in the figure this project appears to be *in control*. Actual hours and costs for both systems analysis and design are less than those projected. If they were greater than expected, the project would appear to be *out of control*.

Table 1–3 provides a breakdown of the initial estimates used in constructing the project planning and control chart. As indicated, data collection and organization together are expected to require the greatest number of systems analysis hours, program design the greatest number of systems design hours, and program and testing the greatest number of systems implementation hours. These high hourly estimates are in line with what analysts and designers experience in practice. Typically, data collection and data organization are the most time-consuming activities, as is computer program design. The high hourly requirements for pro-

TABLE 1–3 Project planning spreadsheet

Systems Analysis	Project Hours	Percent of Total	Costs (in dollars)	Percent of Total
Requirements	120.00	3	4,000.00	4
Data Collection	300.00	7	9,000.00	9
Data Organization	100.00	3	2,500.00	2
Feasibility	200.00	5	6,000.00	6
Logical Design	280.00	7	8,500.00	9
Total	1,000.00	25	30,000.00	30
Systems Design				
Output Design	125.00	3	4,000.00	4
Input Design	150.00	4	4,500.00	4
File Design	200.00	5	6,500.00	7
Program Design	400.00	10	11,000.00	11
Controls Design	75.00	2	2,500.00	3
Technical Design	50.00	1	1,500.00	1
Total	1,000.00	25	30,000.00	30
Systems Implementation				
Program Coding	500.00	13	9,000.00	9
Program and System Testing	1,250.00	31	25,000.00	25
System Installation	250.00	6	6,000.00	6
Total	2,000.00	50	40,000.00	40
Total Project	4,000.00	100	100,000.00	100

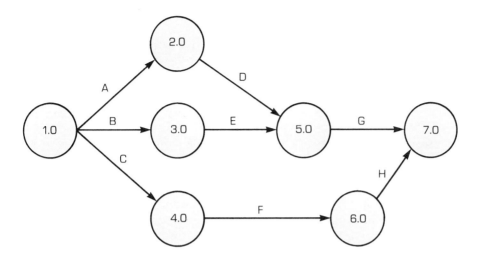

Activity	Node Relationship	Description	Time (hours)
A	1.0→2.0	Design of system output	125
B	1.0→3.0	Design of system input	150
C	1.0→4.0	Design of system files	200
D	2.0→5.0	Design of output programs	75
E	3.0→5.0	Design of input programs	75
F	4.0→6.0	Design of file processing programs	250
G	5.0→7.0	Design of system controls	75
H	6.0→7.0	Completion of technical design specification	50

FIGURE 1–7 Network chart

gram and system testing reflect the time (and expense) associated with producing an error-free software product.

Another useful tool is a *network chart* or a *network control chart*. As shown in **Figure 1–7,** a network control chart consists of small circles (called "nodes") that are connected by arrows (called "activities"). Activity A, for example, has a node relationship of 1.0 to 2.0, is described as the "Design of system output," and requires 125 hours to complete. One advantage of a network chart is that it shows when several activities can be worked on simultaneously. Activities A, B, and C can be worked on concurrently, for example, as can Activities D, E, and F.

Project Organization

Project organization shows how the work of a systems group in an organization is structured and how people are assigned to project activities. Several forms of project organization are permitted in a project management environment. **Figure 1–8** features an *organization chart* to illustrate one of these forms. This "pure" form shows that a hierarchical structure is used to partition the functions and tasks of developmental programs. Within this structure, the *systems manager* is responsible for all systems development work; a *project manager* is responsible for the successful completion of a single developmental project. With a pure form of organization, each project manager defines what needs to be done, determines which resources are required to achieve those ends, and sets the milestones at which actual times and costs are compared against estimated times and costs. Although this form of organization is easy to understand, it is often not realistic. In practice, not all systems work is developmental. Most systems work involves maintaining previously developed systems.

Figure 1–9 shows a modified organizational form known as *matrix organization*. In this structure, two types of managers report to the systems manager. As before, the project manager is responsible for a particular systems development project, such as designing and implementing a new method of processing sales information. The *program manager* is responsible for maintaining previously developed systems. He or she must keep these systems operational, which involves making improvements to these systems (e.g., fixing and modifying these systems as required).

There are several advantages to matrix organization:

- It is flexible. Projects can be started or stopped without disrupting the work of program managers.
- It provides a large number of people with managerial experience.

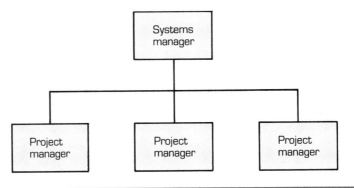

FIGURE 1–8 Pure form of project organization

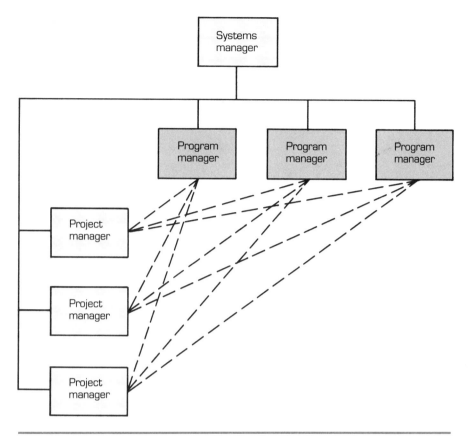

FIGURE 1-9 Matrix organization, showing paths of transfer between operations programs and developmental projects

- It allows analysts to analyze and designers to design. Members of the project staff need not become involved with operational problems.
- It clearly defines project responsibilities. One person, the project manager, is responsible for his or her assigned project.
- It provides temporary assignments. A programmer may be temporarily assigned to a project and then transferred back to a program once the project task is completed (observe the paths of transfer shown on Figure 1-9).
- It provides systems and programming staff with exposure to different types of assignments.

Unfortunately, matrix organization also has some disadvantages:

- The number of people reporting to the systems manager increases.
- Program managers are often reluctant to release key members of their staff, even for short-term assignments.

- Project managers are often reluctant to return key staff members to program assignments.
- The better analysts, designers, and programmers may not always be assigned to systems development activities. Instead, they may be required to perform operational tasks.
- If resources are scarce, project and program managers will struggle to gain the services of the best employees.

A third form of organization is known as *functional organization.* As shown in **Figure 1–10,** systems work can be organized to parallel the work of the organization—namely, marketing, manufacturing, personnel, and accounting. With this form, both project and program tasks are assigned to the same manager. The manager of manufacturing systems, for example, is responsible for developing new manufacturing systems and for monitoring previously implemented manufacturing systems. The advantages of a functional form of organization include the following:

- The responsibility for each functional area of a company is clarified. A single manager is reponsible for his or her area within a firm.
- Communication between systems managers and their functional counterparts is simplified. For example, the manager of marketing systems can communicate directly with members of the marketing staff.
- The better systems analysts can be assigned to systems analysis activities and the better designers to systems design activities.
- Functional organization is easy to explain to others.
- It permits project and program activities to be shared.

Likewise, functional organization has some disadvantages. Compared with matrix organization, it is not as flexible, nor does it provide as wide

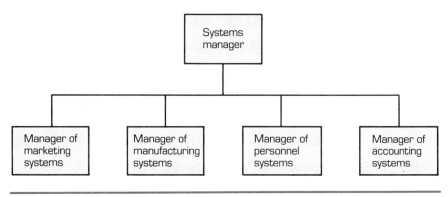

FIGURE 1–10 Functional organization

a variety of people with managerial experience. Perhaps the most serious disadvantage is that developmental projects are often interrupted when development staff is pulled off design work to handle day-to-day operational problems. When this happens frequently, project deadlines slip, making project planning figures worthless.

MANAGING THE SYSTEMS EFFORT

It is the systems department that is responsible for the systems effort. But along with understanding how it is organized, it is important to understand where the systems department lies in relation to the rest of the organization and how the systems effort is administered. In the 1950s and 1960s, it was common to place the systems department, then called data processing, under the larger area of finance and accounting (see **Figure 1–11**). This placement occurred because the primary function of data processing was to handle the processing of routine financial matters. Beginning in the late 1960s, a far more common form of organization was to place all systems activities much higher in the organization. As shown in **Figure 1–12,** some organizations have elevated the systems department to the vice-presidential level. In so doing, the title of "information systems" typically replaces the title of data processing. Systems work and computer software production is currently recognized as a major business function in and of itself.

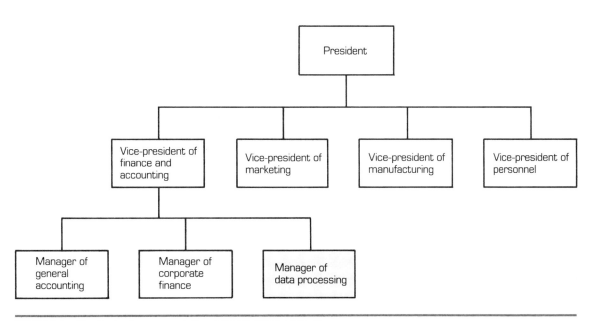

FIGURE 1–11 Early placement of the data processing function

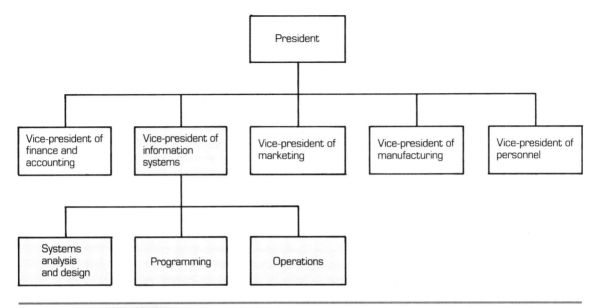

FIGURE 1–12 Organization chart in which the head of information systems is a company vice-president

Systems Administration

Placing the systems department closer to the president elevates systems planning within the organization. Firms experienced in systems matters realize that systems work is best administered when top management is directly involved in systems decision making.

Several organizational approaches currently are used to administer the systems effort in companies. Semprevivo[3] describes five of these: systems initiative approach, steering committee approach, systems clearing-house approach, corporate planning approach, and service center approach.

With the *systems initiative approach*, the systems group is responsible for initiating all new system requests and is, by and large, self-governing in selecting new projects for development. While this approach is very efficient, it often leads to considerable friction between the systems group and other departments. This is especially true when the systems group decides to initiate projects that outside departments and groups perceive as low priority.

The *steering committee approach* places a buffer between the systems group and other departments. As shown in **Figure 1–13,** a steering committee lies between the organization president and such top-level functional areas as marketing and manufacturing. In this "staff capacity," the committee has no direct authority to make decisions, but forwards all recommendations to the president. The purpose of the steering committee is to review new system requests and proposals and to select proj-

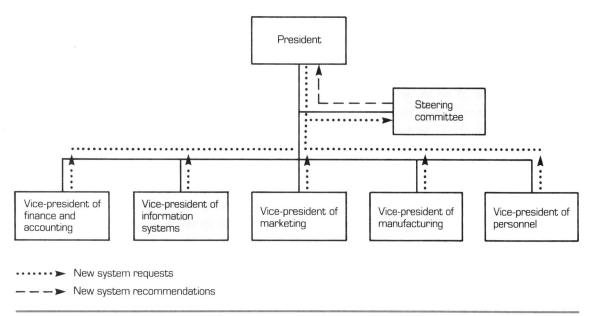

FIGURE 1–13 Steering committee approach

ects that it believes are worthy of additional study. The membership of the committee typically consists of vice-presidents or their representatives and members of the systems group.

Using a steering committee to direct the systems planning effort offers several advantages. The committee helps

- ensure that systems plans are consistent with the organization's long-range plans;
- establish companywide priorities for systems development;
- provide top-level backing to approved projects; and
- make proper use of systems resources, especially the money spent on computing equipment and systems personnel.

Although the steering committee approach is often recommended, its use is not without risk. If committee members lack knowledge about matters relating to computers, for example, the work of the committee tends to be more political than purposeful.

The *systems clearing-house approach* is similar to the systems initiative approach, except that managers of functional areas, such as the vice-president of marketing, submit routine and inexpensive requests for systems work to the systems department, where they are reviewed (and typically cleared). Major requests follow a different route. These are passed upward to top management for its approval (see **Figure 1–14**). Semprevivo calls this a "middle of the road" approach. Its advantages include

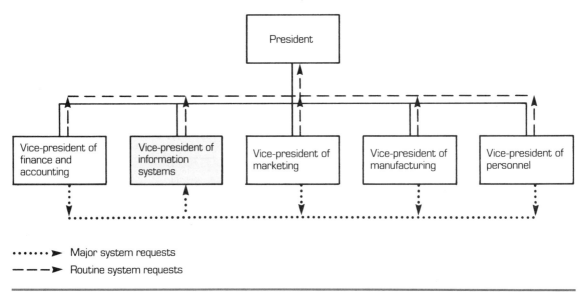

FIGURE 1–14 Clearing-house approach

high efficiency and companywide participation in the systems effort. The major disadvantage is that it can create considerable friction between departments, especially when one department's routine requests are always approved at the expense of other departmental requests.

The *corporate planning approach* changes the position of the systems department within an organization. As shown by **Figure 1–15,** the systems department becomes an organizational staff unit whose primary aim is to provide analysis and consultation to "line" units (units with direct command authority and responsibility for the main functions of the organization). This approach encourages departments to seek out systems consultants to help them with their problems.

The *service center approach* differs in concept from the other four approaches. Here, line units contract with the systems group for systems development and computer services (see **Figure 1–16**). In so doing, the systems group becomes a *cost/responsibility center*—it is responsible for providing a level of service at a specified price. In other words, the systems group becomes a business within a larger business. It is responsible for contracting all work to be performed and for fulfilling the terms of each contract it signs.

Resource Management

Requiring the systems group to provide a specified level of service at an agreed-upon price is consistent with the concepts embodied in *resource management*. With resource management, systems development deci-

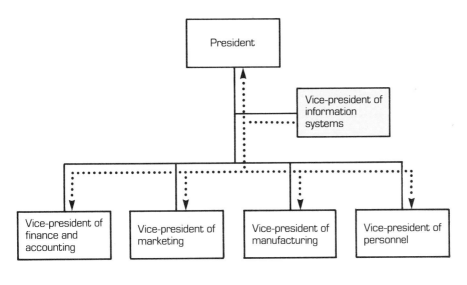

FIGURE 1–15 Corporate planning approach

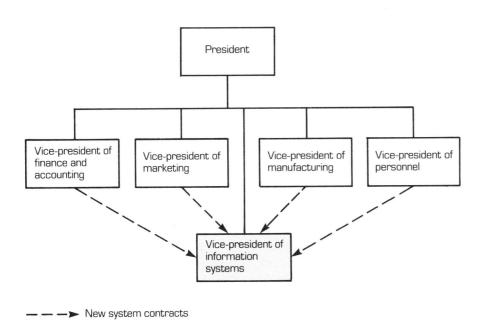

FIGURE 1–16 Service center approach

sions are viewed as investment decisions. Project proposals are evaluated for their capacity to improve overall business efficiency and profitability.

Eliason[4] cites seven opportunities to increase efficiency and profitability in a resource management environment:

1. Eliminate routine tasks.
2. Eliminate duplicate efforts.
3. Improve customer services.
4. Improve managerial reactions to problem situations.
5. Streamline the managerial structure.
6. Improve the use of internal business resources.
7. Improve the understanding of the external business environment.

Of the seven, some lead to *direct savings*. For example, suppose a project is estimated to cost $250,000 to analyze, design, and implement. Once completed, routine tasks and duplicate efforts will be eliminated; fewer people and less equipment will be needed to perform the job. Savings in personnel and equipment are estimated to be $50,000. The *return on investment* for this particular project is set at 20 percent, which is calculated as follows:

$$\text{Return on investment} = \frac{\text{Project savings}}{\text{Project investment}} = \frac{\$50,000}{\$250,000} = .20, \text{ or } 20\%$$

If the return considered average for the company is 15 percent, then this proposed project would be rated as above average.

Besides direct savings (savings that can be measured directly), resource management attempts to estimate *indirect savings*, such as the savings associated with improved customer service or improved managerial response to problem situations. Suppose that a company installs a much improved inventory control system, at a cost of $600,000. Suppose next that the new system is estimated to lead to a 10 percent reduction in inventory plus a 10 percent increase in sales. The reduction in inventory is a direct savings—$120,000 will be saved if inventory is reduced by 10 percent. The increase in sales resulting from improved customer service represents an indirect savings, since it cannot be measured directly.

When evaluating a project, management uses both direct (tangible) and indirect (intangible) savings to project the return on a project investment. Suppose that indirect savings in the example were set at $180,000. With this new figure, the return on the much improved inventory control system would be 50 percent:

$$\text{Return on investment} = \frac{\text{Tangible + intangible savings}}{\text{Project investment}}$$

$$= \frac{\$120,000 + \$180,000}{\$600,000} = .50, \text{ or } 50\%$$

Thus this project would appear to be very favorable.

Besides return on investment, resource management examines the payback associated with each project decision. *Payback* is the time needed by a project to pay for itself. It is the reciprocal of return on investment.

$$\text{Payback} = \frac{\text{Project investment}}{\text{Project savings}} = \frac{1}{\text{Return on investment}}$$

Payback for our last example would be

$$\text{Payback} = \frac{\$600,000}{\$300,000} = 2.00 \text{ years}$$

Thus, this project would pay for itself in two years.

Although a major thrust of resource management is to quantify systems decisions, this is not its only thrust. Another feature is the monitoring of all systems activity to ensure that development and operational work is kept on schedule and within established cost guidelines. This process of monitoring is much like project planning and control monitoring but is now at a higher level in the organization. Using the same project control technique, top managers can determine how various projects are doing; they can identify which projects are ahead or behind in schedule and which cost more or less than initially projected.

The Mansfield, Inc., Case Study

One way to better visualize the work performed by systems analysts, systems designers, and systems maintenance personnel is by studying the work of others. To this end, a brief study dealing with the fictitious firm known as Mansfield, Inc., is included in this chapter and in all subsequent chapters.

Introduction

Mansfield, Inc., is a business wholesaler that buys and sells electronic equipment and supplies. The president of Mansfield, Aaron Gillette, recently decided to form a systems group to streamline all administrative procedures. Aaron remarked, "Since we deal with state-of-the-art equipment, we have to stay ahead of the game. We want our suppliers and customers to recognize that Mansfield has state-of-the-art administrative systems at its disposal."

Company Organization

As shown by **Figure 1–17,** five vice-presidents currently report to the president. David Orring, vice-president for finance and accounting, is very interested in seeing that new systems are designed. "Our customer invoice system, in particular, needs work," he advises. Tony Chung, vice-president of marketing and sales, also appreciates the wisdom of improved administrative systems, adding, "In an electronic age, all administrative systems need to be computer

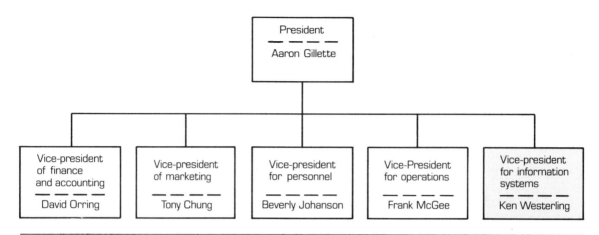

FIGURE 1–17 Top-level organization chart—Mansfield, Inc.

based." Beverly Johanson, vice-president for personnel, was instrumental in getting the newly formed systems group to automate the employee payroll as its first development project. "We just had to do something," she states. "When I came here, our payroll checks were either late or incorrect. I'm surprised the warehouse crews didn't revolt." Frank McGee, vice-president for operations, agrees with Beverly. "I really like the new payroll system," he comments. "I only wish we had a billing and an inventory system that were equally good."

Ken Westerling, vice-president for information systems, while pleased with the new payroll system, is quite aware of the need for other systems. "We just have to bite the bullet," he claims. "We'll see how good we are when we tackle the billing system. The steering committee suggested that we begin work on this system at once."

Organization of the Information Systems Group

Ken Westerling used a modified matrix management approach in organizing his small staff group (see **Figure 1–18**). His major appointments were as follows:

1. Roger Bates was appointed lead systems analyst for all new projects. As lead analyst, Roger could draw on a pool of systems analysts and programmers in putting together an analysis team.
2. John Seevers was appointed lead systems designer for all new projects. As lead designer, John could also draw from a staff pool in putting together a design team.
3. Sandra Cushing was appointed program manager for the employee payroll system. She was responsible for fixing the few bugs that still existed in the design and to make modifications, as required by end users and as approved by the information systems steering committee.
4. Shawn Fortuna was appointed program manager for all other finance and accounting systems, which at the time were limited to three software packages that came with Mansfield's 43301 central computer: a general ledger

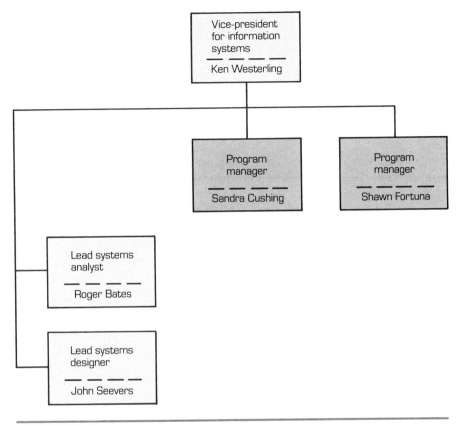

FIGURE 1-18 Organization chart—information systems

package, a sales analysis package, and a general accounting voucher package. "All three are very primitive," Shawn added. "But they are better than nothing at all."

Customer Billing Organization

David Orring, the vice-president for finance and accounting, was charged with preparing all of Mansfield's customer invoices. His comments about the current billing system are quite telling: "Early on, when we were smaller, our manual billing operation worked fine. Today, however, there are times when we employ twelve or more part-time typists to get out the bills. Why, we may run two weeks behind during our peak season."

David continues, "John Havensek, our billing supervisor, has been after me to add additional full-time staff, but I have been holding him off, pushing for a new system. If sales continue to grow to where they reach twenty-four million, we would need to hire twelve additional people simply to type all of our bills. I wish I knew what the systems group was up to. We desperately need a new billing system."

SUMMARY

In the development of new systems, organizations employ systems analysts and systems designers. System analysts specify which types of information should be processed by computer and why they need computer processing. Systems designers determine how the computer will process that information.

The process of systems analysis begins with system requirements analysis and ends with the logical design specification. Through analysis, the systems analyst transforms user requirements into a logical design.

The process of systems design begins with the translation of logical specifications and ends with the technical design specification. The designer sets the hardware, software, procedures, and personnel requirements for the new system.

After the system has been designed, it is implemented. Systems implementation consists of programming, testing, and conversion.

Systems development—namely, systems analysis, design, and implementation—usually is preceded by systems planning. With systems planning, the user specifies what he or she wants the system to do.

Systems development usually is followed by systems maintenance. The purpose of systems maintenance is to keep a developed system operational.

The idea of a project life cycle advances various concepts about how systems development work should be structured and managed. Any large project can be broken down into smaller, well-defined project activities, and in this way systems development can be structured and managed. The project life cycle concept also recognizes that systems development is a birth-death process. As such, its start and stop points can be determined.

Project planning is based on a set of well-defined project activities; it requires the project manager to determine resource requirements for each activity. When this knowledge is combined with an estimate of the time needed to complete each activity, the project manager has the information he or she needs to tell others about the proposed systems development. Project planning and control charts and network charts aid in communicating this information.

Project organization determines how people will be assigned to project activities. It partitions and compartmentalizes the functions and tasks relating to developmental projects and operational programs. Various methods of project organization can be applicable, depending on an organization's needs.

Business firms typically follow a hierarchical structure in partitioning their activities and functions. The more important the function, the more prominent it becomes in the company organization. Today, the function of data processing is quite visible in most modern organizations. Within the past twenty years, data processing has evolved into a major business function.

There are several ways to direct the systems effort in a business. Most firms have seen fit to move away from the systems initiative approach. Many prefer to use either the steering committee or the clearing-house approach, thus involving top management directly in the selection of new projects. Other firms believe that the corporate planning or the service center approach better suits their needs.

A resource management view of data processing maintains that systems development decisions should be treated as investment decisions. Before a project is approved or rejected, its return on investment should be determined. A second task of resource management is monitoring previously approved projects. With monitoring, top management can keep abreast of the resources expended on major systems development projects, much like other large capital investment projects.

REVIEW QUESTIONS

1-1. What is the difference between a systems analyst and a systems designer?
1-2. Name the activities important to the process of systems analysis.
1-3. What types of system requirements must be described logically by the systems analyst and later translated into a software plan by the systems designer?
1-4. How does programming differ from a computer program?
1-5. What is meant by the term *systems development*?
1-6. What principles are advanced by the concept of a project life cycle?
1-7. What is the purpose of project planning?
1-8. What are the advantages of using a network chart compared with a project planning and control chart?
1-9. What is the difference between a project manager and a program manager in a matrix organization?
1-10. List several approaches currently used by companies in administering the systems effort.
1-11. What is meant by the term *resource management*?
1-12. What is meant by the term *return on investment*, and what is the difference between return on investment and payback?

EXERCISES

1-1. Staff manager A is directly responsible for the activities of managers C, D, and E. Correct the organization chart shown on the following page to demonstrate this reporting relationship.

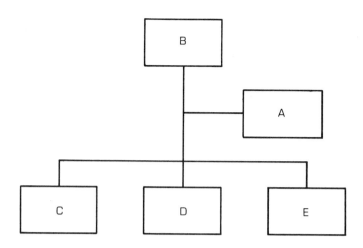

1–2. Prepare an organization chart, given the following information: Line manager A reports to line manager B, who reports to line manager C. Clerical assistant D reports (in a staff capacity) to line manager B.

1–3. Suppose that the implementation of a $1.2 million system leads to tangible cost savings of $200,000 and intangible cost savings of $100,000. What is the return on investment for this project? What is the payback period for this project?

1–4. Design an organization chart to show the following (include the names of the people and their job functions):

- Three managers—Fred Thomas, systems analysis and design; John Mayhew, programming; and Marvin Webster, operations—report to Marsha Williams, director of information systems.
- A user steering committee, consisting of users from marketing, manufacturing, finance and accounting, and personnel, is created to advise Fred Thomas.
- Two supervisors—Norma Whitehead, systems analysis, and Fred Best, systems design—report directly to Fred Thomas.

WORLD INTERIORS, INC.—CASE STUDY 1
DESIGNING ORGANIZATION CHARTS

Introduction

World Interiors, Inc. (WI), is a large direct mail wholesaler and retailer of home interior products. Through the 1970s the company consisted of a single home interiors retail store, selling carpeting, drapery, window blinds, and woven shades. In the early 1980s, following an agreement with a leading manufacturer of window coverings, the company expanded rapidly. Initially, the corporate offices were moved to another state. There the owners began a mail-order business to reach customers nationwide. Finding the mail-order business to be an instant success, the company expanded by opening additional retail outlets. Within five years, company sales grew from $150,000 to an amount in excess of $20 million.

Organization

Rena Logen started WI with her husband, Ray. Although Rena's title was president and Ray's was senior vice-president, in actuality, they both headed the company. At the board-of-director level, the wife and husband worked together to set major goals and policies for the company and to develop strategic plans. At the operational level, they supervised different areas of responsibility. Because of her interests and talents, Rena took charge of all administrative personnel selection, direct sales, and store management. Ray was responsible for supervising all sales and promotional plans, selecting new store locations, and deciding which products to continue selling and which to discontinue.

Besides the Logens, there were three additional top-level managers: Krintine Rinehart, vice-president of direct sales; Ralf Elders, vice-president of accounting; and Gary Meltomon, vice-president of retail store sales. Krintine and Gary reported directly to Ray, who reported directly to Rena. The eastern, western, midwestern, and southern store managers reported directly to Gary.

Direct Sales Department

Personnel reporting directly to Krintine Rinehart included Sheela Robbins, Telephone Sales; Patricia Alexander, Customer Complaints; and Morris Fisher, Customer Order Processing. Sheela supervised six sales associates, who worked the company's direct-line telephones. As the business grew, one or more telephone sales associates were required to be available twelve hours a day, six days a week. Patricia worked alone, handling all customer complaints. Recently, however, she had indicated that her tasks were becoming too difficult for one person to handle. Morris Fisher and his assistant were responsible for mailing promotional materials to customers and for clearing all customer orders once they were received.

Accounting Department

Personnel reporting in a line capacity to Ralf Elders included Mort Foster, accounting, and Marty Rhew, data processing. As chief accountant, Ralf was required to plan WI's cash needs as well as to account for all revenues and expenditures. Mort handled the day-to-day accounting of revenues and expenses. Marty's position was somewhat unique. Most of his time was needed to help coordinate the work of processing customer requests for information and customer orders. The balance of his time was used in keeping records of all customer accounts.

Retail Store Organization

Each retail store required, at a minimum, three employees: a store manager, an inside sales associate, and an installer. The inside sales associate sold ready-made products and took orders for later installation. The installer visited customers at their homes to take measurements, make estimates, and install ordered merchandise.

The Systems Consultant

Recently, WI had contracted for the services of a systems consultant named John Welby. Rena decided to have John serve as a staff adviser and report directly to Ralf Elders in accounting. This would give John the opportunity to study the financial and data processing procedures of the company. Rena also wanted John to report indirectly to her. She felt that an informal reporting relationship was required because Ralf's judgment was not always to be trusted, especially when it came to matters regarding data processing.

CASE ASSIGNMENT 1

Prepare a top-level organization chart of WI showing the titles of the various management positions and, where possible, the name of the person assigned to each position. Even though both Rena and Ray head the company, show Rena as the president. Remember to show all regional store managers.

CASE ASSIGNMENT 2

Prepare an organization chart of the direct sales department, showing the positions and, where possible, the name of the person assigned to each position. Show all sales associate positions.

CASE ASSIGNMENT 3

Prepare an organization chart of the accounting department, showing the positions and, where possible, the name of the person assigned to each position. Include in your chart John Welby's formal and informal reporting relationship.

REFERENCES

1. F. P. Brooks, Jr., *The Mythical Man-Month* (Reading, Mass: Addison-Wesley Publishing Co., 1982).
2. E. Yourdon, *Techniques of Program Structure and Design* (Englewood Cliffs, N.J.: Prentice-Hall, 1975).
3. P. C. Semprevivo, *Systems Analysis*, 2nd ed. (Chicago: Science Research Associates, Inc., 1982), pp. 32–42.
4. A. L. Eliason, *Business Information Systems—Technology, Applications, Management* (Chicago: Science Research Associates, Inc., 1980), pp. 346–348.

Business Systems

INTRODUCTION

SUPPOSE that you are asked to design a new system for a company. As suggested in chapter 1, such an undertaking would first involve devising a logical plan, followed by a detailed systems design. "But, wait a minute," you might reflect. "Before I can design a system, I had better understand what a system is and what a system means in relation to a business. I should also know how to diagram a business system so that others can understand what I am talking about."

This chapter introduces the subject of systems thinking and business systems. When you complete this chapter, you should be able to

- describe what a system is and what is meant by systems thinking;
- describe major subsystems in a business;
- show how to diagram a business system;
- explain what is meant by a business computer system; and
- describe the similarities between a business system, a business computer system, and a computer system.

WHAT IS A SYSTEM?

The term *system* can be interpreted in many ways—as it should. To some, a system is an organized, interrelated set of ideas or constructs. This particular view characterizes an *abstract system*: an intangible but organized set of ideas. To others, a system is an organized, interrelated set of items, such as companies, parts, machines, or procedures. This view corresponds to what is called a *physical system*: a tangible, orga-

33

nized set of items, such as parts or machines. A major difference between these two types of systems is that abstract systems evolve to provide us with a philosophy of how to think about things; physical systems, especially those conceived by people, are purposeful. They are designed to accomplish specific objectives.

Both abstract and physical systems are important to the systems analyst and the systems designer. Abstract systems provide analysts and designers with a way of thinking about problems and suggest ways to solve problems. After working through the exercises contained in this book, for example, you may, ideally, begin to subscribe to a "systems view" in thinking about how to analyze and solve problems. Physical systems, in contrast, provide tangible, useful items. Computer programs, once implemented, represent parts of a larger physical system. Each program is unique to the system; each is written to accomplish a specific objective, such as calculating return on investment or printing the company payroll.

General Model of a System

All systems have eight basic characteristics: goals and purpose, inputs, outputs, boundaries, environment, components, interrelations, and constraints. Because this list is too large to deal with all at once, let's begin with a simple systems model.

Figure 2–1 illustrates such a model. The system has *boundaries*, meaning that the size of the system is limited and that these limits can be defined. Surrounding the system is the *environment*. The environment can influence the system in a variety of ways such as making it more or less difficult for the system to perform. *Inputs* to a system are typically resources, such as capital, labor, information, energy, materials, and machinery. *Outputs* follow from a system. They are usually repre-

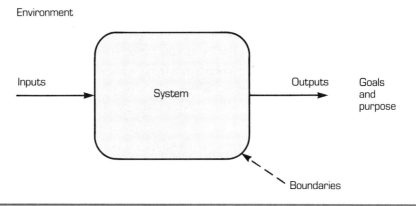

FIGURE 2–1 Simplified systems model

sented as goods or services. A deluxe hamburger, a farm tractor, or a registered student represent different outputs from different systems. Finally, a system is developed for a reason. This reason is expressed as *goals and purpose.* The longer-term purpose of a system might be to open a chain of deluxe hamburger establishments, while a shorter-range goal might be to produce a hamburger that customers are willing to call deluxe.

Figure 2–2 illustrates a more complex systems model. Besides the five characteristics considered in the simplified model, the system consists of components and interfaces. The *components* of a system serve to transform inputs, such as labor and materials, into outputs, such as assembled tractor engines or filled customer orders. The linking together of the components of a system is accomplished by *interrelations*. Interrelations describe how resources pass from one system component to another.

One way to view components and interrelations is to think of components as *processing stations,* and interrelations as *processing channels.* Channels, in this instance, pass partially assembled products, such as information, money, and services, from one processing station to another. **Figure 2–3** illustrates how processing stations and channels work within a system. As shown, customer orders, the inputs to a system, are received by station 1, which is called order processing. This station checks each order for completeness and separates first-time customers from repeat customers (e.g., those customers who have done business

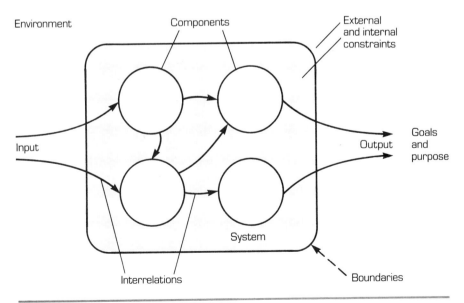

FIGURE 2–2 More complex systems model

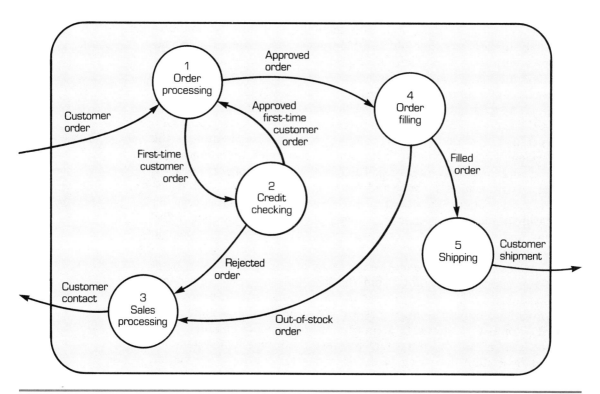

FIGURE 2–3 Order processing

with the company in the past). As indicated next, all first-time customer orders are then sent to station 2, credit checking. At this station, the credit rating of each new customer is either approved or rejected. All approved orders are returned to station 1, where they are combined with repeat-customer orders. Then a split occurs. Each rejected order is sent to the salesperson who wrote the order (station 3, sales processing). Each approved order is sent to station 4, order filling. After an order is filled, it is delivered to shipping (station 5). Shipping, in turn, packs and ships the order to the customer. If an order cannot be filled because of an out-of-stock condition, it is held awaiting the delivery of new stock. Copies of all out-of-stock orders are sent from order filling to the salespersons who wrote the orders (station 3).

As this example demonstrates, processing stations perform a specific function, such as checking credit ratings or filling customer orders. They can accept as inputs products and information; they can produce as outputs products and information. Processing channels pass information (e.g., rejected orders or out-of-stock orders) from one processing station to another, or products (e.g., goods pulled from inventory) from one station to another. They show how one part of an organization can communicate with another. For instance, credit checking, an accounting activity, and order processing, a warehousing activity, both communicate with sales processing, a marketing activity. This communication between different processing stations within an organization is studied carefully by the systems analysis team. By tracing the flow of information between stations, the team can determine how the work of a business is organized and interrelated.

Constraints, the last of the eight characteristics of a system, represent environmental, social, technical, and behavioral factors that typically limit system performance. A regulation imposed on a business by a federal agency is an example of an external or environmental constraint. Rules and procedures, policies, and decisions regarding how resources are to be obtained and used in a company are examples of internal constraints. Returning to the example in Figure 2-3, let's suppose that the processing of new customer orders is limited (that is, constrained) to ten orders per day. Why? Because resources—namely, people assigned to credit checking—may limit the total number of new orders that can be processed. This limitation is an internal constraint—it is imposed on a component within the system.

General Model of a Business System

As discussed in chapter 1, business firms follow a hierarchical structure in partitioning and compartmentalizing their main functions. They use organization charts to depict visually each of these functions and to show how the various functions are interrelated. The main benefits of an organization chart are that it shows the top functional levels, areas of decision making, and formal communication channels.

From a systems perspective, an organization chart is helpful but insufficient in explaining system characteristics. The general structure of a business system must feature different levels of management, each responsible for different types of activities, information processing, and decision making. Any model must also show the relationships between higher-level processing stations and how these stations communicate, through processing channels, with lower-level processing stations.

ANTHONY'S MODEL

Anthony's[1] view of the general structure of a business clarifies some of the differences between various levels of management. His model features three types of planning and control, with a level of management corresponding to each:

1. *Strategic planning*, which determines the longer-term purpose and goals of the organization, the resources needed to obtain those purposes and goals, and the policies and guidelines needed to govern the acquisition, use, and disposition of those resources;

2. Management control, which monitors the use of resources (e.g., people, money, and equipment) to ensure that use is efficient, effective, and consistent with achieving the medium-term goals of the organization; and

3. Operations control, which monitors the daily tasks of various work groups in an organization to ensure that shorter-term goals are achieved.

Anthony's model can be represented as a three-level pyramid (see **Figure 2–4**). At the top, managers are responsible for longer-term strategic planning. This top level typically involves the activities undertaken by the top management in an organization. The middle level characterizes the province of middle management. This middle level serves primarily as a control level—it monitors operational activities and takes corrective action when resource use seems inappropriate. The lower level represents lower-level management. It, too, is a control level. Managers at this level are responsible for monitoring the daily tasks assigned to operational work groups and for making decisions when group performance appears improper.

BLUMENTHAL'S MODEL

Blumenthal's[2] view of the general structure of a business system features the concepts of functional units and management control centers. *Functional units*, such as small work groups, form the basis of the organization. Each unit receives information about levels of resources, makes decisions based on desired outcomes, and takes action to change the rate of flow of resources.

Figure 2–5 indicates that functional units consist of decision centers and activity centers. *Decision centers* receive internal information, infor-

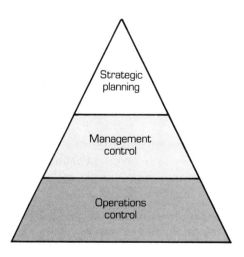

FIGURE 2–4 Three levels of management

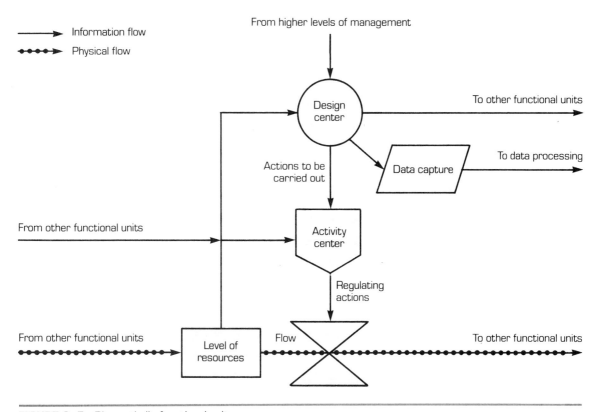

FIGURE 2–5 Blumenthal's functional unit

mation both from higher levels of management and from other functional units. Based on information, decision centers instruct activity centers regarding actions (decisions, activities, tasks) to be carried out. *Activity centers* represent the activating units of an organization. These centers regulate the physical flow of resources between functional units. For example, a warehouse activity center might receive instructions to fill one hundred customer orders. Once filled, the orders are to be placed on pallets and moved to the next functional unit, such as the unit responsible for packing and shipping the picked goods.

Besides instructing activity centers, decision centers capture information and send it either to other functional units or to data processing. Continuing the previous example, suppose that in filling the one hundred orders, six people are required and each takes eight hours to complete his or her assigned activity. The information to be collected—hours worked by employees—would be held by the decision center until hourly figures were required by data processing. Following the transfer of that information, another functional unit, such as the payroll processing unit, would transform the hours worked into employee pay.

Management control centers supervise work assigned to functional

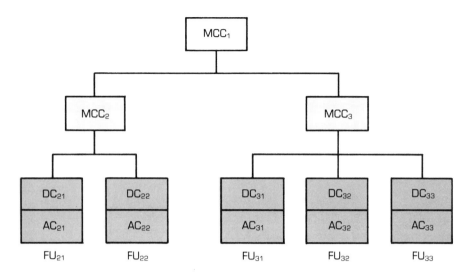

Key:
MCC Management control center
FU   Functional unit
DC   Decision center
AC   Activity center

FIGURE 2-6 Blumenthal's managerial hierarchy

units. As shown in **Figure 2–6,** each management control center acts as a decision center for one or more functional units; each management control center (with the exception of the very top) is supervised by a higher-level management control center. Viewed in this way, any organization consists of a fixed number of functional units linked together by management control centers. More important, information links the various units of an organization. Information produced by one functional unit is passed to other units, or to one or more management control centers. Information is captured for data processing at both the lower and upper levels of an organization.

NEUMANN AND HADASS'S MODEL

Neumann and Hadass[3] combine some of both Anthony's and Blumenthal's thinking in describing the physical structure of information systems within organizations. As shown in **Figure 2–7,** an organization features three main types of information processing systems:

1. Decision support systems. These are largely unstructured systems that examine events external to the organization and compare these events with internal activities. These systems support strategic planning and management control activities.

2. Structured decision systems. These are partially structured systems that examine internal planned events and compare these events with actual events. These systems support management control and operations control activities.

3. Transaction processing systems. These are highly structured systems that process routine, day-to-day internal documents and materials. These systems support operations control activities.

The Neumann and Hadass model helps point out the difference between administrative data processing (ADP) and a management information system (MIS). As the figure illustrates, *administrative data processing* consists primarily of highly structured, transaction-based systems, mixed with partially structured, structured decision systems. *Management information systems*, in contrast, include largely unstructured or partially structured systems. MIS support is directed at middle to upper levels of management.

THE BUSINESS CYCLE MODEL

The business cycle model[4,5] supports the Neumann and Hadass model. It describes two levels of information systems in an organization: transaction-based systems and management-based systems. A *transaction-based system* is organized to process business transactions and provide a written record of processing. Within any business, there exist four main transaction-based systems, with each system containing more than one administrative processing function. These four are the revenue system, the expenditure system, the conversion system, and the treasury (cash management) system (see **Figure 2–8**).

The *revenue system* processes financial transactions associated with the sale of goods to customers and the billing of customers. A sale consists of distributing goods or services to customers in exchange for the promise of future payment. The billing function asks customers to pay for delivered goods and services and involves sending customer invoices and monthly statements.

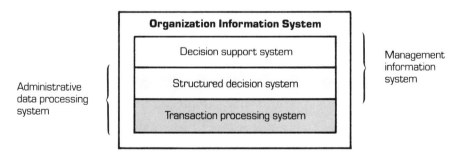

FIGURE 2–7 Neumann and Hadass's information systems model

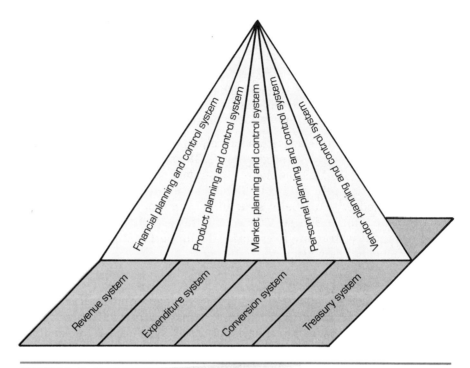

FIGURE 2-8 Business cycle information systems model

The *expenditure system* processes financial transactions associated with the purchasing of property, labor, materials, and services, along with the making of payments. Purchasing involves the obligation to pay for resources obtained from vendors (their goods and services) and employees (their time). Making payments involves the sending of monies (company checks) to vendors and employees.

The *conversion system* processes material, labor, and financial transactions associated with transforming inputs, such as lumber, into outputs, such as furniture. It describes how resources are held, used, and transformed by the various processing stations in a system. It explains how resource levels change within a station and between stations.

The *treasury system* processes financial transactions associated with capital funds received from investors and creditors (such as banks). It determines the cash requirements of a business and records all temporary investments. It shows how funds are to be repaid to investors and creditors.

The second level of the business cycle model consists of management-based planning and control information systems. These systems are designed to draw information from transaction-based information systems. This is not the only difference, however. Unlike other systems, *management-based systems* do not process business transactions. Instead, they are designed to process management *queries:* questions posed by the manager about factual matters of the business. As such, management-based systems are designed to summarize, hold, compare, and translate information. They are built to provide managers with the information they need to make decisions.

Although there are several types of management-based systems, five, in particular, are used in most companies (see Figure 2–8). The *financial planning and control system* reports on the financial soundness of a business. This system is designed to compare planned and actual business revenues and expenditures. The *product planning and control system* evaluates the construction and sale of products offered by a business. This system is designed to compare product forecasts to actual product turnover and sales. The *market planning and control system* evaluates the sale of goods and services to the customers of a business. Besides comparing sales forecasts with actual customer sales, this system examines how the marketplace is affected by economic conditions and trends. The *personnel planning and control system* reports on the labor force employed by a business. This system is designed to compare wage and productivity rates; it is developed to help a business understand what types of people it will need to employ to achieve its objectives. The *vendor planning and control system* evaluates the sources of supply available to a business. This system compares vendor prices and delivery times; it helps explain how supplies are affected by economic conditions and trends.

WHAT IS A BUSINESS COMPUTER SYSTEM?

Before we consider different types of information system designs, we should understand what is meant by a business computer system and how such systems are related to a business enterprise.

A *business computer system* consists of an organized, interrelated set of computer applications, which are designed to process a particular type of business transaction or to prepare a particular type of management information. Thus some business computer systems are transaction based: they are constructed to process the paperwork associated with a particular type of business transaction. Other systems are management based: they report on business conditions and provide detailed and summary information required by managers.

The relationship between a business computer system and its applications can best be understood with an example. **Figure 2–9** uses a systems model to explain the processing of customer bills. The uppermost input to the billing system is a record of what was shipped to the customer. This information is used to prepare customer bills—that is, the amount owed the company by the customer. Preparing customer bills leads to two types of output: the formal business document known as

44   Chapter 2: Business Systems

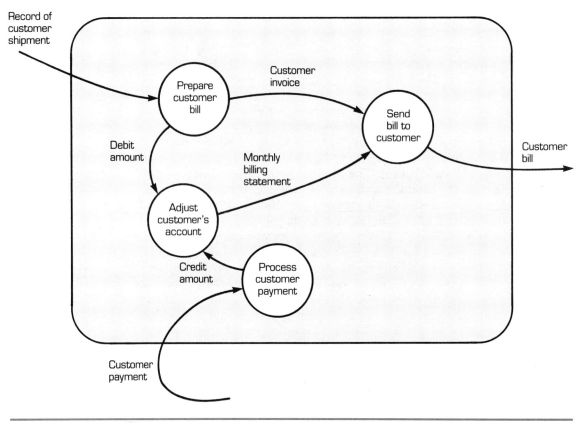

FIGURE 2–9   Billing system

the *customer invoice* and the debit amount shown on the invoice to be added to the customer's account.

The lowermost input to this system is a customer payment. Once received, it is processed by the business, which, for our purposes, leads to adding the payment total (credit amount) to the customer's account. After an appropriate period of time (such as thirty days after the mailing of the invoice), the debit amount is subtracted from the credit amount. If the balance is greater than zero, a second formal billing statement is prepared. This statement is called the *monthly billing statement*.

A business computer system results when processing that was previously done manually is now performed by the computer, and when the products of computer processing are linked together. As **Figure 2–10** shows, a computer-based billing system consists of three components or three *business computer applications*.

"Prepare customer invoice," for example, might also be called the *customer invoice computer application*. The purpose of this application is to transform customer shipments into customer charges, print the cus-

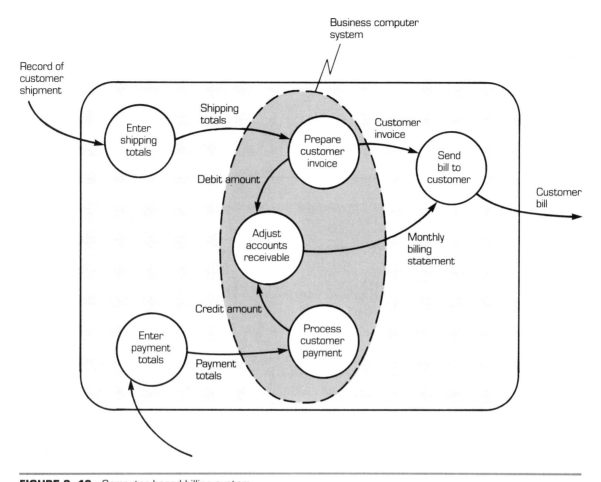

FIGURE 2–10 Computer-based billing system

tomer invoice, and transmit debit amount totals to adjust the customer's account.

"Process customer payment" could also be called the *customer payment computer application.* This application processes payments received from customers and transmits credit amount totals to adjust the customer's account.

"Adjust accounts receivable" is perhaps better known as the *accounts receivable computer application.* This application adds debit amounts and subtracts credit totals, maintains a record of all customer accounts, and prints monthly statements whenever account balances remain above zero.

Figure 2–10 illustrates two other features of a business computer system: a business computer system exists within a larger system, and it often adds new steps to a system. Consider the first of these two features.

The billing system, considered as a whole, consists of processing performed by the computer and processing done manually. Next, consider the difference between a manual and a computer-based system. If we compare Figure 2–9 with Figure 2–10, we can see that two separate steps are needed to enter data into processing and that both steps lie outside computer processing. These manual steps are *data entry activities*. New steps in processing must be studied carefully by the systems analysis team, primarily because they often limit system performance. Data entry activities, for example, may limit the number of transactions that can be successfully handled by a business.

The Structure of a Business Computer System

We can learn more about the structure of a business computer system by paying closer attention to its components—namely, business computer applications. By definition, a business computer application is an organized set of processing procedures, organized as computer programs and designed to handle a specific type of processing function. **Figure 2–11** shows the types of programs used by the customer invoice computer application. One program is designed to get valid shipping totals. It is written to receive data and to verify that these data are correct. Another program calculates customer charges and is designed to convert shipping totals into dollar charges. If ten units are shipped at a cost of $5 per unit, the program computes a dollar charge of $50. Still another program is written to print customer invoices. This program produces the formal business document known as the customer invoice. A final program, to debit accounts receivable, is written to alter a customer's receivable account. The account is changed to show the current dollar amount now due from the customer.

The arrangement of processing procedures in a business computer application typically follows an input-processing-output (IPO) sequence.

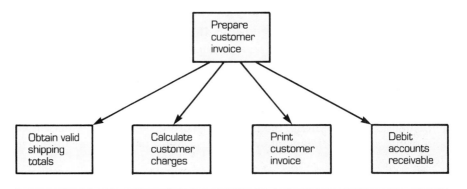

FIGURE 2–11 Application programs

Later, in chapters 3, 9, and 12, we will learn how to construct program structure charts and IPO diagrams. For now, it is important to understand that each business computer application can be viewed as a subsystem within a larger system, and that each application consists of one or more computer programs.

Interrelations of a Business Computer System

Another way to learn more about a business computer system is to study its interrelations—the ways in which business computer applications are linked together and with other manual activities. Once again, let us examine the work flow of customer invoicing. As shown in **Figure 2–12,** three interrelations exist—two that tie the processing of invoices with manual activities and one that ties processing to the accounts receivable application.

The study of interrelations combined with components required in processing points out three principal functions of information systems design: data transmission, data transformation, and data storage. *Data*

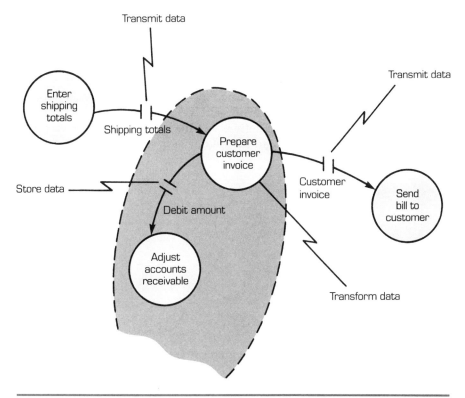

FIGURE 2–12 Business system interrelations

transmission brings data to and from a business computer system. Figure 2–12 shows that shipping totals must be transmitted to invoice processing and that customer invoices must be transmitted to an outside station to be mailed. *Data transformation* transforms data from one form to another and is accomplished by system components. In the example, shipping totals are transformed into customer charges. *Data storage* is the holding of data, generally for a short period of time. The link between invoice processing and accounts receivable processing is accomplished by the storage of debit totals, for instance. These totals are held for use by the accounts receivable application. Data storage acts as a buffer: it permits data to be built up by one part of a system and to be released for use by another part.

The System Trinity

In the study of business computer systems, the concept of the system trinity must be kept in mind. As characterized by Churchman,[6] the system trinity consists of three parties who must work cooperatively: the client or user, the decision maker, and the designer (see **Figure 2–13**). Each of these parties tends to view the system differently. The *client*, for example, is interested in system performance. He or she will be interested in how a system works and how well the system can satisfy the user's needs. The *decision maker* is interested in the resources required by the system. He or she will want to know system costs as well as which resources (e.g., people, materials) will be consumed by its operation. The *designer* is interested in how the system will be developed and maintained. He or she will want to satisfy the client's design requirements and personal design objectives, while keeping resource requirements within the range specified by the decision maker.

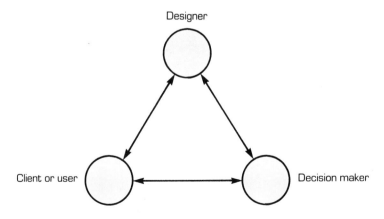

FIGURE 2–13 The system trinity

Even though the system trinity seems clear enough, in some instances it is difficult to separate the individuals who make up the three corners of the trinity. For example, suppose you decide to bake and later eat a cake. You would be the designer and the client in that situation. Since you decided to bake the cake, thus deciding to use your personal resources, you would also be the decision maker.

Suppose next that you are asked to define the client, the decision maker, and the designer for a college or university bookstore. You are told that faculty committees determine which books are to be ordered; however, the bookstore manager is responsible for ordering all books, stocking the bookstore shelves, and making payments to publishers.

What conclusion did you reach? If you determined that the client is the student, the decision maker is the faculty committee, and the designer is the bookstore manager, then you understand the concept of the system trinity.

You may wonder why such an understanding is important to systems analysis and design. Consider the following: In logically defining and designing a new system, the analysis and design teams must determine the needs, preferences, motivations, and perceptions of the client and the decision maker. In this way, the features important to all parties (including the designer's) can be incorporated into the logical and the technical design specifications. (For a more detailed discussion of the roles of the three parties see note 7.)

WHAT IS A COMPUTER SYSTEM?

A computer system is much like other systems discussed thus far. It comprises the following five components: computer equipment, which is often called *hardware*; computer programs to operate the computer, which are called *system software*; other programs written for a business computer system, called *application software*; *procedures*; and *data processing personnel*. Of these five, system and application software contain the instructions needed to make the computer operational. Procedures are required to explain how data are to be prepared for processing, how processing steps are to be monitored, and how the results of processing are to be distributed. Data processing personnel are needed to keep hardware, software, and procedures in good working order.

Computer Hardware

The hardware of a computer performs several basic functions: input, output, data storage, transmission, and processing of data. These functions form a hierarchy (see **Figure 2–14**). At the bottom of the hierarchy are input and output functions (I/O functions). Before data can be processed, it usually must be read, with the reading of data accomplished by some type of *input device*. Common types of input devices are display termi-

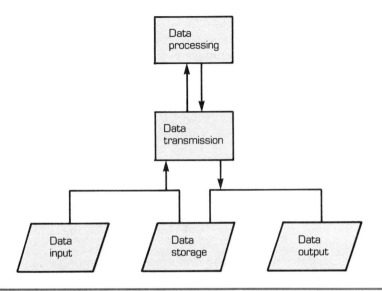

FIGURE 2–14 Functions of computer hardware

nals, hard-copy terminals, optical scanners, and card readers. Likewise, after data are processed, they usually must be written in a form that can be readily understood by people, with the writing performed by some type of *output device*. Common types of output devices are line printers and display terminals.

Data storage involves the capability to store data and computer programs in *machine-readable form*—a form that can be read directly by the computer. The storing of data is accomplished by some type of *data storage device*, such as a magnetic disk or magnetic tape unit. These units act as storage bins, serving as both input and output to processing. For example, they hold data and programs until they are needed in processing; and they accept data and programs following processing.

Data transmission controls the flow of data to and from I/O and data storage devices. Imagine 300 or more computer terminals and twenty or more data storage devices trying to send messages to a processing unit. Coordinating these messages so that none is lost or mixed requires some type of *data transmission device*. On large computers, such a device might be called a communications front-end, or a network processor.

Data processing transforms data into a desired form. Because data processing represents the central part of the hierarchy, this function is accomplished by a *central processing unit* (CPU). The CPU of a computer comprises three parts: an arithmetic/logic unit, an internal storage unit, and a control unit. The *arithmetic/logic unit* manipulates data until the transformation into a desired form is complete. The *internal storage unit* (commonly called *memory*) is similar to a data storage device: it stores data and programs transferred to it by an I/O device, a data storage device, or the arithmetic/logic unit. The *control unit* is similar to a data transmission device; it directs the flow of data and programs to and from the arithmetic/logic unit and the internal storage unit.

Figure 2–15 shows how a microcomputer incorporates the functions of data input and output, data storage, and data transmission and processing. The computer keyboard is used to enter data into processing; the line printer is used to show processed results. In addition, the display screen shows both data keyed to processing (input) and processed results (output). The particular data storage medium illustrated is a floppy disk, which is inserted into a magnetic disk data storage device. A disk medium that is easy to bend is called a *floppy disk*; a disk medium that is not easy to bend is called a *hard disk*. Data transmission

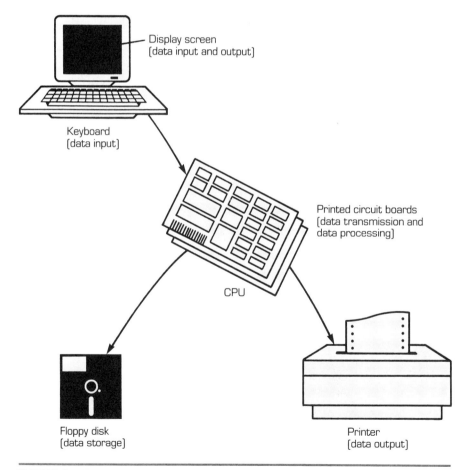

FIGURE 2–15 A microcomputer system incorporating the functions of data input and output, storage, and transmission and processing

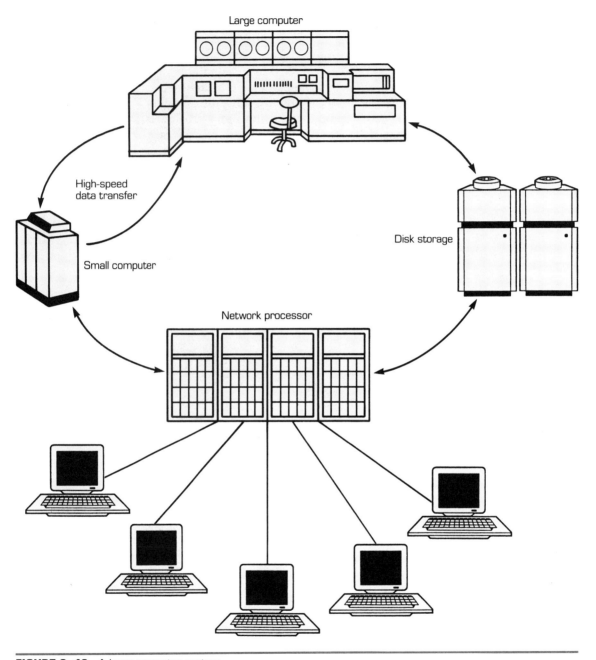

FIGURE 2–16 A large computer system

and processing are accomplished through a variety of silicon chips, seated on printed circuit boards. These chips provide timing, control use of memory, control transmission of data to and from processing, and coordinate input and output to the CPU.

Figure 2–16 shows a larger computer system, with the same functional parts. In this instance, numerous computer terminals transmit input and receive output. In addition, several magnetic disk units are connected to the system. Data transmission, as shown, is coordinated by a network processor. With many large computers, a network processor is a small computer dedicated to controlling the flow of data to and from the larger computer that it serves. Data processing is accomplished by the large computer. Once processed results are available, they can be released to the network processor for distribution.

Perhaps an easier way to understand the functions of hardware is through a practical illustration. **Figure 2–17** provides another view of the

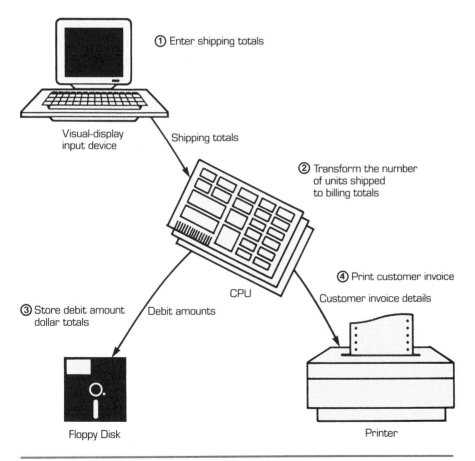

FIGURE 2–17 A microcomputer used to process customer invoices

work flow associated with customer invoicing. As indicated, the entry of shipping totals is accomplished by a visual display input device. After totals are entered, the CPU transforms the number of units shipped into billing totals (e.g., the dollar amount that is due). At that point, the CPU must perform two other tasks: it must transmit debit amount dollar totals to a floppy disk, where they are to be stored, and transmit customer invoice details to the computer printer.

System Software

System software is often as important as the CPU of a computer. This type of software permits the hardware to work optimally. An *operating system* is an example of system software. Designed most often by the hardware manufacturer, the operating system allocates the various resources of the computer (e.g., disk storage space, internal memory space, and so forth) so that application software can be processed efficiently.

An operating system consists of two types of computer programs: control programs and processing programs. *Control programs* supervise the running of application software, which includes routing data from various input/output devices to memory, allocating processing time, and scheduling processing tasks. *Processing programs* consist of language processors and service or utility programs. A COBOL compiler is an example of a language processor. The compiler is designed to transform statements written in the COBOL language into machine-readable processing instructions.

Application Software

Application software consists of computer programs designed to perform common processing tasks, such as processing an employee payroll or customer invoices. The design of this type of software is of immediate interest to the systems analyst and designer, for in writing application software, hardware and system software specifics must be considered. A systems designer may seek a printer speed of 600 lines per minute, only to be constrained by a printer with a maximum speed of 300 lines per minute. Likewise, a designer may seek to write all computer programs in the "C" language, only to discover that a "C" language processor is not available. Or a systems analyst may want to enter data into processing from a remote plant location only to discover that the existing computer hardware does not have the data transmission capabilities to support such a design requirement.

Procedures

Another task of importance to the systems analyst or designer is the writing of procedures. *Operational procedures* indicate what is to be done, along with by whom, when, and from what location it is to be done. A line from a procedure might read:

1. Select the tape, serial #1003.
2. Attach a read-only ring.
3. Mount on tape unit 006.

Other information included in operational procedures answers the questions of by whom, when, and from what location the task is to be performed. The procedure for the example shown before should tell us that the instruction is to be performed by a computer operator, completed before the customer invoice register is printed, and done at central computing.

Data Processing Personnel

Staff members are required to perform several tasks in conjunction with the computer. Some of these tasks concern the analysis of a system and the subsequent design of an improved system. Other tasks include systems maintenance and computer programming. Still others deal with preparing data for processing, scheduling jobs for processing, and distributing processed results.

Within a company there are a variety of ways to organize the data processing function. Some companies separate systems analysis and design from programming and from computer operations. **Figure 2–18** illustrates this method of organization. The manager of systems analysis and design is responsible for all systems development work, with the

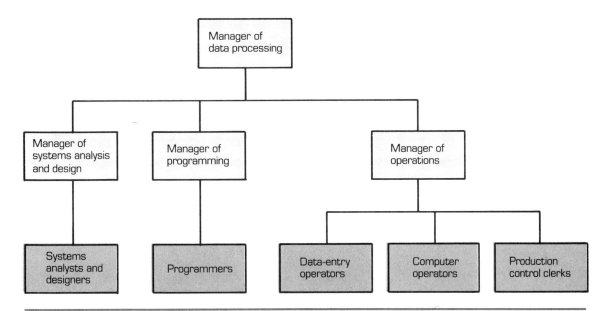

FIGURE 2–18 Organization of a medium-size data processing function

exception of programming. Responsibility for the writing of all application software is assigned to the manager of programming. The manager of operations is responsible for three types of activities: data entry, computer operations, and production control. In this environment, data processing personnel are needed to prepare data for processing (data entry activities), to operate the computer by following operating procedures (computer operations activities), and to schedule computer use and distribute the results of processing (production control activities).

The Mansfield, Inc., Case Study

Aaron Gillette, president of Mansfield, Inc., is well aware of David Orring's cry for a new billing system. "We want the billing system to be as trouble-free as the payroll system," remarked Aaron. "This means that we will not be able to deliver the new system yesterday, as David might wish, for when we do deliver, we want it to be a quality system."

Project Steering Committee

In consultation with the president, Ken Westerling formed a project steering committee to help frame the logical requirements for Mansfield's new billing system. The members of the team consisted of:

- Aaron Gillette, president
- David Orring, vice-president for finance and accounting
- Ken Westerling, vice-president for information systems
- Roger Bates, lead systems analyst
- John Havensek, billing supervisor
- Shawn Fortuna, program manager
- Marvin Wilson, budget officer

Ken explains: "I have staffed the committee in such a way as to represent the three parties of the system trinity" (see **Figure 2–19**). When Roger Bates asked for clarification, Ken added: "We need Aaron or one of his representatives to help us come up with the resources needed to complete the project. I didn't want to ask Aaron to set aside thousands of dollars unless he understands why we need such a large sum of money."

Project Organization

Roger Bates moved quickly in forming a systems analysis team. Shawn Fortuna agreed to the transfer of two members of her staff, Carolyn Liddy, systems specialist, and Neil Mann, programmer/analyst, to help Roger prepare the logical specification. John Havensek was also asked to join the team on a part-time basis. "John should be able to provide eight or so hours each week to help with the design," David Orring commented. "The only period in which he will not be of much help will be in six months, when we enter our peak season. However, I would hope that by then the new system would be up and running."

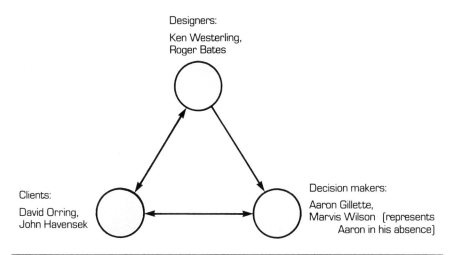

FIGURE 2-19 System trinity for the new billing system

SUMMARY

By now, you should have a better understanding of what constitutes a system and what is meant by a systems view. You should also have an appreciation of the major subsystems in a business. At lower levels, transaction-based systems are designed to process matters concerning revenues, expenditures, conversion, and treasury. At upper levels, strategic planning and the planning and control of lower-level activities become more important. Systems that respond to management queries are needed to provide managers with information needed for decision making.

Even though multiple subsystems exist within any organization, each subsystem can be examined through its basic characteristics. The eight system characteristics are: goals and purpose, input, output, boundaries, environment, components, interfaces, and constraints.

People who are new to systems analysis often overlook the fact that the computer is only part of a larger system. That business computer systems are but a part of larger business systems is important to remember.

There are different levels of planning and control within an organization. Anthony separates strategic planning from management control and operations control. Blumenthal separates management control systems from functional units.

Different types of information processing systems are needed for different types of planning and control. Decision support systems, structured decision systems, and transaction processing systems provide support to different levels of management; transaction- and management-based systems must be designed differently in order to handle different organizational processing requirements.

A business computer system consists of an organized and interrelated set of computer applications that are designed to process a particular type of business transaction, or a particular type of management information. Again, such a system is part of a larger system.

Through systems analysis, the analysis team decides which part of a business system will be improved with the use of the computer. Following that determination, the team must determine which computer applications are needed to realize those improvements. In so doing, the team must specify data transmission, transformation, and storage requirements.

The concept of the system trinity recognizes that the systems analyst and designer constitute just one of the parties important to the systems development process. The other two parties are the client, (the person who will use the completed system) and the decision maker (the person who determines the resources that will be spent on the development of the new system).

A computer system is configured to process business computer applications. Such a system employs I/O and data transmission devices to transmit data to and from a business computer application, a CPU to transform data (as directed by the operating system and application software), and data storage devices to hold data preceding and following processing.

The systems analyst must determine data transmission, transformation, and storage needs. By knowing such needs, the analyst can specify computer system requirements. The speed of I/O devices, for example, is determined in part by data transmission needs; the speed of the CPU is determined in part by transformation needs; and the size of the data storage device is determined in part by storage needs.

A computer system is more than just hardware and software. In order for a computer to be used properly, procedures are needed to indicate what is to be done, as well as by whom, when, and from what location work is to be done. In larger work environments, various people are required to deal with the data processing function. Some are responsible for designing and programming systems; others are responsible for preparing data for processing, for operating the computer, and for scheduling and controlling processing.

REVIEW QUESTIONS

2–1. How does an abstract system differ from a physical system?

2–2. Name the basic characteristics of a system.

2–3. According to Anthony, what are the different types of planning and control?

2–4. According to Blumenthal, what is the difference between a decision center and an activity center?

2–5. According to Neumann and Hadass, what are the main types of information processing systems?

2–6. What is the difference between a transaction-based system and a management-based system?

2–7. What is a business computer system?

2–8. How does a business computer application differ from a business computer system?

2–9. Name the principal features of information systems design, as pointed out by a study of the interrelations of a business computer system.

2–10. What are the main components of a computer system?

2–11. What components make up the CPU of the computer?

2–12. How does systems software differ from application software?

EXERCISES

2–1. What is wrong with the following system?

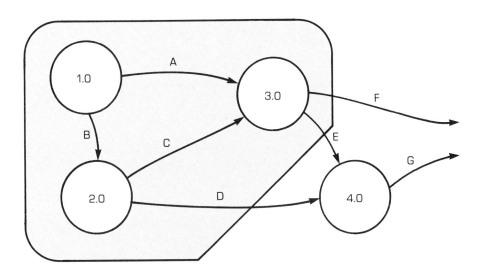

2–2. Suppose we know the following about processing deluxe hamburgers (the system diagram shown on page 60 describes the process):

- When a customer order is received, it is placed on a spindle.
- The customer order remains on the spindle until it is removed by the chef. The chef instructs the kitchen helper to prepare the customer's plate while the hamburger is being broiled.
- Once the hamburger is cooked, it is set on the customer's plate. The plate is set on the counter.
- The plate remains on the counter until the waiter picks it up to serve the customer.

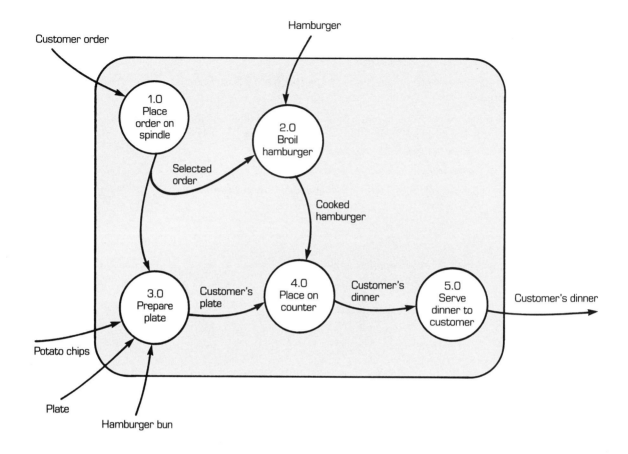

Answer the following:
1. What are the purpose and goals of this system?
2. What are the inputs and outputs to the system?
3. List the components of the system.
4. List the internal interrelations.
5. Describe the system boundary and the environment.
6. Describe possible internal constraints.
7. Describe possible external constraints.

2–3. Complete the general system model shown on page 61. You are required to show that a business computer system is part of a larger business system and made up of an interrelated set of computer applications.

The labels needed to complete the diagram are as follows: inputs, outputs, business computer system, business system, distribute outputs (use two times), computer application (use three times), enter data, components, environment, and interrelations.

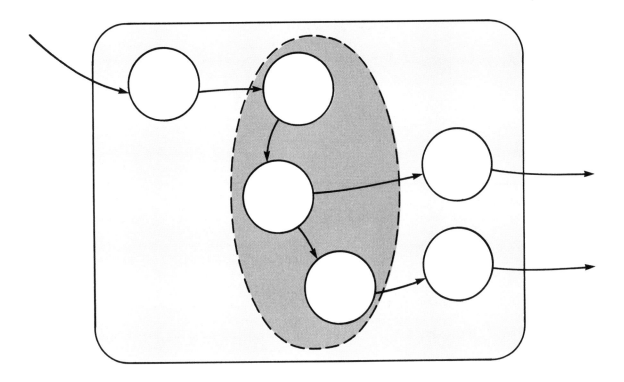

2–4. Suppose that we know the following about the procedures followed by a psychologist and a receptionist. Each time a client arrives at the psychologist's main office, the receptionist pulls the client's folder from the files and asks the client to take a seat. When the psychologist is ready to talk with the client, the client is asked to step into the psychologist's private office; the receptionist hands the psychologist the client's folder.

Before entering into a discussion with the client, the psychologist takes two sheets of specially marked paper. The date and time of day are entered on each of the two sheets. Following this, the psychologist begins the session with the client, which consists of talking, listening, and taking notes on the specially marked paper.

At the end of the session, the psychologist asks the client to leave. Once the room is empty, the psychologist makes final notes and places them in the client's folder. When the next client arrives, the receptionist hands the psychologist that client's folder, and picks up the previous client's folder. (The receptionist is responsible for placing the previous client's folder back in the files.)

Diagram the main components and interrelations of this system. Include the tasks required of the receptionist and of the psychologist; show input and output to the system.

2–5. Prepare system trinity diagrams for the following situations. In each case identify the client, the decision maker, and the designer.

- A college or university library, where a new acquisitions manager asks faculty members to specify which new books to acquire.
- A police department, where voters decide on the city budget and the police department's budget is part of the larger city budget.
- A toy store, where the store is family owned and operated.

WORLD INTERIORS, INC.—CASE STUDY 2
THE SYSTEM TRINITY

Introduction

When John Welby took on the job of system consultant, he realized that he would be the man in the middle. "It will be difficult gaining the trust of either Ralf Elders or Marty Rhew if they believe that everything they say will be passed upward to Rena," he muttered. Then, too, he was not clear about how communications were passed between Rena and Ray. If he confided in Rena on matters for which Ray was responsible, would she keep Ray informed of their discussions? If so, the findings of his study would probably be translated from one end of the company to the other.

Marty would probably represent one of John's special problems. Rena and Ray advised him that Marty was not capable when it came to the design of a new system. He could barely operate and maintain WI's existing computer system.

System Planning

As far as John could determine, WI had no strategic plan, much less a clear idea on what was needed in the way of management control or operations control. "We just grew too fast," remarked Rena. "We realize we need a well-designed system of controls for both our mail-order and our in-store sales." Ray agreed: "When we were small, we could do everything by hand. At about the $100,000-per-month level we ran into trouble. We had to let go and have others help us. In looking back, we were not very good at letting go. Even today, we are in trouble. We desperately need a new customer mail-order and payment system."

Customer Mail-Order and Payment System

WI's existing mail-order and payment system had been revised three years earlier by an outside software company. After a two-month study of WI's operations, the software company representatives convinced Rena to acquire a small personal computer, with the software needed to process all incoming mail orders and all payments to WI suppliers.

"The installed system never worked as expected," Ray commented. "We were told that the new system would print out all orders with a minimum of effort by members of our staff. As it stands today, Marty spends 90 percent of his time trying to keep the system operational. Each time he corrects one problem, another one pops up. It is no wonder that Morris Fisher, in order processing, and Mort Foster, in accounting, are upset. They have to live with an awful processing system."

System Steering Group Committee

In an attempt to resolve some of the real and potential problems at WI, John decided to form a system steering group committee (see chapter 1 for a description of a steering group). John stated, "The committee will serve several purposes. It will help resolve communications problems, improve the quality of system planning, and determine the requirements of a new customer mail-order and payment system. The only difficult task is to decide on the membership of the committee." In pondering the issue of membership, John prepared a long and a short list as follows:

- Long List:
    - Rena Logen
    - Ray Logen
    - Ralf Elders
    - Krintine Rinehart
    - Marty Rhew
    - Gary Meltomon
    - Mort Foster
    - Morris Fisher
    - John Welby
    - Jim Sponskowski, store manager, western region
- Short list:
    - Rena Logen
    - Ray Logen
    - Ralf Elders
    - Krintine Rinehart
    - Gary Meltomon
    - John Welby

CASE ASSIGNMENT 1

Prepare a system trinity diagram showing the decision maker, the client, and the designer of WI's existing customer mail-order and payment system.

CASE ASSIGNMENT 2

Prepare a system trinity diagram showing the decision maker, the client, and the designer of WI's proposed customer mail-order and payment system.

CASE ASSIGNMENT 3

Help John define the membership of WI's system steering group committee. Why might John favor the short list rather than the long one? Why might he favor the long list rather than the short one? What would you propose regarding the make-up of the committee?

REFERENCES

1. R. A. Anthony, *Planning and Control Systems: A Framework for Analysis* (Boston: Harvard University Graduate School of Business Administration, 1965).
2. S. C. Blumenthal, *Management Information Systems: A Framework for Planning and Management* (Englewood Cliffs, N.J.: Prentice-Hall, 1969).
3. S. Neumann and M. Hadass, "On Decision-Support Systems," *California Management Review* 22 (Spring 1980): 77–84.
4. *A Guide for Studying and Evaluating Internal Accounting Controls* (Arthur Anderson, January 1978).
5. A. L. Eliason, "Administrative Processing Cycles and the Design of MIS," *Proceedings of the Ninth Annual Meeting of the American Institute of Decision Sciences*, ed. J. D. Stolen and J. J. Conway, 1977, 647–649.
6. C. W. Churchman, *The Design of Inquiring Systems: Basic Concepts of Systems and Organizations* (New York: Basic Books, 1971).
7. N. Ahituv and S. Neumann, *Principles of Information Systems for Management*, 2nd ed., (Dubuque, Iowa: William C. Brown Publishers, 1982), pp. 102–104.

3
Tools of Structured Analysis and Design

INTRODUCTION

UNDERSTANDING the process of systems analysis and design, as discussed in chapter 1, and the make-up of business systems, as discussed in chapter 2, can help clarify how to approach the systems design effort. It does little, however, in explaining how design is done. Using the tools of structured design, this chapter begins such an explanation.

At this point you might ask, "Well, what do you mean by structured design and why are structured design tools important?" *Structured design* involves the use of a set of tools and techniques that allow us to take a complex problem and break it down into simpler terms. The primary reason for using structured design is that it works.

Why does it work? According to Page-Jones,[1] structured design

1. allows the form of the problem to guide the form of the solution;
2. provides a means of subdividing a system into "black boxes" (i.e., modules that perform specific processing functions) and into hierarchies suitable for systems implementation;
3. features a variety of tools that simplify systems understanding;
4. offers a set of strategies for evolving a design solution; and
5. offers criteria for evaluating the quality of a design solution.

In short, structured design provides us with a well-defined set of methods to be used over the entire project life cycle. These methods help the systems analysis team logically define the features of a system. Equally important, these methods provide continuity from the logical to the technical design of a system.

With this listing of reasons supporting the case for structured design—compared with the more traditional methods of studying systems—the objectives of this chapter can be summarized. When you complete this chapter, you should be able to

- describe the main tools important to structured systems analysis;
- describe the main tools important to structured systems design;
- use tools important to structured systems analysis and design; and
- use tools important to traditional systems analysis and design.

TOOLS OF STRUCTURED ANALYSIS

Three groups of tools are used by systems analysts and designers to make complex systems more understandable: tools of structured analysis, of structured design, and of traditional systems analysis and design. The first of these concerns the logical description of a system. This group consists of tools of structured analysis, which include data flow diagrams, a data dictionary, data structure diagrams, and structured English.

In this section, we will review each of these tools. To add realism to this review, we will let Roger Bates, the lead systems analyst at Mansfield, Inc., explain the way in which the systems analysis team logically began to define the payroll system. Roger and his colleagues used the data flow diagram (DFD) as their principal tool of analysis. As you will soon appreciate, the DFD is a major problem-solving tool. It allows complex systems to be broken down into increasingly smaller subsystems. In this way, the subsystems of a business system can be isolated and studied with regard to how information must be transmitted, transformed, and stored.

Data Flow Diagram

A *data flow diagram* graphically depicts the information flows and points of processing in a system. These diagrams are used to show how data flow in a system, where data are transformed by a system, and where data are stored by a system. The techniques associated with data flow diagramming (also known as bubble and activity charting) are not new. However, only recently have these techniques been recognized as important to systems analysis and design. Leading this movement have been Yourdon and his colleagues,[2,3,4] as well as former associates of the Yourdon group.

Figure 3–1 illustrates the basic form of a DFD, or bubble chart. Its construction resembles both the network chart shown in chapter 1 and the system flow diagrams shown in chapter 2. In actuality, DFDs differ from network charts in various ways. Each bubble appearing on a DFD, for example, changes the form of data. This transformation can occur in a variety of ways, as the next three examples illustrate. In **Figure 3–2a**,

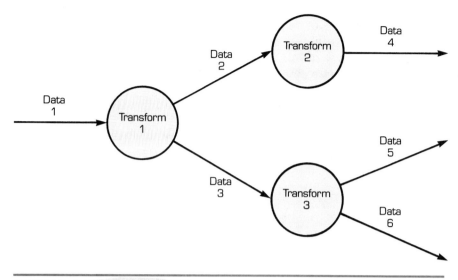

FIGURE 3–1 Bubble chart: basic form

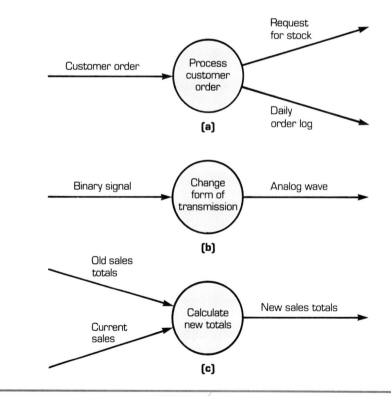

FIGURE 3–2 Examples of data transforms

the customer order data flow is transformed into two subsequent data flows, request for stock and daily order log. In **Figure 3–2b,** the data flow simply changes its form, from binary signal to analog wave. In **Figure 3–2c,** data are merged. Old sales totals are combined with current sales to form new sales totals.

Observe that a DFD does not indicate how transformation takes place. For that, another tool is required. Likewise, as Figures 3–1 and 3–2 indicate, a DFD makes no attempt to assign times or costs to the various data paths and transforms (the nodes, or places where data are changed by a process). The intent is to show information flow rather than to show how to control sets of related activities.

There is considerable similarity between the basic form of the DFD and systems flow diagrams, such as Figures 2–9 and 2–10. These two figures are also DFDs in that they show how data flow through a system.

CONTEXT DIAGRAM

The construction of a set of DFDs begins with the design of a context diagram. A *context diagram* frames the essence of a system. It represents the top-level diagram in a leveled set of DFDs (see Page-Jones for a discussion of a leveled set of DFDs[5]). Roger Bates explains the context diagram prepared for the employee pay system at Mansfield, Inc.

"We began our logical analysis of the employee pay system by showing how the system works in relation to our organization. In other words, we described the context of the system. Consider **Figure 3–3.** In preparing this context diagram, we had to identify all *data flows* to and from the employee pay system and all major *sources* and *sinks*. Let's review each of these terms.

"A data flow might be compared with a pipeline that carries data from point A to point B. As shown, an employee's payroll record represents a data flow. The data are carried from the payroll department, point A, to the system, point B.

"The context diagram also shows the names of the major data flows, which either feed data to the system or carry data produced by the system. For example, data contained on the employee's timecard are used to feed the system while data contained on the employee's paycheck are produced by the system.

"Let's consider next the terms used in constructing a context diagram. A source, such as a person or machine, is a provider of data to a system, while a sink is a receiver of data from a system. I should mention that the same item, person, or object may be both a source and a sink. The employee indicated in Figure 3–3, for example, is a source (he or she provides timecard information to the system) and a sink (he or she receives an employee paycheck). The payroll department illustrated is also a source and a sink. This department provides the employee's payroll record and receives the paycheck register. The employee's

Chapter 3: Tools of Structured Analysis and Design

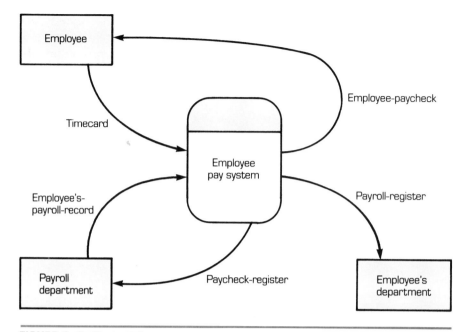

FIGURE 3-3 Context diagram

department, meanwhile, is a sink, since it only receives the payroll register."

Roger concludes his description of a context diagram by adding, "In designing the context diagram, we attempted to identify all sources and sinks to the system and all primary data flows. We also used a different symbol in representing the system. Instead of a circle, we have replaced it with a rounded rectangle. This symbol is consistent with the DFD notation suggested by Gane and Sarson.[6]"

LEVEL-0 DIAGRAM

Once the context diagram is finished, a *level-0 DFD* is prepared. A level-0 DFD describes the major functions, or transforms, of a system. In addition, it shows how the major functions of a system are linked together. We let Roger continue his discusson by explaining the level-0 DFD prepared for the payroll system.

"We started by identifying the three main functions of payroll processing," he began. "These functions were

1.0 Prepare record of hours worked.
2.0 Prepare payroll register.
3.0 Prepare employee paychecks.

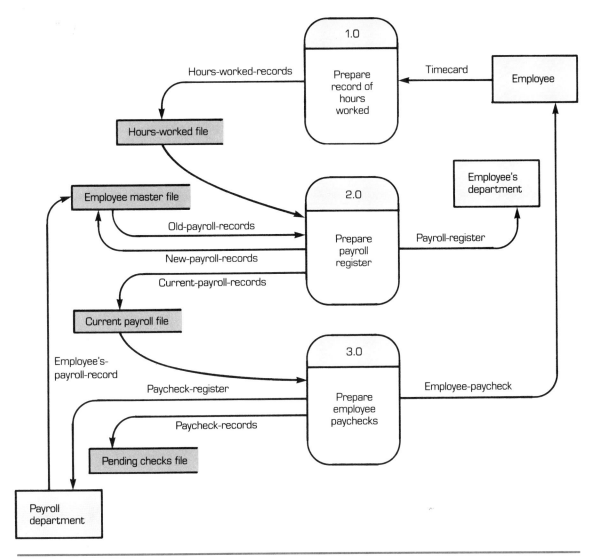

FIGURE 3-4 Level-0 data flow diagram

These functions tell us what our system must be designed to do.

"Once we were satisfied with these definitions, we identified the major data flows to and from each function and where the system could be expected to store data," Roger continued. "**Figure 3-4** shows our completed level-0 DFD. Let's examine the three symbols used in the makeup of this diagram, which incidentally are the symbols recommended by Gane and Sarson.[7]"

The transform symbol:

"The most prominent symbol in a DFD is the transform symbol. It shows where each major transform occurs in a system, tells what the transform is all about, and indicates at which level the transform occurs. Each transform describes a function of a system. The description itself begins with an action verb and ends with an object. 'Prepare' (the verb), for example, is followed by 'record of hours worked' (the object showing what has been prepared). Several examples of action verbs followed by the word *what* are shown below. After the verb, ask yourself a question using the word *what*. For instance: Prepare what? The response: Prepare employee paychecks.

Prepare (what)	Enter (what)	Move (what)
Get (what)	Edit (what)	Display (what)
Print (what)	Compute (what)	List (what)
Merge (what)	Total (what)	Verify (what)

"Each transform is also numbered. The level-0 notation is 1.0, 2.0, 3.0. If, however, the DFD contained five functions instead of three, the notation would be 1.0, 2.0, 3.0, 4.0, and 5.0.

The data store symbol:

"The second most prominent symbol is the data store symbol, which shows where a repository of data exists in a system. A data store is another term for a file and can be created either by people or by the computer. Thus, a data store may be a disk, tape, or tub file; however, the specific type of file is never shown.

"Data can be placed in or on a data store as an output of a transform or placed in a data store by a source. As Figure 3–4 shows, we determined that the payroll department (a source) placed an employee record directly in a data store.

The source/sink symbol:

"The rectangle identifies each source and sink in a system. The value of showing where data come from and where data go is that important environmental factors can be recognized. For example, timecards at Mansfield are prepared by employees; they are not prepared by each employee's department.

Data flow connecting lines:

"Besides the three graphic symbols, a DFD uses connecting lines to show major data flows in a system. These single- or double-directional lines are labeled. Each label attempts to describe the type of data flowing to and from each transform, data store, source, or sink. We select each label carefully. For example, if a timecard is found in a system, we need to label it as such. We try to avoid using meaningless labels such as data, detail, notes, records, messages, or stuff."

LOWER-LEVEL DIAGRAMS

Each major function described on the level-0 diagram can be described in terms of its subfunctions. Dividing a system into its component parts is known as *decomposition*. Roger Bates explains once again.

"Decomposition is the process of breaking down a system into smaller and smaller pieces until each transform shown on a DFD becomes clear enough to require no further explanation. When a transform reaches this state it is called a *functional primitive*. DeMarco[8] states that a functional primitive requires no further decomposition.

"In decomposing the payroll system, we began by preparing an outline of functions and subfunctions. An abbreviated outline is as follows:

1.0 Prepare record of hours worked
2.0 Prepare payroll register
    2.1 Sum hours worked
    2.2 Calculate current pay
    2.3 Calculate year-to-date pay
    2.4 Print payroll register
3.0 Prepare employee paychecks

"This outline shows that function 2.0, prepare payroll register, can be decomposed into four subfunctions: sum hours worked, calculate current pay, calculate year-to-date pay, and print payroll register. Collectively, these four subfunctions define the steps important to preparing the payroll register. They identify the names of the subfunctions to be used in preparing a lower-level DFD.

"**Figure 3–5** illustrates the decomposed view of function 2.0, prepare payroll register. As indicated, information from the hours-worked data store is used initially in the summation of hours worked. Once total

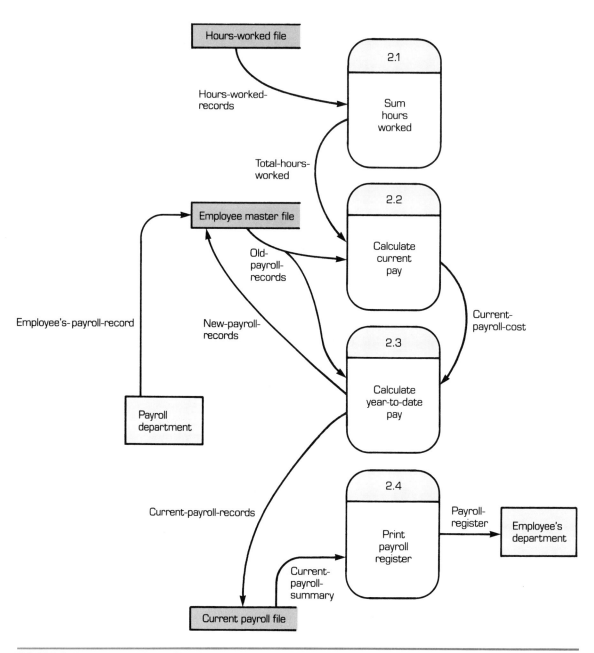

FIGURE 3–5 Lower-level data flow diagram

hours worked are known, the evaluation of current (employee) pay is performed, followed by the calculation of year-to-date (employee) pay. Following all calculations, the payroll register is printed and sent to the employee's department. The register shows the details of processing. It shows how much each employee in a department is to be paid."

(In chapter 5, we will return to this lower-level DFD to explain its properties in more detail. At this point it is important to understand that each function shown on the level-0 DFD can be decomposed into an integrated set of subfunctions.)

Data Dictionary

A *data dictionary* is prepared by the analysis team to define, at the very minimum, the meaning of each data flow shown on a DFD. When all people understand the meaning of terms, it becomes easier to design logical and physical systems. Once again, we will let Roger explain.

"Consider the data flow labeled current payroll cost, which is shown on Figure 3–5," Roger commented. "In defining this term, we had to ask ourselves the question, What does current payroll cost consist of and how can it be subdivided into smaller terms, much like DFDs are partitioned?

"Our analysis led us to conclude that current payroll cost consisted of several smaller terms or data flows. We were able to define current payroll cost as follows:

   current payroll cost = current gross pay, where
   current gross pay = regular pay + overtime pay

"We must decompose each term in the data dictionary until it is said to be *self-defining*," Roger continued. "For example, since neither regular pay nor overtime pay is very clear, we can say that neither is self-defining. We can further write:

   regular pay = regular hours worked × hourly pay rate
   overtime pay = overtime hours worked × overtime pay rate

"If the terms appearing on the right-hand side of these two equations are not self-defining, then we would have to define what we mean by regular hours worked, and so forth."

Data Structure Diagrams

Data structure diagrams are prepared to show the contents of data stores. Typically, a data structure diagram (DSD) provides two types of information: the list of items contained in a data store and the key or path that provides access to the data store. As an example, consider how Roger instructed his staff to prepare DSDs for Mansfield's payroll system.

"We had to design a number of DSDs in designing the payroll system at Mansfield," he remarked. "Let's examine the data store entitled employee master file to explain what we mean (see Figure 3–5). As illustrated, this file is important in supplying the old payroll record and in storing the contents of the new payroll record. Consider **Figure 3–6** next. In describing the contents of the employee master file, we initially identified the key fields. A *key field* determines how access to a record is possible. With employees, we decided that access should be controlled in two ways: either by knowing the employee number (a unique number assigned to each employee) or by knowing the employee name. The remainder of the employee record contains other fields, such as the employee's wage code. These other fields are known as *nonkey fields*.

"**Figure 3–7a** shows how we organized the contents of the nonkey fields for the employee master file," Roger added. "Initially, we grouped together common employee information categories, such as employee address, personnel, wage history, and so forth. Following this, we identified the specific fields required for each category. Under the category entitled personnel, for example, we decided to store six specific types of information (see **Figure 3–7b).** For each employee, we stored his or her department number, date started employment, date of birth, sex, marital status, and race."

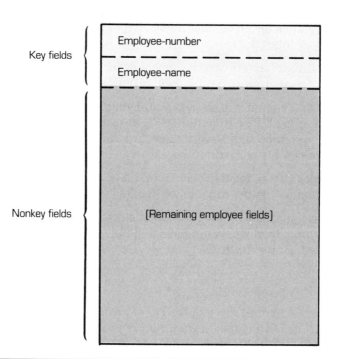

FIGURE 3–6 Data structure diagram (DSD) showing the key fields in the employee master file

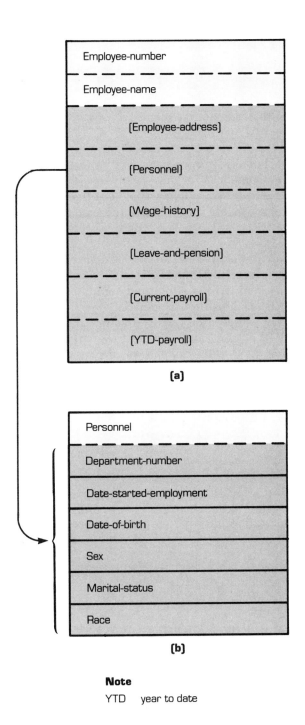

Note
YTD   year to date

FIGURE 3–7 DSD showing (a) employee information categories and (b) contents of Personnel category

Structured English

Structured English is a narrative technique used to describe the inner workings of the transforms shown on a DFD. The term *structured English* implies a highly restricted subset of the English language, which embodies rules that are similar in many ways to the rules of programming logic.

Although there are several versions of structured English, there is no official version. For our purposes, we will show structured English as short imperative sentences, beginning with a strong verb such as *get*, *move*, *put*, or *multiply*. We might write: "Get the employee record from the employee master file" or "Multiply regular-hours-worked times hourly-pay-rate to get regular-pay." Both of these sentences illustrate structured English.

Besides short imperative sentences, structured English contains sequence, decision, and repetition logic constructs, which are identical in form to those used in writing structured computer programs. Chapters 6 and 14 discuss these constructs in detail.

Let's consider a larger example to illustrate the use of structured English. We will let Roger Bates explain how structured English helped explain the internal workings of the transforms within the payroll system. Roger begins:

"We set out to write structured English for all lower-level transforms shown on our DFDs. In this way, we were able to describe the innermost workings of payroll processing. Consider the following description prepared for transform 2.1, sum hours worked (see Figure 3–5):

```
REPEAT for each employee
   Read hours-worked-record
   IF daily-hours exceed 8.0
       THEN add excess hours to overtime-hours-worked
         and add eight to regular-hours-worked
       ELSE add daily-hours to regular-hours-worked
   END-IF
   Sum overtime- and regular-hours-worked to
     determine total-hours-worked
UNTIL an employee's record is read
```

"This particular structured English example shows how we were able to determine the value of the data flow total-hours-worked shown on the DFD. It also shows that structured English is made up of sequential statements (statement 1, statement 2, and so forth), loops (REPEAT-UNTIL), and decision constructs (IF-THEN). Why do we indent when we write structured English? To make the language easier to interpret and comprehend. What we are doing is using a structured notation to describe our DFDs."

TOOLS OF STRUCTURED DESIGN

Once the data flows associated with a system have been fully diagrammed and documented, a systems designer or design team must decide which parts of the system should be processed by computer and, where they will, to propose a plan of computer processing. In this section, we will begin to examine how a logical design depicted by a set of DFDs is translated into a technical design description. A design team usually discovers that the process of translation ranges from quite easy to very difficult, depending on the nature and scope of the design problem. Fortunately, design teams have at their disposal various design tools and techniques, called the tools of structured design. The specific tools important to this second group are

- a structure chart,
- pseudocode,
- input and output layouts,
- file and database layouts, and
- processing controls.

Our discussion in this section is limited to a review of each of these tools. Chapters 9 through 13, which deal with the subject of systems design, will provide more complete coverage of these and other design topics.

Structure Chart

As its name suggests, a *structure chart* is a graphic picture of the internal structure or organization of a computer-based system. Much like the development of DFDs, the design of structure charts follows a top-down design. Beginning with the uppermost level of a system, the designer breaks down processing into a series of lower-level modules. A *module* represents a collection of coded computer program statements and consists of four basic functions: input, output, processing function, and internal data.

Before we continue, let's answer a commonly raised question. Many people ask, "Isn't writing a computer program an art, and, since programmers differ so much in organizing their coded instructions, isn't it impossible to show how a program should be organized?"

Consider the following response: "Programming style does differ from one person to another, and, if left to their own devices, programmers will vary greatly in how they design a program." However, this response also indicates why a structured approach is well advised. Such an approach permits others to study and comprehend a designer's thinking; it helps clarify how the designer organized the structure of the system he or she was assigned to work on.

THE PAYROLL SYSTEM EXAMPLE

Let's consider once again the payroll system designed by the staff at Mansfield, Inc., to illustrate how structure charts can be used in systems design. **Figure 3–8** illustrates the internal structure of the employee pay system. Compare this figure with the logical description of the system (see Figure 3–4). Similar to the level-0 DFD, the structure chart depicts a system that is divided into three primary modules: Update hours worked, Print payroll register, and Print employee paychecks. This particular structure chart is also called a level-0 chart. Why? Because it illustrates the top or uppermost level of a complex system.

The figure contains one special feature: a diamond-shaped symbol. This symbol means that it is possible to invoke—that is, to call or perform—any one of three lower-level modules. The uppermost module can update the hours worked, or print the payroll register, or print employee paychecks. After one of these options is processed, control is returned to the employee pay system—the uppermost module, which directs and controls the execution of all lower-level modules.

Figure 3–9 shows a partially completed level-2 structure chart. This chart begins to organize the steps important to module 2.0, Print payroll register. As indicated, four lower-level modules make up module 2.0.

This process of beginning with an upper-level module, subdividing it into two or more second-level modules, subdividing all second-level modules into third-level modules, and so forth illustrates once again the process called decomposition. As in the development of DFDs, decomposition is required to show the details of processing.

The figure shows a horizontal arrow in addition to four downward-pointing arrows. A horizontal arrow means that an iteration is to be added to processing. That is, modules 2.1, 2.2, and 2.3 are to be processed repeatedly until a signal is received to indicate that processing should be stopped. Once that signal is received, module 2.4 is invoked.

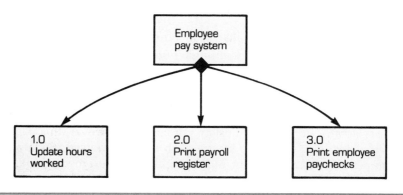

FIGURE 3–8 Level-0 structure chart

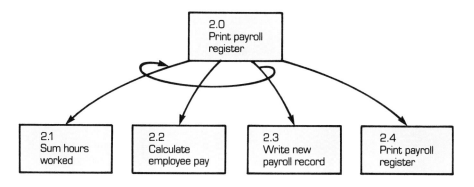

FIGURE 3–9 Level-2 structure chart (partial)

Consider the following structured English form, which describes how this iteration works:

```
REPEAT the following:
    Sum hours worked
    Calculate employee pay
    Write new payroll record
UNTIL a signal to stop is received
THEN write the payroll register.
```

SYMBOLS AND CONVENTIONS

Before we can proceed, we need to define some of the symbols used in designing structure charts. The higher-level module in a structure chart (see **Figure 3–10**) is the *calling module*, while the lower-level module

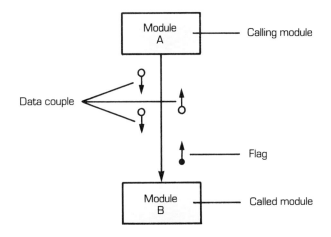

FIGURE 3–10 Connecting conventions

is the *called module*. This is similar to a boss-worker relationship, and we will treat it as such. In this instance, the boss asks each worker to perform some small task (such as add a list of figures or print a report). Each boss may be responsible for several workers; however, each worker reports to a single boss.

The figure shows two types of connecting symbols: a data couple symbol and a flag symbol. A *data couple symbol* is indicated by a small arrow with a white tail. This symbol is used to define the type of information to be passed from the boss to the worker or from the worker back to the boss. As shown in **Figure 3–11,** the boss might send total hours worked to a worker and expect the worker to convert this information into regular hours, overtime hours, and gross pay, and to send these data back once processed.

Besides the data couple symbol, the *flag symbol* is a small arrow with a black tail. A flag symbol can be used to describe other data (a *descriptive flag*) or to tell the boss to take some action (a *control flag*). **Figure 3–12** illustrates both types of flags. The hours-in-range flag is a descriptive flag. It describes the state of processing and lets the boss decide whether additional action is necessary. The print-out-of-range flag is a control flag. It tells the boss to take some action and for this reason is also known as a *nag flag*. A control or nag flag is one to avoid. As in other work settings, the worker should bring matters to the attention of his or her boss; however, the worker should not attempt to control or nag the boss. We will return to this concept once we begin to design computer programs.

Figure 3–13 shows a nearly finished level-2 structure chart. We will let Roger explain the organization of processing.

"To begin, we send the hours worked record to sum the hours worked

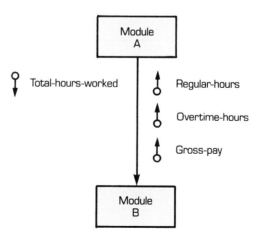

FIGURE 3–11 Passage of data couples

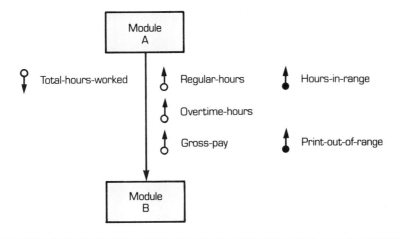

FIGURE 3–12 Passage of flags

(see module 2.1), which when completed sends three data couples—employee number, regular hours worked, and overtime hours worked—to the boss (see module 2.0, Print payroll register). The boss then sends all three data couples in order to calculate employee pay (see module 2.2). The first thing this second lower-level module does is check the accuracy of the employee number. If the number is incorrect, a flag is re-

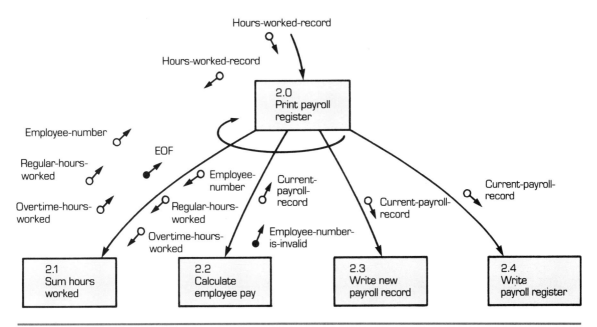

FIGURE 3–13 Level-2 structure chart

quired to tell the boss that the employee number is invalid; the boss is not notified if the number is valid. Following this, the module computes employee pay and returns the current payroll record. The third lower-level module (see module 2.3), entitled Write new payroll record, receives the current payroll record and writes it into a file. Following this step, processing loops back once again to sum hours worked. This iterative process continues until an end-of-file (EOF) flag is received. Once this flag is received, the boss sends all current payroll records to the fourth lower-level module (see module 2.4) to produce the payroll register."

Pseudocode

Pseudocode is a high-level language used to specify the processing functions to be completed by a module shown on a structure chart. While similar to structured English, pseudocode contains a level of detail that is more akin to computer code than English. As an example of pseudocode, let's consider the pseudocode written to describe module 2.1 in Figure 3–13. The DFD told us that input to this module was the number of hours an employee worked, while the output was regular hours worked plus overtime hours worked. The pseudocode written for this module is as follows:

```
Employee-count = 0
Stored-employee-number = -99
Regular-hours-worked = 0
Overtime-hours-worked = 0
Open hours-worked-file
FOR each hours-worked-record in the hours-worked-file DO
   Read hours-worked-record
   IF stored employee-number ≠ employee-number read
     from file and is > 0
       THEN send stored-employee-number
            send regular-hours-worked
            send overtime-hours-worked
            employee-count = employee-count +1
            regular-hours-worked = 0
            overtime-hours-worked = 0
       ELSE stored-employee-number = employee-number
            read from file
   END-IF
   IF daily-hours > 8.0
       THEN add (daily-hours - 8.0) to overtime-hours-
            worked
            add 8.0 to regular-hours-worked
       ELSE add daily-hours to regular-hours-worked
   END-IF
UNTIL EOF
Close hours-worked-file
```

Observe the difference between pseudocode and structured English (see page 76). With pseudocode, the specific details of processing—the opening and closing of files, the initialization of variables, the use of formulas, and so forth—capture the complete logical description of a module's inner workings. On receipt of pseudocode, a programmer should be able to transform the logic into program code. Read again the pseudocode shown above. Could you transform this logical description into computer code?

Input and Output Layouts

Input and output layouts concern the actual design of inputs to and outputs from computer processing. The objective is to provide mockups of all major input and output media. These mockups typically include representative samples of all source documents (e.g., documents prepared before processing, visual display screens [VDSs], computer-printed reports and documents, and special types of input or output such as optically scanned input data forms).

Figure 3–14 illustrates a mockup of a VDS. The user initially is presented with descriptive text before being told to enter the number of one choice. When the user desires no other characteristics, the letter *N* is keyed.

```
A program to be evaluated can be defined
by up to four different characteristics.

Please enter the number of one choice.

         1)  Age
         2)  Sex
         3)  Race
         4)  Admittance date
         5)  Family type
         6)  Mother's educational level
         7)  Family income
         8)  Program
         9)  Identifying condition
        10)  Impairment level

         N)  No other characteristics
```

FIGURE 3–14 Layout of an input screen

File and Database Layouts

Similar to input and output layouts, *file and database layouts* provide mockups of files (or databases) to be implemented by the systems design team. The objective here is to provide a *file specification:* the documentation that specifies how a file is to be organized, the contents of each record stored on the file, the keys for each record, the size of the file, how the file is to be modified (e.g., updated), and how the file is to be reset, backed up for security reasons, and purged to remove obsolete records.

Figure 3–15 illustrates a partial description of a file concerning course description. This file contains records of courses offered by a department. The record description includes the specification of each field name, the field type, and the field length. Thus, the field name COURSENUM (meaning course number) consists of four characters,

Course Description Record

Field Name	Type	Length
TLN	digits	4
Dept	char	4
Coursenum	char	4
Title	char	16
Time	digits	11
Day	char	3
Room	digits	3
Bldg	char	3
Instr	char	12
Notes	char	7
Totopen	digits	3
Curenro	digits	3
Priority1	digits	3
Priority2	digits	3
Priority3	digits	3
Total		82

Indexed by TLN

Indexed by Dept

FIGURE 3–15 Layout of a course description record

while the field name ROOM (for room number) is restricted to three digits.

The figure shows that a course description record consists of fifteen different fields and that each record requires eighty-two characters or digits. In addition, the file specification indicates that two key indexes will be available for course description records. First, records will be arranged by term-line number (TLN), a unique number assigned to each course offered by the department. By entering the appropriate TLN, a single record can be retrieved from the file. Second, records can be accessed by department. By entering a department number, a user can access all courses for that department from the file.

Processing Controls

The last tool important to structured design involves the specification of processing controls. *Processing controls* help verify that all coded program statements work correctly. They are especially helpful during the implementation and operation of a system.

Generally, a design team provides for four types of processing controls: source-document controls, input controls, program controls, and output controls. *Source-document controls* consist of matching counts and amounts recorded on source documents to counts and amounts entered into processing. In the Mansfield payroll example, a count of the number of timecards would be made. This count would be compared against the number of timecards processed by the computer. *Input controls* (including file input) check the accuracy of data transmission. Note that in writing pseudocode for summing hours worked an employee count is kept. Why? This count can be printed to verify that all employees have been accounted for. *Program controls* include written procedures to trap data in error or data suspected of being in error. Suppose that the room number field on Figure 3–15 is always less than 800. A program control might be written into a file processing program to check the accuracy of all room numbers. The code might read:

```
IF room > 800
    THEN display error message "room number is too
    large"
END-IF
```

Finally, *output controls* check the accuracy of data processing and transmission. The payroll register used in the payroll example is one type of output control. Roger Bates explains this printed document: "The payroll register produced by the system is nothing more than a listing showing what employee paychecks will look like once we go ahead and insert the blank check forms in the printer. Before printing actual checks, however, we print the register and inspect it carefully, noting anything suspicious—such as two or more checks for the same employee. Only when we have conducted a thorough audit of the register do we go ahead and print employee paychecks."

TRADITIONAL TOOLS OF SYSTEMS ANALYSIS AND DESIGN

Besides the tools important to structured systems analysis and design, other, more traditional tools help depict the logical features of computer-based systems. Three of these tools are *system flowcharts, program processing menus,* and *program flowcharts.* System and program flowcharts trace the flow of data through a system; however, compared to DFDs, (which tend to be quite abstract), both system and program flowcharts provide a concrete portrayal of processing. Program processing menus show the specific programs or job steps to be installed in a computer system and, as such, provide a high-level overview of the organization of a system design. Let's consider each of these more traditional tools in more detail.

System Flowcharts

System flowcharts attempt to show the major steps in processing and, much like a level-0 DFD, are designed to show the primary inputs to and outputs from each main step in processing. To simplify documentation, special symbols are used. (Refer to any standard flowcharting template if you are not familiar with these symbols.)

Figure 3–16 illustrates the general form of a system flowchart, which can be summarized as follows:

1. Data are keyed to processing, as indicated by the *document symbol* and the *manual input symbol.* The *communications link symbol* shows that after data are keyed, data transmission occurs.

2. Data are read from a file to processing, as illustrated by the *online storage symbol.* Thus, there are two types of input to the process: keyed input and stored input.

3. Data are transformed in processing, as shown by the *process symbol.*

4. The results of processing are written to a terminal screen, as indicated by the *display symbol.* The communications link symbol shows that following processing, data transmission occurs.

5. The results of processing are also written into a computer file, as shown by the online storage symbol. Thus, this process leads to two types of output: displayed output and stored output.

Figure 3–17 provides another view of a system flowchart. The purpose of the process here is to verify data. Similarly to a DFD, the name of the process begins with a verb and ends with the object to the verb. "Verify what?" "Verify data." Inputs to this verification process consist

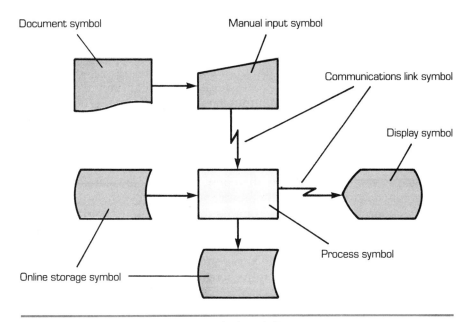

FIGURE 3–16 System flowchart: standard symbols

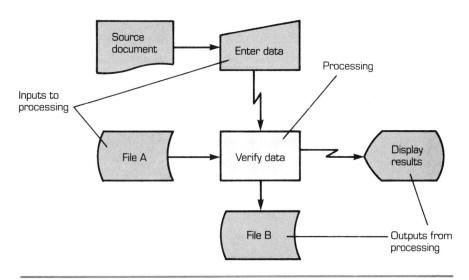

FIGURE 3–17 Interactive processing

of data recorded on a source document and keyed into processing, and data read from file A. Outputs, meanwhile, are made up of displayed results and data written into file B. Thus this flowchart suggests that all verified data are written into file B. Finally, observe the organization of a system flowchart: inputs in processing are placed at the top and to the left-hand side of the process; outputs are positioned at the bottom and to the right-hand side. The flowchart itself is read from top to bottom.

Besides showing input to and output from a process, a systems flowchart demonstrates whether a process is an interactive or a batch processing step. With *interactive processing*, transactions are entered one at a time. This feature permits the computer terminal operator to enter into a conversation with the computer. Figure 3–17 illustrates such a processing arrangement. For example, the process of verifying data might initially require the operator to enter a social security number. Following entry and transmission of the number, the process would check to determine whether the keyed number appeared to be correct. If not, a message would appear on the screen such as

SOCIAL SECURITY NUMBER 639-3W-4082 IS INCORRECT.
 PLEASE REENTER.

Figure 3–18 shows a batch method of processing. With *batch processing*, transactions are processed as a complete set. As illustrated, data from file A are combined with data from file B to produce printed results and an output file, file C. In this environment, operator communications take place before a processing run begins and after all transactions have

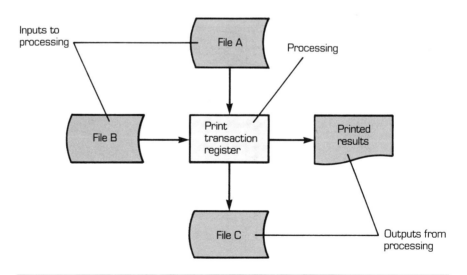

FIGURE 3–18   Batch processing

been processed. If errors are discovered during a run, the process may suddenly end, or the type of error discovered may be included as part of printed results.

Figure 3–19 illustrates how system flowcharts are related to both DFDs and structure charts. Three system flowcharts are needed to show the three DFD processing steps: preparing the record of hours worked, preparing the payroll register, and preparing employee paychecks; these same flowcharts correspond to the modules shown on the level-0 structure chart (see Figure 3–8). Again, Roger Bates explains each of these payroll processing system flowcharts.

"1.0 is 'Update hours worked.' This interactive processing step accepts time card hours keyed to processing and updates the hours-worked file. Processing consists of adding the old hours worked record to the current hours worked record and of writing this new record to the hours-worked file. The old and new hours-worked totals are displayed for visual verification.

"2.0 is 'Print payroll register.' This batch processing program accepts hours-worked totals read from the hours-worked file and employee payroll data read from the employee master file. Processing consists of summing the hours worked, calculating current pay, and calculating year-to-date pay. All payroll totals are printed on the payroll register and written into the current payroll file. All changes to a employee's record are written to the employee master file.

"3.0 is 'Print employee paychecks.' This second batch processing program accepts payroll data stored on the current payroll file and prints employee paychecks. A register of processing is also printed; a summary of all checks issued is written to the pending checks file."

Program Processing Menus

Besides system flowcharts, program processing menus are designed to show the organization of processing. These menus identify the major computer programs to be designed for a system, in which each program follows from a process shown on a system flowchart. In addition, some program processing menus separate interactive from batch processing programs. This separation helps show how a software design is to be organized.

Figure 3–20 shows the expanded employee payroll computer system designed for Mansfield, Inc. Four interactive programs are shown. In addition to updating the hours-worked file, an interactive update program is needed to change the contents of the employee master file. Two display programs are also indicated. The first permits a user to review hours-worked figures; the second allows a user to review an employee's payroll account.

The figure also shows that the batch portion of processing is greatly expanded. Two additional registers are required: the hours-worked reg-

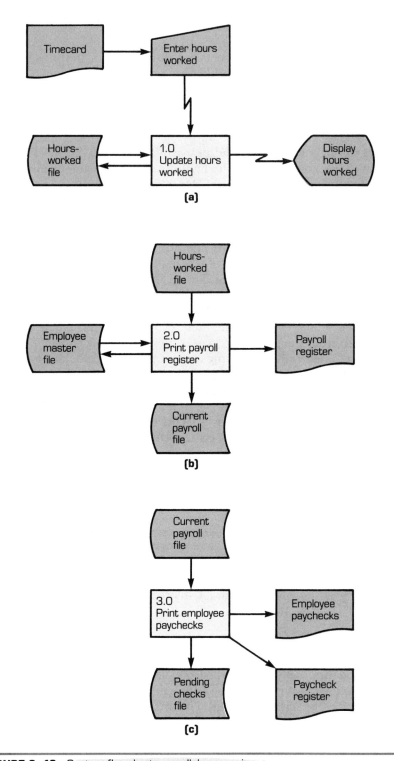

FIGURE 3–19 System flowcharts: parallel processing

```
I.1    UPDATE HOURS WORKED
I.2    UPDATE EMPLOYEE MASTER FILE
I.3    DISPLAY HOURS WORKED
I.4    DISPLAY EMPLOYEE PAYROLL ACCOUNT

B.1    PRINT PAYROLL REGISTER
B.2    PRINT HOURS WORKED REGISTER
B.3    PRINT EMPLOYEE MASTER REGISTER
B.4    PRINT EMPLOYEE PAYCHECKS
B.5    PRINT PAYROLL JOURNALS
B.6    PRINT QUARTERLY REPORTS
B.7    PRINT YEAR-TO-DATE REPORTS
B.8    BACKUP FILES
B.9    RESET FILES
<10>   EXIT
```

FIGURE 3–20 Program processing menu

ister and the employee master register. Likewise, three types of payroll report summaries are indicated: payroll journals, quarterly reports, and year-to-date reports. Finally, two types of file processing activities are indicated: backup files and reset files. These programs permit the user to make a backup copy of important computer files and to reset certain processing parameters, such as account totals at the beginning of each quarter or fiscal year.

Program Flowcharts

Program flowcharts provide a graphic representation of program logic. They differ from system flowcharts in that they trace the flow of a set of processing instructions. **Figure 3–21** illustrates a program flowchart to depict the procedure described earlier using pseudocode. In this instance, the diamond symbol represents a decision or branch point. The rectangle represents a process.

Let's review this procedure once again. The flowchart indicates that following the initialization of variables, hours-worked records are read until an end-of-file signal is received. After reading a record (and finding no EOF), the first condition tested is whether the record contains a new employee number. If so, figures on total hours worked are sent for the old employee number, and hours-worked sums are initialized (set to zero) for the next employee number. The second condition tested is whether the employee's reported time is greater than eight hours. If so, overtime and regular hours are summed; if not, only regular hours are

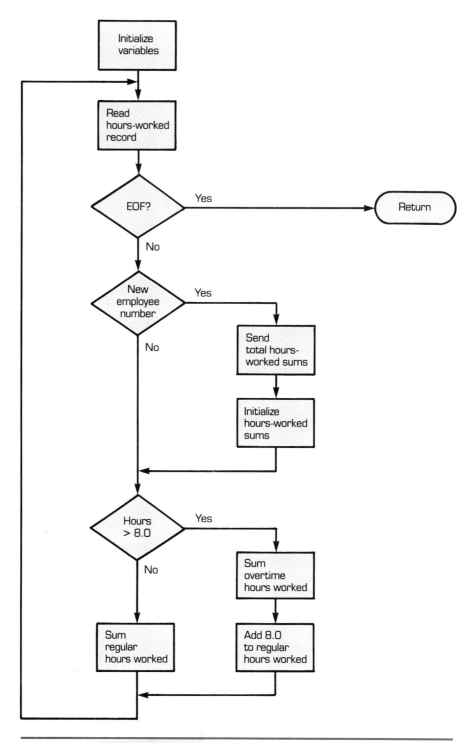

FIGURE 3–21 Program flowchart

summed. Once hours are summed, the procedure loops back to read another hours-worked record.

After reviewing this program flowchart, you might wonder which tool is best, pseudocode or program flowcharts. While most practicing designers prefer pseudocode, both tools contain both positive and negative features. Pseudocode is much easier to write (and to repair), is closer to actual program code, can describe complex procedures, and is more specific. Program flowcharts, however, can visually convey what pseudocode cannot. Thus, a designer's personal preference comes into play in determining which tool is best.

The Mansfield, Inc., Case Study

Roger Bates began his study of the billing system by preparing two diagrams: a context diagram and a level-0 diagram. "These diagrams will help all members of the team to visualize what we mean by a billing system," Roger commented. He added, "I stress the word *all!* John Havensek should be able to understand what we in systems are up to. If he does not understand, how can he communicate his satisfaction with the system to David Orring? I consider it important to gain the confidence of the finance and accounting group early on."

Context Diagram

Figure 3-22 illustrates the context diagram that describes the billing system. "Let me explain the context of the system," Roger began. "As you can see, the system must be able to produce two types of output: customer invoices and monthly statements. Customer invoices are mailed to the customer following the shipment of merchandise from one of our warehouses. If a customer (e.g., a retail store) decides not to pay the amount shown on the invoice, we print a summary of the outstanding charge later on. The monthly statement shows these summary charges.

"To appreciate how the system works, let's trace the processing of one transaction," Roger continued. "The system is activated by the receipt of customer shipping information. Shipping information tells us who the customer is and what was ordered and shipped. What the system must do is translate this information into customer charges."

"How does the system do this?" Neil Mann, a programmer/analyst, questioned.

"We begin by getting a customer record from marketing and one or more product records from manufacturing," Roger explained. "A customer record provides us with important customer information, such as the customer's bill-to address. This address is usually different from the customer's ship-to address. Likewise, a product record provides us with important product information, such as the price of the product. If the customer orders a dozen different products, we have to pull a dozen product records."

Roger then concluded, "What the billing system must do is to combine cus-

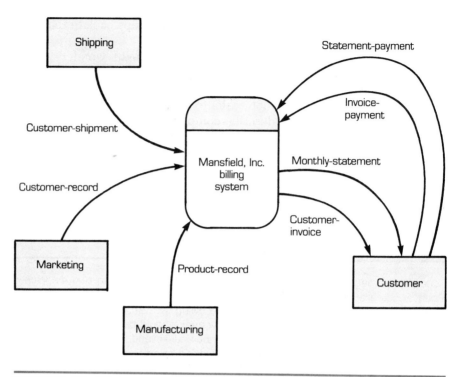

FIGURE 3—22 Billing system context diagram

tomer shipment, customer record, and product record information to create a customer invoice. We send the invoice to the customer, expecting in return an invoice payment. Failing to receive a payment, we send the customer a monthly statement, expecting in return a statement payment. Still failing to receive a payment leads to another monthly statement, and so on, until the outstanding bill has been paid."

Level-0 Diagram

Once Roger was confident that all members of the team understood the context of the billing system, he turned his attention to the level-0 DFD.

"The level-0 DFD shows the internal functions of the billing system," he stated. "As indicated [see **Figure 3—23**], there are three main processing steps; preparing the customer invoice, processing the customer payment, and preparing the monthly statement. Let me explain each of these steps.

"Step 1.0, 'Prepare customer invoice,' is, obviously, where we prepare the customer invoice," Roger explained. "Here, we combine shipping information with customer and product records. One output shown is the customer invoice, and another output is called the invoice summary."

"What's the purpose of the summary?" questioned Carolyn Liddy, systems specialist.

"The invoice summary shows which products were shipped and to whom, as well as the total balance-due charge to be paid by the customer," Roger added.

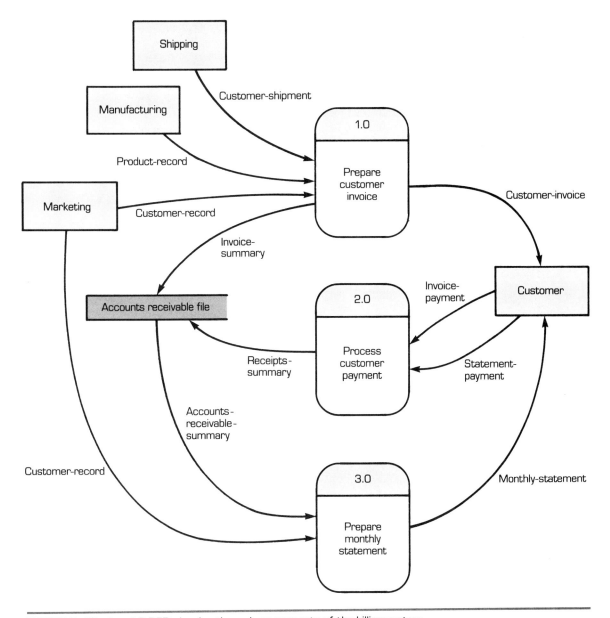

FIGURE 3–23 Level-0 DFD showing the main components of the billing system

"As indicated, we place this invoice summary information in the accounts receivable (A/R) file."

Roger continued, 'Process customer payment,' step 2.0, is where we process invoice and statement payments. In each case, we must produce a receipts summary, which, as you might expect, shows the amount paid by customer invoice number. We add this payment information to the A/R file."

Neil interrupted. "Doesn't the third step match invoice charges outstanding against customer payments received?"

"Yes, it does," Roger remarked. "'Prepare monthly statement,' step 3.0, is where we match the invoice summary against the receipts or payment summary (e.g., the accounts receivable summary). If the match leads to a zero balance due, this tells us that the customer bill is paid in full. If the balance due is greater than zero, however, we show this balance on the monthly statement. Or if the balance due is less than zero, we show a credit amount on the monthly statement. Before printing a monthly statement, we obtain the customer record once again. We have to check to determine whether the customer bill-to address is still valid."

SUMMARY

Structured analysis and design feature a set of tools and techniques specifically designed to break down complex problems into simpler terms. The tools of structured analysis (data flow diagrams, data dictionary, data structure diagrams, and structured English) were developed to help with the logical description of a system. The tools of structure design (structure charts, pseudocode, input and output layouts, file and database layouts, and processing controls) were developed to help with the physical design of a system. To round out the main tools of analysis and design, system flowcharts, program processing menus, and program flowcharts help describe the features of a complex system.

A principal tool of systems analysis is data flow diagrams. DFDs show how data flow in a system, how data are transformed by a system, and where data are stored in a system. The construction of a set of DFDs begins with the design of a context diagram (to place the system into context), followed by the design of a level-0 DFD (to identify the main functions of the logical system).

The data dictionary defines the meaning of each data flow shown on a DFD, while a data structure diagram defines the contents of each data store shown on a DFD. Structured English is used to provide a written description of the inner workings of each transform shown on a DFD.

A principal tool of systems design is the structure chart. This type of chart shows the internal structure of a complex design by subdividing a procedure into smaller modules. Much like DFDs, structure charts are graphic, able to be partitioned, and easy to understand. They embody data couples and flags to show how data are passed from one module to another.

Pseudocode is a high-level language used to specify the processing steps within each module shown on a structure chart. Input and output layouts and file and database layouts provide mockups of all input and output data. Processing controls help verify that coded program statements work as intended.

System flowcharts represent a more traditional tool of systems analysis and design. They are designed to show the main input to and output from each main step in processing, in which the steps are defined by a program processing menu. Program flowcharts are more detailed; they provide a graphic representation of program logic.

REVIEW QUESTIONS

3–1. What is meant by the term *structured design*?
3–2. How is a data flow diagram used by a systems analyst?
3–3. What is the difference between a context diagram and a level-0 DFD?
3–4. What is a data dictionary and why is it useful?
3–5. What is a key field and of what importance are key fields in designing a data structure diagram?
3–6. How is structured English different from everyday English?
3–7. What is a structure chart designed to show?
3–8. What is the difference between a calling module and a called module? Between a boss module and a worker module?
3–9. How does pseudocode differ from structured English?
3–10. What is a file specification?
3–11. What is the difference between an input processing control and a program processing control?
3–12. What are system flowcharts designed to show?
3–13. What is the difference between interactive and batch processing?
3–14. How do program flowcharts differ from system flowcharts?

EXERCISES

3–1. What is wrong with the following context diagram and level-0 DFDs that were designed for the same system?

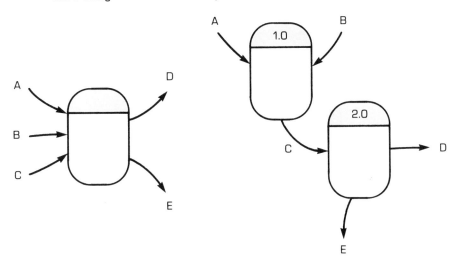

3-2. Rewrite the following paragraph in structured English.

"Each day we receive customer payments in the mail. There are several payments. We do several things with each payment. First, we read the check (to make sure it is acceptable). If not, we return it to the customer. Next, we compare the check amount with the payment stub amount. If the amounts are different, we change the stub amount to be the same as the check amount. Last, we write the check number and amount on the bank deposit slip."

3-3. The accounts payable (A/P) department of A-Bacon Company, Inc., receives a *vendor invoice*, a formal billing statement that tells A-Bacon what it owes from a previous purchase. The A/P staff compare the billing information shown on the vendor invoice with the information shown on a *material receipt slip*, a slip prepared by the receiving department when materials from vendors are received. If A/P employees find that the charges are accurate, and if they have enough *cash on hand*, they produce a *vendor check* to be mailed to the vendor. The amount of cash on hand is tabulated daily. This information is sent to the A/P department by the corporate finance department.

Prepare a context diagram describing A-Bacon's payables processing system.

3-4. With customer invoicing, the shipping department of A-Bacon sends one set of *shipping papers* to the billing department and a second set to the customer. The billing department multiplies the number of units shipped by the unit price and produces a *customer invoice*, a formal billing statement that tells the customer what he or she owes. Once printed, the original copy of the invoice is sent to the customer; a second copy is sent to the accounts receivable (A/R) department. If the customer finds the charges on the invoice to be acceptable, a *customer payment* is made. The payment is received and processed by the A/R department.

Prepare a level-0 DFD describing the main parts of A-Bacon's billing system.

3-5. Suppose that you are told the following about a marshmallow toasting system:

1. Start the fire
2. Repeat
   2.1 Get a marshmallow
   2.2 Toast the marshmallow
   2.3 Eat the marshmallow
       Until you have had enough.
3. Put out the fire.

Draw a structure chart to describe this system.

3-6. Add data couples and flags to the following structure chart; before starting, we are told:

1. We will always rent a bike unless there is none in stock. When the bike is rented, the clerk will hand us a bike ticket.

2. We will always ride the bike unless there is an equipment failure.
3. We will always return the bike and hand the clerk the bike ticket, unless the bike shop is closed.

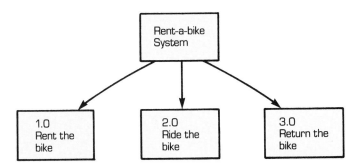

WORLD INTERIORS, INC.—CASE STUDY 3
DESIGNING TOP-LEVEL DATA FLOW DIAGRAMS

Introduction

The rapid growth experienced by WI was in large part the result of a once-in-a-lifetime opportunity. In the early 1980s, a leading manufacturer agreed to let WI begin a mail-order business to feature its products. Before that time, the manufacturer had used distributors who operated as middlemen. These distributors took orders from retailers, who in turn sold directly to customers. No distributor other than WI, however, sold directly to customers.

The mail-order success experienced by WI surprised everyone—WI, the manufacturer, and especially the distributors. Charging unfair competition, the distributors convinced the manufacturer not to open additional mail-order outlets or to license firms such as WI to sell their home interior products directly. The manufacturer agreed to these demands, with one exception: it honored its agreement with WI. In so doing, WI was given the exclusive rights to market goods directly to the customer; its retail cost of goods thus was lower than any other retail store in the nation.

Direct Sales Processing

Even though it appeared complicated—because of the large number of orders processed daily—WI had an easy-to-understand system of processing customer orders received in the mail. Krintine Rinehart explained:

"All customers are required to send two items with their order: a completed customer order form and a personal check for full payment of the merchandise ordered. What we do by telephone is tell customers about our products. We obtain their names and addresses and send them, at no expense, product information and customer order forms. We might also tell them how to measure correctly or explain different product options. Our forms are so simple that we do not compute the cost of the product selected. We expect the customer to be able to do this. The customer is expected to return the order with a personal check.

"Once we receive a customer order, we clear it. Clearing the order consists of making sure the information shown on the order is complete and correct and that payment is in full. We place the cleared order in an accepted-orders file. Daily, we pull orders from this file to process all customer invoices. During this stage, we use the computer for processing. The customer invoice is needed to tell the manufacturer which goods are needed. Each manufacturing plant requires that we follow slightly different order processing procedures. Consequently, we use a variety of customer invoice forms. Once we have completed the customer invoice we send a copy to the manufacturing plant, send a copy to the customer, and store a copy in our pending customer invoice file."

Accounts Payable Processing

Ralf Elders helped complete the description of order processing. He explained the steps important to paying the manufacturer for goods shipped to customers as follows.

"The system designed to process factory invoices is more complicated than the system for processing customer mail orders," he said. "Initially, the customer invoice is compared with the factory invoice (i.e., the invoice mailed to us by the manufacturer) to obtain a match and to determine whether the manufacturer's charges are correct. As you might suspect, this matching can take considerable time. If we find a match and correct charges, the approved factory invoice is placed in our accounts payable file. We use the due dates shown on the factory invoice to determine when to pay. The process of payment consists of removing the factory invoice from the accounts payable file and using the information recorded on it to prepare a vendor-voucher check. The voucher portion of the check explains what led to the writing of the check and includes our customer invoice number and the factory invoice number. We send the completed vendor-voucher check to the manufacturer. This completes the accounts payable cycle. The only remaining steps are to keep a register of all outstanding checks and to resolve any problems found on disapproved factory invoices."

CASE ASSIGNMENT 1

Prepare a context diagram showing the major sources and sinks important to WI's mail-order and payment system. Include in your diagram all primary data flows. Label all sources, sinks, and data flows. Show how the filled customer order is shipped to the customer. Label this data flow "Filled customer order."

CASE ASSIGNMENT 2

Prepare a level-0 DFD showing the principal parts of WI's customer mail-order processing and payment system. Include in your diagram all primary data flows, main transforms, sources, and sinks. In preparing this DFD, limit your analysis to two transforms. Label one transform "Process customer orders." Label the second transform "Pay manufacturer's charges." Remember to label all data flows and all data stores.

REFERENCES

1. M. Page-Jones, *The Practical Guide to Structured Systems Design* (New York: Yourdon Press, 1980).
2. E. Yourdon, *Techniques of Program Structure and Design* (Englewood Cliffs, N.J.: Prentice-Hall, 1975).
3. T. DeMarco, *Structured Analysis and System Design* (New York: Yourdon Press, 1978).
4. C. Gane and T. Sarson, *Structured Systems Analysis* (Englewood Cliffs, N.J.: Prentice-Hall, 1979).
5. M. Page-Jones, *The Practical Guide to Structured Systems Design.*
6. C. Gane and T. Sarson, *Structured Systems Analysis.*
7. Gane and Sarson, *Structured Systems Analysis.*
8. T. DeMarco, *Structured Analysis and System Design.*

System Requirements

INTRODUCTION

BEFORE an analyst can begin to take a complex problem and break it down into its component parts, he or she must first identify and understand the nature of the problem. Only then can an analyst start to define system requirements. The organization of this chapter follows this line of reasoning. It begins with an examination of problem solving and looks at a number of problem-solving tools. It ends with a description of the process leading to preparing a statement of system requirements. When you complete this chapter, you should be able to

- describe why problem solving is important to the work of systems analysis;
- identify the most common types of system problems that frequently occur in organizations;
- use tools important to identifying which factors lead to system problems;
- show how changes to a system will affect service levels;
- identify the issues important to the determination of system feasibility; and
- prepare a project specification to finalize system requirements.

Before we continue, however, you might ask, "Why is problem solving important in determining system requirements? After all, don't managers assign projects to systems analysts and instruct them to carry them out?" A response to these questions is more difficult than it may appear. In practice, managers do assign projects to analysts and ask them to carry them out. However, it becomes the analyst's responsibility to determine

- which problems the project is designed to resolve,
- which resources will be required to solve those problems, and
- whether the project should be done at all.

This final determination (telling a manager that a project should or should not be done, based on a preliminary study) is difficult, especially when the manager has a different set of expectations. Nonetheless, this situation occurs more frequently than might be expected. Consider the following example.

A manager of a medium-size manufacturing firm assigns a project to an analysis team, asking it to design an automated set of procedures for work-in-process inventory. After a preliminary study, the team decides that it would be unwise to automate the procedures at this time. Why? Because any change would be too short-lived as a result of the changing nature of the work-in-process inventory system and the efficiency of existing manual procedures. In its study of system requirements, the team concluded that a computer-based approach would have little to no cost advantage. Moreover, any new system would be too difficult to maintain relative to the benefits it was expected to return.

PROBLEM CLASSIFICATION AND DEFINITION

One of the most difficult tasks of systems analysis is developing a clear, in-depth understanding of the problem being investigated. Most experienced analysts take considerable time in handling this task. They realize—often as a result of a painful experience—that without a clear understanding of the problem, it becomes impossible to specify the requirements for a new system with any accuracy.

Several questions should be posed before any new project assignment is initiated. These include:

- What is the problem?
- How complex is it?
- What are its likely causes?
- Why is it important that the problem be solved?
- What are possible solutions to the problem?
- What types of benefits can be expected once the problem is solved?

Defining the Problem

It takes considerable skill to determine the problem, its complexity, and its likely causes. The analyst might begin this task of definition by attempting to determine whether the problem can be classified according to one or more problem types. That is, because there are various system problems that occur frequently, the analyst often can define the characteristics of the problem through examination of its attributes. An example can illustrate this finding.

A manager comments, "We need a new budgeting system. Our current one seems to vary in quality from one month to the next. Besides, reports are often late, incomplete, and contain misleading information. Why, we must spend a fortune simply trying to keep the system up and going."

Careful analysis of this statement suggests a number of different problems: the problem of *reliability* (the system varies in quality from one month to the next), the problem of *accuracy* (there are too many errors), the problem of *timeliness* (reports are often late), the problem of *validity* (reports contain misleading information), and the problem of *economy* (the system is costly to keep up and running).

Besides the problems of reliability, validity, accuracy, economy, and timeliness, the problems of *capacity* and of *throughput* also arise. Capacity problems occur when a component of a system is not large enough. Two people attempting to do the work of six illustrates a capacity problem. Throughput problems deal with the efficiency of a component of a system. Six people doing the work of two represents a problem of throughput. Let's consider each of these seven problems in more detail.

THE PROBLEM OF RELIABILITY

A system suffers from the problem of reliability when procedures work some but not all of the time, or when use of the same procedure leads to different results. Users will contend—and rightfully so—that a system is unreliable if it is not dependable or cannot be trusted. Imagine how you would feel if a friend of yours said he would call you each evening at 8:00 but often fails to call.

Analysts must work continually to improve the reliability of systems. They strive to do this by running software tests to document that two runs of a computer program lead to identical results, by selecting equipment with low failure rates, and by monitoring processing schedules to ensure that results are produced and delivered on time. When we study the subject of software testing, we will give more attention to this problem of reliability.

THE PROBLEM OF VALIDITY

Systems that produce invalid results are often the most troublesome to users and system managers. These systems draw incorrect conclusions. A report might show that demand is increasing and that additional stock should be ordered for inventory. In actuality, suppose that demand is decreasing. Ordering additional stock in this instance would be an incorrect decision.

Maintaining software validity is a troublesome design problem. The objective in design is to produce a flawless product, one that will hold up over time. Determining whether this objective has been achieved, however, is often difficult because many software products are poorly tested and often designed according to a set of design assumptions. An example will help illustrate.

A software manager believes that her design staff has built a flawless market analysis software product. After a month of testing, the product works as expected, malfunctions only once in a while, and meets the expectations of marketing managers—who have exerted considerable pressure on the software manager to complete the product. Shortly after the introduction of the product, the software begins to report invalid results. The marketing managers demand an explanation. Further testing reveals that the estimate of future sales was based on the assumption that demand followed a sine wave with an assumed wave frequency. As long as actual demand followed the same frequency, the software worked as expected. When the frequency changed, the software continued to report results; however, those results were no longer valid.

THE PROBLEM OF ACCURACY

The problem of accuracy is similar to the problems of reliability and validity. A system is inaccurate when processing is error-prone. For example, assume that several people are required to post company expense transactions against departmental budget numbers. If the posting procedure is complex and the number of transactions large, a fair number of errors may occur (e.g., 1 percent of all transactions). Because of inaccuracy, the entire budget system might be viewed as unreliable and often invalid. However, these are symptoms of the real problem—namely, the problem of accuracy.

Routine, transaction-based manual procedures are likely candidates for conversion to computer-based methods of processing because the computer is far more accurate than human beings. Provided that software is written properly—a major problem in some instances—the problem of accuracy all but disappears. Eliminating the problem of accuracy helps explain why computer technology has been accepted so widely by so many organizations.

THE PROBLEM OF ECONOMY

Besides improving processing accuracy, organizations seek to improve processing economy. A system suffers from the problem of economy when existing methods of transmitting, processing, and storing information are costly. An organization might discover that the cost of handling the paperwork associated with each purchase order is $25. This high cost is viewed as a problem of economy. After the installation of a new method of processing, the cost per purchase is substantially reduced—from $25 per order to $8 per order.

Projects with clear-cut savings are likely candidates for conversion to computer-based methods of processing. Much like the problem of accuracy, the problem of economy is relatively easy to identify. The danger with this area is the naive assumption—by both users and system managers—that computer processing will eliminate the problem of economy. Ask budget managers whether this assumption is true. They will report that some projects cost far more than they return. Thus, before moving ahead on a project assignment, the analyst must ask, "Is the project worth doing?" A partial answer to this question follows from determining the return on investment expected from the project (see chapter 1). If the return is not sufficient, more economical projects should be selected.

THE PROBLEM OF TIMELINESS

The problem of timeliness relates more to the transmission of information than to the processing or storing of information. A system suffers from the problem of timeliness if information is available but cannot be retrieved when and where it is needed. As people become more familiar with information systems and how they function, they generally realize how much easier it is to process and store information than it is to retrieve it. For example, visualize your method of processing and storing canceled personal checks. Visualize next how you would retrieve a single check returned by the bank three months ago. If the number of checks is small, say one hundred or less, the search time to find the check would not be too great—say, five minutes or less. However, if the number is large—say, 10,000 or more—the time needed to find the check becomes considerable.

Organizations have committed extensive resources to handle the problem of timeliness in recent years. Fingertip access to information has been the desired objective. The findings to date show that only modest success has been achieved in improving this problem area. Only when retrieval problems are small and well defined has the overall success rate improved. For example, most stock market price retrieval systems work well. Ask any stockbroker for the latest price of a stock. The broker should be able to display the current price, the price at the beginning of the day, and the high and the low price for the day. This information should not take the broker minutes to retrieve. Rather, he or she should have nearly immediate fingertip access to such information.

THE PROBLEM OF CAPACITY

The problem of capacity occurs when a system component is not large enough. On holidays, for instance, it is often difficult to place a long-distance telephone call. Capacity problems are especially common in organizations that experience peak periods of business. During peak periods, inadequate processing capacity, transmission capacity, storage capacity, staff capacity, and the like may all exist. Capacity problems are also evident in rapidly growing organizations. With growth, smaller-capacity equipment must be replaced with larger-capacity equipment; smaller staff groups must be expanded into larger staff groups. In either case, some reorganization is needed to handle the increasing volume of business.

Many system problems are directed at solving capacity problems. Because it is often difficult to justify the purchase of new equipment or the hiring of new staff, people tend to put off such decisions until the very last moment. Consequently, when the systems group is contacted, the problem of capacity is easy to spot; the difficulty, however, lies in knowing how to handle the problem. Generally, the analyst suggests a short-term solution to the problem in order to gain time to permit the formulation of a longer-term solution. For instance, an analyst might recommend: "Let's hire five part-time employees to help us get through the peak period." When a short-term approach fails, the analyst may be tempted to implement a quick-fix computer-based solution. Unfortunately, this solution carries with it the associated danger of creating an even more severe system problem in the near future.

THE PROBLEM OF THROUGHPUT

The problem of throughput may be viewed as the reverse of the problem of capacity. Throughput deals with the efficiency of a system. If system capacity is high and production low, a problem of throughput occurs. Consider the following example.

Five programmers are assigned to a fairly straightforward programming assignment consisting of 10,000 lines of computer code. After thirty days of coding, the programming team is evaluated. It is discovered that they have completed 6,000 usable lines of code. Now, if each programmer worked eight hours a day, a total of 1,200 hours would have been expended on the project. Calculated differently, the average production rate for each programmer would be 5 lines of code per hour (6,000 lines divided by 1,200 hours). These findings might lead the analyst to conclude that there is a problem of throughput.

Similar to the problem of capacity, the problem of throughput may be much easier to spot than to treat. When repeated equipment breakdowns lead to low rates of production (and when the equipment has been purchased and cannot be returned), an organization can badger the vendor into fixing the equipment but can achieve little more short of legal action. Likewise, when groups of people exhibit low rates of production, such as the five-person programming team, the problem becomes even more complicated. Badgering and threats may not work at all. Rather, a manager must be able to determine the root cause of the problem in order for any improvement in throughput to take place.

Evaluating the Problem

Once a problem area has been identified, the evaluation can begin. Problem evaluation consists of asking the following questions: Why is it important that the problem be solved? What are possible solutions to the problem? What types of benefits can be expected once the problem is solved? There will be times when an analyst will recommend that nothing be done to solve the problem. Suppose that an analyst discovers that the real problem lies with the supervisor of an area. Because of mistakes by this one person, the throughput rate is 20 percent less than had been expected. However, suppose next that the supervisor is new to the job, is smart enough to realize where mistakes were made, and knows how not to repeat them in the future. Given this situation, the analyst might close the book on this project, recommending that no action be taken at this time.

Consider a different set of circumstances. Suppose that an analyst determines that a problem of low throughput can be traced to a computer printer. Suppose further that the problem must be corrected. Once the problem has been identified, the analyst would prepare a solutions table to list possible problem solutions and the expected benefits from each. **Table 4–1** illustrates such a table. Although the three solutions all have apparent benefits, the best solution is not at all evident. At first glance, it would appear that the final solution would depend on the availability of money to spend on improving the computer printer.

In this section, we have spent considerably more time examining how an analyst identifies a problem compared with how the problem is evaluated. In practice, this uneven split also occurs. As a general rule, analysts spend 75 percent of the system requirements phase of analysis defining the problem and 25 percent evaluating and documenting their findings. Note also that we have limited our discussion to seven major types of system problems. Because of this limitation, you might ask, "What about the problems of communication? Of group conflict? Of management? Of system security? Are these problems as well? Are these types of problems also evaluated by the analyst?" By all means, yes. Although our discussion has been restricted to more technical system

TABLE 4–1 Solutions table

Possible Solutions	Expected Benefits
Replace with new printer	Lowest expected maintenance Longest service contract Less noise Lowest ribbon cost
Replace with used printer	Lower expected maintenance Lower out-of-pocket costs Longer service contract Less noise
Repair existing printer	Lowest out-of-pocket cost Greatest product familiarity

problems, individual or group problems also occur in a systems environment and require identification and evaluation.

Still another limitation is the coverage given to determining the feasibility of taking some action to solve a problem. The concept of *feasibility* entails the joint questions of "Can something be done?" and, if so, "Should it be done given a particular set of circumstances?" For example, is it possible to climb a mountain when we have at our disposal only a forty-foot rope, and, if so, should we attempt such a climb given the size of our rope? We will examine this concept in more detail at the end of this chapter.

PROBLEM IDENTIFICATION AND EVALUATION TOOLS

Various tools are used routinely by systems analysts in their attempts to identify and evaluate different types of system problems. As discussed in chapter 3, one of these tools is the data flow diagram (DFD) and in particular the context diagram and the level-0 diagram. By drawing the context diagram, the analyst can conceptualize the important sources and sinks associated with a system. The level-0 DFD shows the major processing steps and the data flows important to each step.

Other tools are also important to problem identification and evaluation. Three of these are causal analysis, marginal analysis, and functional analysis. Causal analysis deals primarily with problem identification, while marginal analysis is a problem evaluation tool. Functional analysis is a system requirements tool and is solution directed in its orientation. Collectively, these tools reinforce (rather than diminish) the importance of DFDs. They help define the goals and purpose, boundary, main components, and interrelations between components of the system being studied.

Causal Analysis

Causal analysis (or cause-and-effect analysis) is a bubble charting activity designed to identify the factors that cause a problem to occur. Because people tend to concentrate on the effects and clues of a problem rather than its root cause or causes, analysts need a method to help them separate factors that appear to be the problem from factors that actually cause the problem.

Figure 4–1 shows the basic form of a causal analysis diagram. An analyst is required to separate cause from effect by asking a series of questions beginning with "why" or "what." Consider the following. Warren is in a terrible mood. He is difficult to talk to, and he is clearly angry. Joyce finds the courage to ask, "Warren, what's wrong? You seem troubled by something." Warren snaps, "I'm sorry if I appear to be a bit low today. I just didn't sleep well." After hearing this, Joyce leaves, confident that she has identified the problem. "Warren's problem," she states, "is the result of a poor night's sleep."

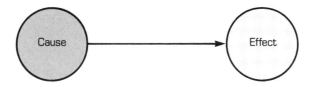

FIGURE 4–1 Cause-effect relationship

Unfortunately, Joyce's analysis did not go deep enough. She did not identify the root cause of Warren's problem. After Warren said he didn't sleep well, Joyce should have asked, "Why didn't you sleep well?" The "why" question in this instance would have made Warren think about what led to his sleep-disturbed night. He might then have stated, "I have been having trouble with my lower back. I may have pulled something yesterday moving some heavy boxes. I should see a doctor about it before it turns me into an emotional wreck."

As this example demonstrates, a causal analysis tends to identify the clues to a problem before it identifies a problem's true cause. Let's return to a previous example. Gayleen remarks, "We need a new budgeting system. Our current one seems to vary in quality from one month to another. Besides, reports are often late, incomplete, and contain misleading information. Why, we must spend a fortune trying to keep the system up and going." Jeffrey, a systems analyst, identifies several potential system problems based on Gayleen's statements. There appear to be problems of system reliability, accuracy, timeliness, and economy. In Jeffrey's mind, however, these are clues. They identify the different classes of problems that must be examined. What he must do next is to determine why these problems exist. This is necessary if he is to uncover the true causes of the problems. He begins to construct a causal diagram, placing the effects on the right-hand side, the clues running down the middle of the diagram, and the causes of the problem on the left-hand side. **Figure 4–2** illustrates the partially completed diagram. Jeffrey then asks, "Why are there too many errors?" His investigation uncovers the following:

1. Staff members responsible for entering data into the budget system are not well trained. The employees responsible for data entry are new and do not fully understand how to assign charges to departments.

2. The computer programs written for the system contain no built-in audit procedures. Lacking audit checks, incorrect expenditure totals are never spotted as part of processing.

3. The computer programs written to edit budget data are faulty. Budget account numbers are never edited for correctness. This permits charges belonging to one department to be posted to another.

Jeffrey might carry his analysis one step further. He could ask, "Why is there a lack of trained staff?" or "Why are there not built-in audits?"

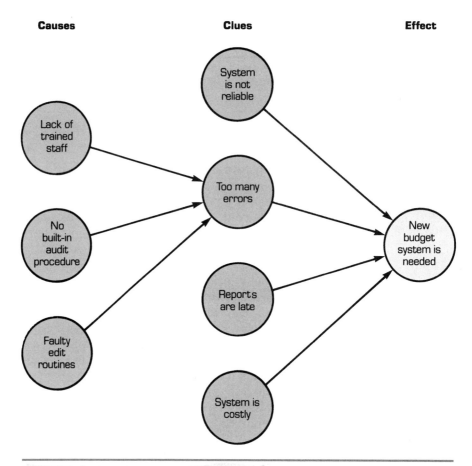

FIGURE 4-2 Causal analysis—an example

or "Why are there faulty edit routines?" These questions, however, will reveal little new information, other than to fix the blame on one or more individuals. Consequently, Jeffrey is satisfied. Of the three causes, all can be corrected. He recommends that staff be trained and that computer programs be modified to properly edit and audit budget system transactions.

Marginal Analysis

Marginal analysis is another graphic technique designed to help the analyst understand a problem environment. This method often picks up where causal analysis leaves off and is used more for problem evaluation than problem identification. Let's return to the previous example by assuming that Jeffrey is only partially successful in convincing others of the need to rewrite some of the budget system edit and audit procedures. He is now faced with the problem of recommending how much to spend

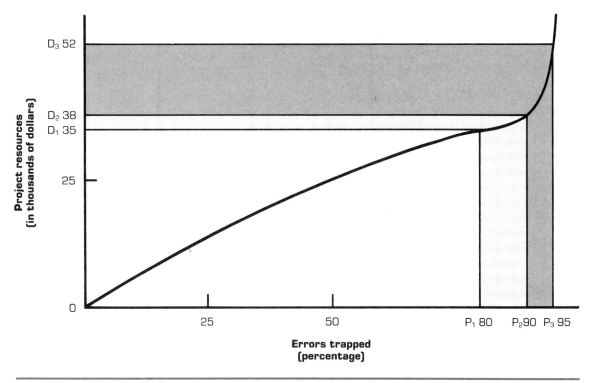

FIGURE 4–3 Project cost curve

on the project. To help him in his thinking, he constructs a cost curve, showing the percentage of errors that would be trapped at various dollar levels (see **Figure 4–3**). He then decides to compare the marginal effects of trapping errors at three levels: the 80 percent level (P_1), the 90 percent level (P_2), and the 95 percent level (P_3). He computes the following:

$$\frac{D_2 - D_1}{P_2 - P_1} = \frac{\$38,000 - 35,000}{90\% - 80\%} = \frac{\$3,000}{10\%}$$
$$= \$300 \text{ per one percentage change}$$

$$\frac{D_3 - D_2}{P_3 - P_2} = \frac{\$52,000 - 38,000}{95\% - 90\%} = \frac{\$14,000}{5\%}$$
$$= \$2,800 \text{ per one percentage change}$$

Because of the relatively little change in cost between D_2 and D_1, Jeffrey concludes that the 90 percent level can be justified. Nevertheless, he cannot recommend moving beyond the 90 percent level, because of the sharp increase in cost.

Figure 4–4 shows the two basic forms of the marginal analysis model. Generally, the analyst is asked to draw a cost curve to compare the resources (dollars, people, equipment) required of some activity with the

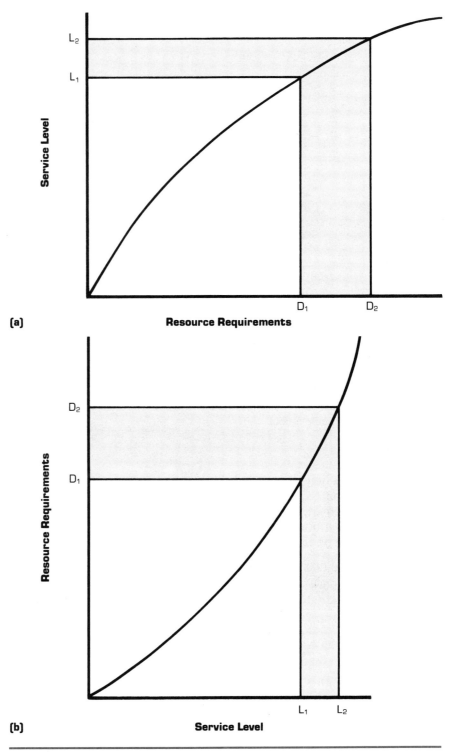

FIGURE 4–4 Marginal analysis—two basic forms

service level to be achieved. Defined early on, the *service level* becomes a standard to be incorporated into the system analysis and design project specification. For example, consider Figure 4–4 once again. A company might decide to increase the staff size from ten to twelve (D_1 and D_2). In so doing it projects increased production of from 100 units per day to 125 units per day (L_1 to L_2). The new service level (L_2) becomes the standard to be built into other company forecasts.

In practice, service levels can take a wide variety of forms, including the number of errors trapped, the number of lines printed per minute, the number of records translated, and the number of hours expended on product testing. During problem evaluation, the analyst attempts to define the service level to be achieved as a result of the successful implementation of a project. The resources needed to achieve this level are also determined. Both service-level projections and resource requirements are essential parts of any written document that defines system requirements.

Functional Analysis

Functional analysis is another graphic technique, though it is more solution directed than problem directed in its orientation. The primary purpose of functional analysis is to identify and describe the functions to be added to or modified within a system, and to show how those changes will affect existing service levels. A secondary purpose of functional analysis is to determine whether changes in function lead to higher or lower system resource requirements.

Figure 4–5 illustrates the general model associated with functional analysis. The analyst is required to study the old effect, as measured by

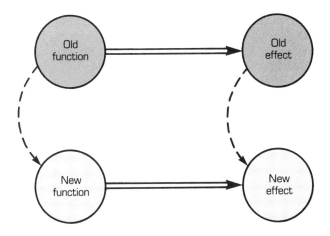

FIGURE 4–5   Functional analysis—general model

the existing service level, and to determine the system function or functions that bring about this effect. Following this, a new function must be prescribed. This new function produces a new effect, which, it is hoped, represents an improved service level.

As indicated by the general model, functional analysis is an extension of causal analysis. Let's examine two examples to show further how functional and causal analysis are related in practice. The first example is when Jeffrey determined (using causal analysis) that a lack of trained staff led to excessive errors in the budget system. In this instance, the effect is too many errors, the system function is data entry, and the existing level of service is the percentage of transactions entered into processing that are free from error. Jeffrey decides that he must somehow modify or change the system function in order to create a different effect—one that will improve the level of service.

Figure 4–6 illustrates how Jeffrey decides to change the functions of the system. First, he adds two new functions to monitor the types of data entry errors and to determine data entry training requirements (see Figure 4–6a). These new functions identify two types of data entry training needs: training for special problems and training for all newly hired personnel. Second, Jeffrey adds these two new data entry training functions to his model (see Figure 4–6b): new-hire training and special problems training. He reasons that a better-trained staff will create a new effect—allowing fewer erroneous transactions to enter into processing.

Figure 4–7 shows a second functional analysis diagram to describe how Jeffrey proposes to improve the budget system's faulty edit routines. He begins by stating his objective: to devise a plan that will improve the computer editing of budget system transactions. Following this, he decides to monitor editing problems (a new function) and to determine edit requirements (another new function). His study of edit requirements leads to the design of three edit procedures (three additional functions): an account number edit procedure, an account name edit procedure, and an account cross-check edit procedure. Jeffrey reasons, "Once added to the budget system software, these improved edit procedures will significantly improve the existing level of service. The effect will be to trap any erroneous budget number entered into processing, and to keep charges belonging to one department from being posted to another."

Sometimes an analyst will place small plus or minus signs beside functional analysis components. Plus signs are used to show net increases to system performance, while minus signs show net decreases to system performance (generally, higher costs). **Figure 4–8** illustrates the effect of making functional changes to the budget system. First, the various costs of systems study are noted: $1,000 to monitor editing problems and $5,000 to determine edit requirements. Second, the costs of adding procedures to the system are estimated; $2,500 to design the account number edit procedure, $2,500 to design the account name edit procedure, and $4,000 to design the account cross-check edit procedure.

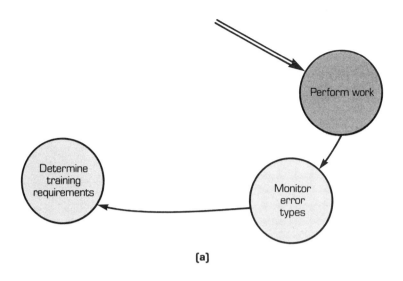

(a)

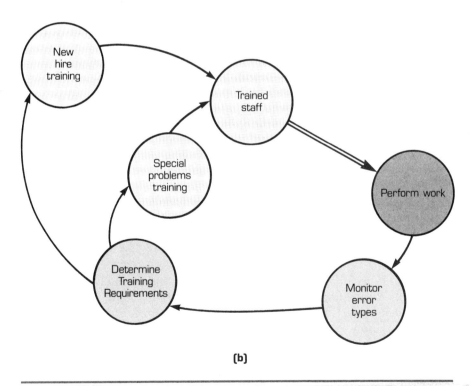

(b)

FIGURE 4–6 Functional analysis (a) to determine new requirements and (b) steps to create desired effect

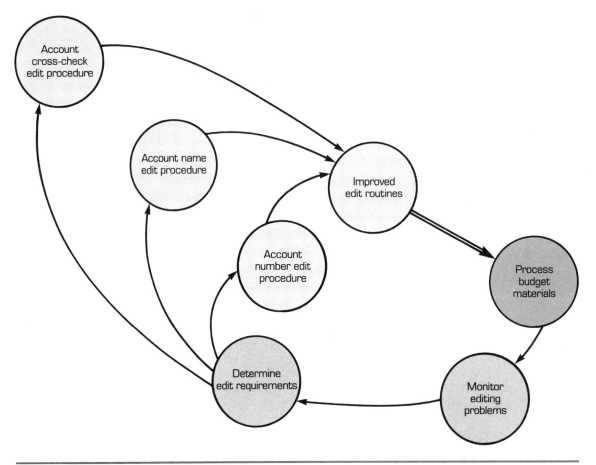

FIGURE 4-7 Functional analysis—another example

Third, the benefits of each new procedure are indicated; $10,000 in savings from the new cross-check edit procedure, $6,000 from the new account name edit procedure, and $8,000 from the new account number edit procedure. Finally, the overall benefit is shown: a $9,000 savings in the processing of budget materials.

The value in identifying the costs of system improvements and where they occur, as well as the system benefits and their dollar value, is that it simplifies projecting the return on investment for a project. **Figure 4-9** illustrates how such a return is calculated. Initially, the total system investment is entered in the table. Next, system savings are determined where, as shown, savings result from an improved service level. Consider the savings that follow from reducing the number of errors from sixty to twenty per month. As shown, the total cost of resolving these errors will be reduced from $3,000 to $1,000 per month. The net effect

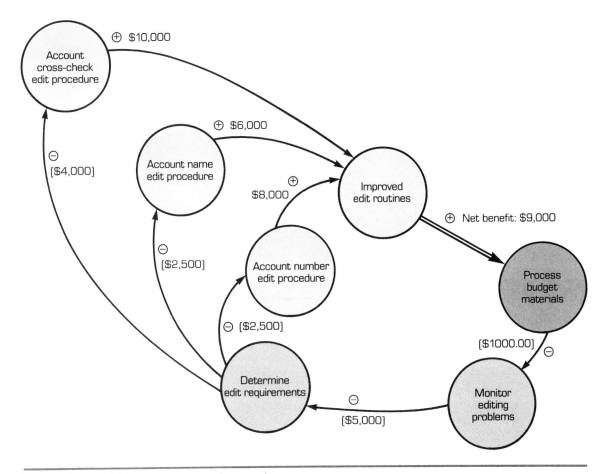

FIGURE 4–8 Adding system costs and system investment

is a monthly savings of $2,000. On an annual basis, system savings of $24,000 ($2,000 × 12), divided by the system investment of $15,000, leads to a projected return on investment of 160 percent; the payback period for this project is .63 years, or approximately thirty-three weeks.

We should keep in mind that causal analysis, marginal analysis, and functional analysis are problem identification and evaluation tools. As tools, they may or may not be appropriate in a particular situation. Moreover, they can be used in conjunction with other tools, such as DFDs and organization charts. With proper use, however, their purpose should be clear: they help clarify the true nature of a problem situation and help show what can be done to fix or control the situation.

Each of these tools is used informally by the practicing systems analyst. Marginal analysis, for instance, might be derived conceptually on the back of an envelope. Most analysts generally do not spend countless

System investment	$15,000
System savings	24,000
Return on investment	160%
Payback period	.63

	Before Improvements	After Improvements
Cost/error	$50	$50
Number of errors	60/month	20/month
Total cost	$3,000/month	$1,000/month
Net dollar change		$2,000/month
Annual dollar change		$24,000/year

FIGURE 4-9 Determining return on investment

hours attempting to draw carefully balanced causal or functional diagrams. Their diagrams tend to be personal sketches, drawn to better understand the system they have been asked to investigate. The diagrams are carefully drawn only when they are to be presented to the client or decision maker—to prove a point or to explain the logic of a situation.

If you compare and contrast the three methods of analysis presented here to the design of a DFD, you might conclude that a functional analysis diagram is similar to a DFD. This is correct. Without too much difficulty, we could have transformed Figures 4–7 and 4–8 into DFDs. For now, remember that the tools of systems analysis and design are not selected at random. Rather, certain tools complement one another. Collectively, they greatly assist the analyst in finding solutions to systems problems.

FINALIZING PROJECT REQUIREMENTS

Following the definition and evaluation of a problem, the analyst must determine whether it is feasible to take some action. The issue of feasibility asks the joint questions of "Can something be done?" and, if so, "Should it be done given a particular set of circumstances?" Once the questions of feasibility have been answered, the analyst can go ahead and prepare the *project specification*. Such a specification is used to finalize project requirements. It describes how a system is to be modified, based on satisfying a defined set of business needs, and how these modifications will change existing service levels in the organization.

Determining Project Feasibility

The analyst or analysis team must conduct a preliminary feasibility study before any project recommendations can be made. The purpose of such a study is to indicate which course of action has the best chance for success and, conversely, which action should not be undertaken at this time. The analyst must always remember that even the best possible projects may need to be rejected or delayed, given a particular set of circumstances. Let's return to our previous example one final time. Suppose that Jeffrey's feasibility study indicated that the budget system required extensive modification and, based on a partial cost analysis (see Figure 4–9), the risk of such a project, measured in economic terms, was slight. However, let's suppose next that Jeffrey decides not to recommend that the project be scheduled for development at this time. He reports, "It is not feasible to modify the budget system at this time." Why, you might ask. Jeffrey adds, "Members of the data processing department quite agree that there are problems with the budget system. However, at present, they have too great a work backlog to be able to turn their attention to this project. Hiring additional staff will not help, either. No one has the time to train new employees."

When examining the issue of feasibility, the analyst must consider five areas: economic, technical, social, management, and legal. Each type is briefly considered next.

1. Economic feasibility—an evaluation of the economic risk associated with a new project. This evaluation must compare one-time developmental and recurring system costs with cost reductions and benefits. A return-on-investment analysis helps the analyst determine whether a project is feasible from the standpoint of economics.

2. Technical feasibility—an evaluation of the technical possibility and the risk associated with completing a new project. This evaluation must examine both equipment and staff constraints; it must determine whether the proposed system can be implemented fully within the projected time frame.

3. Social feasibility—a determination of whether a proposed project will be acceptable to people employed by the organization. This determination typically examines the probability of the project being accepted by the parties directly affected by the proposed system change.

4. Management feasibility—a determination of whether a proposed project infringes on established company policy and, as such, whether the project will be acceptable to management. If management does not approve a proposed project, it will be viewed as infeasible.

5. Legal feasibility—a determination of whether a proposed project infringes on known federal, state, or local laws as well as any pending legislation. Although in some instances the project might appear sound, on closer investigation it may be found to infringe on several legal areas.

Of these five types of feasibility, technical feasibility generally is the most difficult to determine. Consider the following illustration. A medium-size tool and die manufacturer ($5 million annually in sales) decides to develop a job costing and order processing software product, working cooperatively with a leading computer vendor. The vendor agrees to write, test, and implement all software in six months, for $50,000. Three years later, the money is spent; however, the plans for the software remain on the drawing board. The president of the company explains, "The vendor went through a reorganization and we got trapped in the middle."

Meanwhile, the vendor, still believing in the proposed software, assigns a new programming team to the project. This team determines that the company's computer is too small and requires an upgrade to complete the job. The cost of the upgrade is set at $30,000; the additional cost to complete the software product is set at $60,000. In repayment for the mistakes of the past, the vendor gives the company a credit of $50,000. The company, once again optimistic about the project, agrees to continue the development.

Examples such as this can warn beginning analysts to expect the unexpected when estimating cost and time figures for a new project. Generally, more time should be spent defining the project before deciding to begin any design or implementation. Moreover, feasibility determinations typically occur more than once—early on, in deciding whether or not to begin the logical design, as well as later, in deciding whether or not to begin the technical design. For this reason, this early determination of feasibility is often called the *preliminary feasibility study*, or simply the *project feasibility study*.

Project Specification

The project specification is prepared in order to record the results of the system requirements analysis. It describes the problems to be resolved and the results of the project feasibility study. **Table 4–2** shows the major parts of this planning document. In practice, the format for this specification varies somewhat from one organization to another. Even so, the intent of the document remains the same: it should identify which system problem was studied, describe the different types of analyses performed, show the findings of these analyses, and explain the recommended set of actions.

Let's briefly examine each major part of the specification outlined in the table.

The *executive overview* summarizes the problem investigated and the course of study; it lists the major findings of the study, followed by the recommended set of actions. There are several reasons for summarizing the purpose and results of the study before beginning to consider any details. First, managers, pressed for time, may read only the executive summary. Second, some managers will read the entire proposal only if

TABLE 4-2 Project specification outline

Executive Overview

- Which problem was investigated?
- How was it studied?
- What were the findings of the study?
- What is the purpose of this report?

System Description

- What are the main parts of the system?
- How are they tied together?
- Which functions do they perform?
- Why is there a problem with performance?

Project Description

- What improved functions are being proposed?
- How will these improvements change existing service levels?

Feasibility

- What are the implications of the new system?
- Are these implications acceptable?

Planning Assumptions

- What assumptions were made during the preliminary investigation?

Alternatives

- What alternative courses of action were considered?
- What additional functions are possible?

User Support

- What resources will be required by the user during the development period?
- What types of documentation are to be prepared for the user?

Detailed Schedule

- How long will it take to analyze, design, and implement the new system?
- How much will it cost?
- What are the various resource requirements?

it attracts their interest. If the recommendations drawn by the study are significant, the chances improve that the manager will read the entire report. Third, managers are familiar with a reporting format that provides an executive summary, followed by pages of supporting detail. Well-designed computer-printed reports, for instance, begin with a summary page followed by the report details. Fourth, managers contend, and often rightly, that if the findings of a report cannot be summarized in one page (or at most two), the body of the report is probably not well thought out.

The *system description* identifies the main components of the existing system and shows how these components are interrelated; it also defines the main functions performed by the system and explains why there is a problem with its performance. This section describes the system and the system environment and defines the problems discovered by the analyst. A context diagram and a level-0 DFD can be extremely useful in describing the main components and important data flows within a system. A causal analysis diagram may also help the analyst define the problems with system performance. Regardless of which tools are used, this section must document the nature of the problem, stating clearly, its causes and effects.

The *project description* outlines the changes to be made to the existing system, specifies why these changes are recommended, and projects how they will modify existing service levels. This section is often the most difficult to prepare. It must describe in sufficient detail the set of actions to be taken and why those actions are recommended. In preparing the project description, the analyst may use marginal and functional analysis diagrams. In particular, he or she must show how changes to system functions will lead to improved levels of service.

The *feasibility* section describes why the proposed project is feasible from economic, technical, social, management, and legal standpoints and describes the implications of the fully implemented project. This part summarizes the findings of the project feasibility study. It states why the proposed course of action is most likely to succeed. It identifies the risks associated with the proposed project and indicates whether these risks are slight, moderate, or large. It summarizes the implications of the fully designed and implemented project and specifies whether these implications are acceptable.

The three-step sequence of *actions* (to be taken), *risks* (to be borne), and *implications* (to be realized) is a familiar decision-making format. Besides defining what is feasible, the analyst clarifies the risks and implications of the decision to proceed. In this way, decision makers have a means of comparing one project with another.

The *planning assumptions* section describes any assumptions made during the course of the study, along with why those assumptions were necessary. Generally, the analyst must make a variety of assumptions when attempting to define system requirements. These might be hardware related ("Ample computer time will be available"), staff related ("Existing staff will design the proposed system"), or even strategy or policy related ("The organization encourages decentralization of computing").

On the one hand, planning assumptions serve as a protective measure. By stating these assumptions, analysts can protect (at least partially) their conclusions. On the other hand, assumptions identify additional risks of the proposed system. Decision makers need to know which parts of the study were based on assumptions, and they need to evaluate those assumptions in terms of their soundness.

The *alternatives* section describes other possible courses of action and explains the basis for determining the scope of the proposed project.

Besides proposing one best course of action, as described in the project description and feasibility sections, the analyst is expected to provide evidence that other, less favored ways of solving a problem have been considered. Some managers refuse to accept a proposal if alternative courses of action have not been considered.

Along with describing different courses of action, the analyst is also expected to be able to defend the scope of the proposed project. The *scope* determines the extent to which features are to be added to a new system. For example, the analyst might propose a *bare-bones system*. As the term suggests, such a proposed system is nothing fancy but will get the job done. Conversely, the analyst might enlarge the scope of the project by proposing a *fully equipped system*, which not only gets the job done but also handles both current and projected managerial reporting needs. Although the fully equipped system would be more expensive and more difficult to design than the bare-bones system, it would be expected to do more. In practice, the analyst generally is advised to begin with a bare-bones design, adding functions (or enhancements) to the system as time and cost allow. The alternatives section of the proposal describes these additional functions.

The *user support* section specifies the resources required by the user during the analysis, design, implementation, and operation of the new system and explains the type of education required for the user to gain the full benefits of the new system. During the shuffle to get a favored project approved, the analyst may be tempted to bypass the user, structuring the project proposal to highlight the work to be done by the systems group. Likewise, managers often assume that users are fully aware of their responsibilities and that they will be able to give full support to a new system without needing additional resources. When these circumstances occur, the users feel left out of a discussion in which they should play an important role.

To avoid bypassing the user, the analyst should include a separate section describing user resource requirements and education needs. This section identifies user resource needs (dollars, staff, equipment) that are a direct result of the proposed project. It should outline educational requirements for users. The analyst is also expected to indicate, as well as possible, the kind of training to be conducted (and who will conduct it) and the kind of user documentation to be prepared, with preliminary descriptions of its contents.

The *detailed schedule* section provides the preliminary cost and time estimates and establishes the resource requirements of the project. This final section could also be called the project management section. It most often consists of a project planning chart and a network chart (see chapter 1). These charts project the number of labor hours and dollars required for system development over the estimated life of the project. They identify the main project activities and show estimated project milestones.

The Mansfield, Inc., Case Study

The systems analysis team of Roger Bates, Carolyn Liddy, Neil Mann, and John Havensek took two weeks to define the problems with the current billing system. Carolyn discovered that it took up to five days to print invoices following shipments during peak periods. Neil was concerned with the problems of posting receipt summaries to the accounts receivable (A/R) file. "Sometimes we bill customers after they have paid," he remarked. "We are awfully slow in posting payments."

Roger asked John to study the procedures used in updating customer and product records. John's findings were similar to Carolyn's and Neil's. "It takes up to two weeks for marketing to change a customer record," added John. "Manufacturing is better with posting new prices, but I suspect that because of the delay in revising product records, we sometimes have billed our customers with old price information."

Roger made some discoveries of his own. "I have found problems with A/R processing," he commented. "Sometimes we cannot match a set of invoice charges to a set of customer payments. As a consequence, the system continues to bill customers, even though they may have paid. If this is not bad enough, when the irate customer calls to complain, we lack customer-complaint procedures. It is quite clear that our current system is causing our customers to be very upset with the way Mansfield conducts its business."

Causal Analysis

Roger drew a causal analysis diagram to show the analysis team which factors contributed to problems with the current A/R system (see **Figure 4—10**). He went on to explain his diagram. "Customers are upset for two reasons," he said. "The current system produces bills that are inaccurate, and we lack procedures to handle customer complaints. As to why the accounts receivable system is not accurate, my preliminary investigation led to three possible causes. First, we make frequent account number changes. Each month we must process one hundred or more changes. These changed accounts are the ones we seem to have problems with. Second, we lack adequate system update procedures. We might change customer account numbers showing balance-due amounts, only to later forget to change the account numbers for payments to be applied to the old number. This leads to the situation Neil discovered, that we sometimes bill customers after they have paid. Third, we attempt to match charges to payments only at month end. Since 1,000 accounts must be processed in a day or two, it is little wonder that we make mistakes."

Roger went on to address some of the root problems. "These are items that must be improved with any new system," he continued.

"Customer account numbers are tied to the customer's postal code," he explained. "The ramifications of our approach are that whenever a customer changes his or her address, we have to modify our account number sequence. Moving three blocks away can change a customer account number from 97401CAPP to 97403CAPP. In addition, customer update procedures were never integrated. Consequently, if we change a customer account number, we

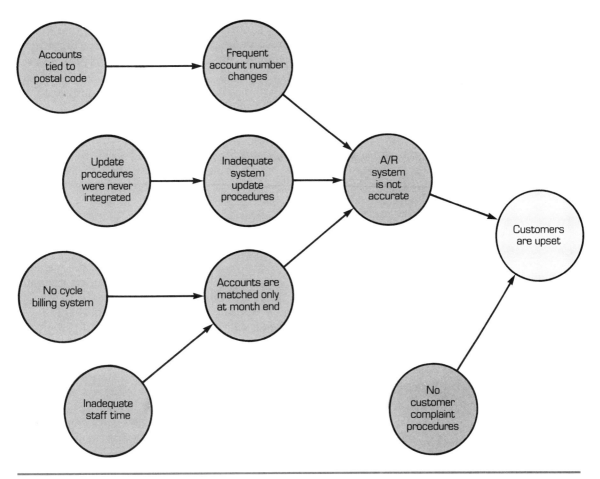

FIGURE 4-10 Causal diagram tracing reasons for customer complaints

usually fail to tell the staff processing customer payments of this change. When this happens, an incoming payment may be applied to a nonexistent account.

"We need to consider changing over to a cycle billing system," Roger continued. "With such a system, monthly statements for groups of accounts are produced at different times during the month. For example, we might decide to divide our accounts receivable accounts into four groups and produce monthly statements four times each month."

"Finally," Roger concluded, "inadequate staff time is another root problem. Our small receivables staff is one of the reasons why we currently use month-end billing rather than cycle billing. At present, we would be overwhelmed if we were told to produce statements at four different times during the month. The staff would probably tell us something like, 'Hey, no way. That's impossible!'"

SUMMARY

When determining system requirements, the systems analysis team must define the problem to be resolved, evaluate the problem, determine whether it is feasible to do something about it, and, providing it is, prepare the project specification.

Defining the problem is often a difficult and time-consuming task. In practice, most analysts look for common system problems, including the seven types discussed in this chapter (the problems of reliability, validity, accuracy, economy, timeliness, capacity, and throughput). However, other problems arise as well, including accounts tied to postal codes or such cumbersome business practices as month-end billing for all customers rather than some form of cycle billing.

The analysis team often uses graphic representations to define and evaluate system problems. Causal analysis helps analysts define the root causes of a problem, while marginal analysis helps them evaluate the problem environment. Functional analysis, a third technique, helps analysts identify the functions to be added to or modified within a system. Functional analysis also determines whether changes in functions lead to increased or diminished system resource requirements.

Following the definition of system problems and their evaluation, project requirements can be finalized. The major product of this activity is the project specification: a written document that records the results of the system requirements analysis. Besides describing the problems to be resolved, the specification summarizes the results of the project feasibility study.

Project feasibility is the determination of which course of action is most likely to succeed. When examining the issue of feasibility, the analyst must consider five types: economic, technical, social, management, and legal.

The contents of the project specification draw heavily on the findings of the feasibility study. The heart of this document is the project description, which describes the changes to the existing system and explains why these changes are needed and how they will modify existing service levels. Immediately following this section is one entitled "Feasibility." This section discusses both the feasibility of the project as well as the implications of the fully designed and implemented system.

Completion of the project specification provides the setting for the next stage of analysis, which is the detailed examination of user requirements. At this point, the project specification has documented the findings brought about by preliminary fact-finding: the definition of the problem, the determination of whether it is feasible to proceed with the project, and the preliminary view of the functions of the proposed system. The next step is further study. The analysis team must collect additional information to determine the full extent of user requirements and the degree to which a new system can satisfy those requirements.

REVIEW QUESTIONS

4–1. When does a system suffer from the problem of reliability?

4–2. How does the problem of validity differ from the problem of accuracy?

4–3. How does the problem of capacity differ from the problem of throughput?

4–4. What should be done once a problem area in a system has been identified?

4–5. What is causal analysis designed to accomplish?

4–6. What is meant by a "service level"?

4–7. What is the purpose of functional analysis?

4–8. What are the types of feasibility that the analyst must consider?

4–9. How does the analysis team finalize project requirements?

4–10. What is the difference between a bare-bones system and a fully equipped system?

4–11. What are the major parts of a project specification?

EXERCISES

4–1. What is wrong with the following causal analysis diagram?

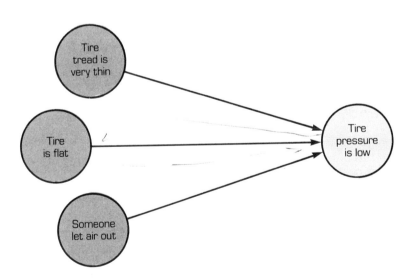

4–2. How would an analyst interpret the following functional analysis diagram?

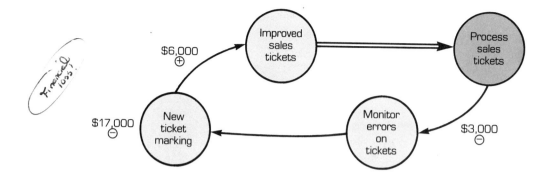

4-3. The Bohemia Elementary School asked the district's systems analyst, Jill McCone, to help determine the number of computers needed for classroom use. Jill estimated that each computer would cost $2,000. She determined that one hundred children would need access to a computer; however, not all of the children would use the computer room at the same time.

Jill estimated that if twenty-five computers were purchased, 75 percent of the students would have access to a machine when they needed one. This percentage would increase to 80 percent if thirty computers were purchased, to 85 percent if thirty-six computers were purchased, to 90 percent if forty-four computers were purchased, and to 95 percent if fifty-nine computers were purchased.

First, prepare a marginal analysis chart for the principal of the Bohemia Elementary School showing how resource requirements change as the service level improves.

Second, how many computers should be acquired if the principal decides not to pay more than $3,200 for improving the service level by 1 percent? (Construct a table to determine your answer.)

4-4. The president of the A-Bacon Company, Inc., was bothered by the current billing system. "We take too long to get out customer invoices," he muttered. "Some invoices are sent out anywhere from seven to ten days after we ship goods to customers."

George Falsted, A-Bacon's systems specialist, is asked to examine the customer invoicing problem. He discovered that part of the delay between shipping and invoicing is to be expected. "We have to transfer unit prices from the price book to the invoice, compute the extended price, and determine the dollar amount for each invoice," remarked Guy Williams, head of the billing department. "Besides," he added, "we still have to type each invoice."

George discovered that part of the delay was abnormal. "Shipping papers are incomplete," he stated. "Sometimes freight and insurance charges are missing; at other times, the customer number is missing. We even run into some papers where the number of units shipped is missing. This gives our billing clerks a fit!"

George then questioned Martha Mathews, shipping supervisor, to clarify several items. Martha remarked, "We don't always check the shipping papers before sending them to billing. Our shipping foreman tries, but he's not perfect. It really gets hectic on the floor with all shipments headed off at once."

Prepare a causal analysis diagram to show the clues and causes of A-Bacon's billing problem.

4–5. George Falsted, systems specialist at A-Bacon Company, Inc., needed to determine whether it would be worthwhile to improve the paperwork processing procedures of the shipping department (see exercise 4–4). He estimated that it would cost $500 each month for a clerk to check all shipping papers before copies were sent to the customer and to the billing department. The advantage of this checking would be to send out customer invoices earlier.

George estimated that once prompt billing procedures were in place, the savings to the company would be substantial. "Accounts receivable dollars would be reduced by $100,000," George indicated. "Since we have to borrow money to finance our receivables, we should be able to save a bundle. Our current cost of capital (the interest charged A-Bacon on its borrowed money) is currently averaging 12 percent."

Prepare a functional analysis diagram to show the change in A-Bacon's system. Show increases and decreases in system performance and the overall effect of making the functional change.

4–6. Jeffrey, the systems analyst discussed earlier, decided to estimate the return on investment to be expected from improvements in data entry procedures (see Figure 4–6b). He estimated that with improved training, the number of data entry errors could be reduced by sixty per month (from one hundred to forty). He estimated next that the cost to the company for each error was $20.

Jeffrey also came up with training cost estimates. "It will run $500 per month for training new employees and $400 per month to provide training on special problems," he indicated. "Training is not cheap."

Finally, Jeffrey determined that it would cost $100 per month to monitor each error type and $200 per month to determine new training requirements.

Set up a table similar to Table 4–2 to show the return on investment resulting from the improved data entry system.

WORLD INTERIORS, INC.—CASE STUDY 4
DESIGNING CAUSAL ANALYSIS DIAGRAMS

Introduction

Rena and Ray Logen decided to employ John Welby, the systems consultant, because they realized they needed help. It seemed that as the business grew, so did the number of business problems. "We are out of control," remarked Rena. "We need someone to help us define our problems."

John took on his new task at WI with great enthusiasm. He was quick to grasp the way in which WI operated and was able to understand why Rena believed the company to be out of control. What he did not understand immediately was how to solve WI's data processing problems. He decided that what he needed to do first was gather information and study WI's current procedures. "I'd better define WI's problems before I go much further," he said.

Retail Store Problems

John's preliminary discussions with Rena and Ray, and later with the four regional retail store managers, indicated that improving retail store management was an important near-term goal of the company. Ray summarized: "We need a new system. This is our biggest retail store problem. And we must improve store management before we can open any additional stores."

Rena went on to elaborate: "Currently, we have no retail sales dollar controls. Some of our managers could be robbing us blind and we could only suspect what was going on." Ray added: "We do not know where we are making, or losing, money. We ask store managers to supply their regional managers with weekly and monthly store sales figures, but this information tells us little about which products are selling well and nothing about the profit we are making on our various product lines. We simply have not had the time to write retail store procedures for handling these matters. These procedures should tell us which types of data we need and how we should process them."

John's discussions with regional store managers led to similar responses. Here is a sampling of the regional managers' comments:

"We need some direction from top management."

"Perhaps we could install some type of point-of-sale system."

"We need to know which products to stock and to keep track of which products customers buy from us."

"We need a computer to help us manage our data."

"We need to know the effects of our various advertising and promotional programs. Currently, we do not have this information."

"I think the key to improving our information lies in the acquisition of a new computer system."

"We need a management training program."

"We need someone to help us write procedures for our retail stores. We do not know how to write procedures. We do not know where to begin."

Direct Sales Processing Problems

John's discussion with members of the direct sales staff led him to conclude that the retail stores were not the only problem area for WI. Krintine Rinehart in direct sales stated, "We have order processing, computer, employee, customer, and interoffice problems. Where do you wish to start?" As an opener, John obtained permission to interview Krintine's staff.

"We need a new computer," remarked Morris Fisher. "This so-called custom software doesn't work. It has bugs. It can't be maintained. We get into a sort routine and, at times, can never get out."

Patricia Alexander stated, "We cannot keep track of our problem orders. We should be able to enter the details of problem orders directly into the computer, but right now we can't. Marty Rhew tells us that he's too busy to make changes. Frankly, the system is so fragile, I don't think he dares."

"We need a better method of helping customers with their questions," Sheela Robbins indicated. "We could save so much time if we could retrieve information from the computer directly. For example, a customer might call to find out about an order. If the computer is busy—as it always is—we have to take the customer's name and number, indicating that we will call back. Why is the computer always busy? It's too small."

WI's Computer System

The computer system currently used by WI consisted of an off-brand microprocessor, with 64K bytes of memory, a 10-megabyte hard disk, a single floppy disk drive, and a standalone, dot matrix printer. When WI purchased the computer, it also contracted for and obtained an order processing software system. The custom-made software was designed by a small data processing consulting firm. Members of the firm spent several months working directly with Rena and her staff in developing the design. The software permitted WI to keep information on all customer orders on line. In addition, the software featured the printing of customer invoices, schedules of all orders placed with WI's suppliers, and mailing labels for any or all of WI's direct mail customers.

John's discussions with Ralf Elders and Marty Rhew suggested that the small computer had been inadequate from the very beginning. Although it was large enough to handle customer orders for direct sales, it was far too small to handle both retail store and direct sales orders. Moreover, it could not be used to process the company's accounts payables or to analyze direct mail and retail store sales. Marty Rhew indicated that he was wary of changing the existing computer programs. "The off-brand hardware and the custom software are giving us fits," he complained. "First, it's almost impossible to find someone qualified to maintain the machine. Second, the software was designed to be turnkey. We were told to turn the computer on and to use it. The software development team never provided us with documentation or even source code. Frankly, I think we purchased a piece of junk."

CASE ASSIGNMENT 1

Prepare a causal analysis diagram showing which factors have contributed to Ray's statement that the retail stores need a new system. Based on your analysis, what are the underlying problems? What, if any, are the intermediate problems? What needs will be satisfied by resolving these problems?

CASE ASSIGNMENT 2

Prepare a causal analysis diagram showing which factors have contributed to the general feeling that WI needs a new computer system. Based on your analysis, what are the underlying problems? What are the keys to resolving those problems?

5
Data Collection and Analysis

INTRODUCTION

ONCE the systems analyst has determined that a problem exists and has obtained permission to do something about it, the process of data collection and analysis can begin. At first glance, the work associated with data collection and analysis appears to be relatively straightforward, especially when system problems have been well defined in advance. Nevertheless, this view can be misleading. In practice, it is difficult to determine which types of information should be collected and analyzed. Consider the following:

Suppose that a group of students is asked, "What type of information would you like to have about a new school?" One member of the group might state, "I would like to see the artist's drawing of the school." Another might add, "I would like a summary of why we need the school." Still another might ask, "I would like to see a breakdown of the new school costs."

What does this example suggest? It suggests that people differ widely when it comes to their perceived need for information. Some people want more information than others, while some want information that is presented differently. The example also suggests that data collection and analysis can have political implications, especially when the analyst must decide how to satisfy everyone's needs and when all those needs are different.

In this chapter, we will use broad definitions to depict the process of data collection and data analysis. *Data collection*, the gathering of systems information, includes two main parts: tracing data flows and conducting interviews, and determining which types of decision-related materials are important to the system. *Data analysis* involves making sense of the collected data. The purpose of data analysis is to show how data logically flow within the system.

We will begin by looking at data flow diagrams (DFDs) once again. The systems analyst constructs DFDs to determine the characteristics of a system and to specify the types of information to collect and to analyze. Next, the analyst considers using personal interviews and questionnaires. By asking questions, the analyst can pinpoint user expectations and fears, system requirements, and system features that would be nice to have. Finally, the importance of written procedures and internal and external forms will be reviewed. This evidence helps show how processing is organized. In many instances, procedures and forms show the decision-making requirements of a new system.

When you complete this chapter, you should be able to

- describe why DFDs are important to data collection;
- decompose a top-level DFD into ever-smaller pieces;
- identify the functional primitives of a system and determine whether the system is in balance;
- understand the differences between a structured and a nonstructured personal interview;
- summarize the advantages and disadvantages of questionnaires; and
- explain why secondary sources are important to data collection and analysis.

DATA COLLECTION AND DATA FLOW DIAGRAMS

Data collection is not a random activity. Before it is started, the analysis team must have a good idea of who should be interviewed and why. Team members need to know which forms, documents, reports, and other materials are to be collected and how each item relates to the system under study. How, then, does the team devise a data collection plan? In large part, the question of where to begin and how to organize the work of data collection is answered once the context diagram and level-0 DFD have been finalized. Besides using these diagrams to show the main data flows within a system, the analysis team can add additional information to clarify and identify system properties. For example, the team can identify the names of departments, individuals and processes that send or receive information in a system (sources and sinks) and the departments, individuals, and processes, both human and machine, that transform information in a system.

Let's examine once again the context diagram prepared for the employee pay system (see **Figure 5–1**). This diagram focuses on three things about data collection: people to be interviewed outside the employee pay system, important inputs to the system, and the main outputs of the system. For example, people to be interviewed outside the employee pay

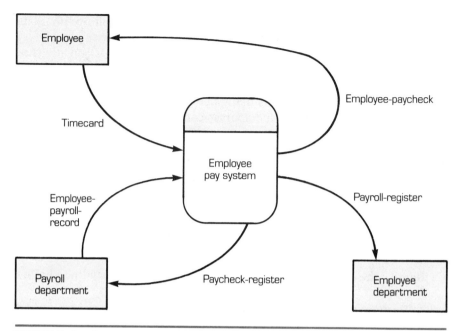

FIGURE 5–1 Context diagram

system include employees, managers of departments, and members of the payroll department. Important inputs to the system include the employee's payroll record and timecard. Samples of this input will have to be collected and studied. The main outputs of the system include the payroll register, the employee paycheck, and the paycheck register. Samples of these documents must also be collected and studied.

The level-0 DFD provides considerably more information about the employee pay system and the data collection effort than the context digram (see **Figure 5–2**). Besides planning to interview people and collect materials outside the system, the analyst can use the level-0 diagram for help in determining who should be interviewed and which forms to collect within the system. As illustrated, the analysis team must speak with people who prepare the record of hours worked, the payroll register, and the employee paychecks. In addition, the team must determine the contents and structure of several data stores. Specifically, the team must determine the contents of the hours-worked file, the employee master file, the current payroll file, and the pending checks file. Finally, the team must determine, by using personal interviews and studying procedures and forms, how transforms change data flows. For example, the team must discover how the hours-worked record and the previous payroll record are combined (see transform 2.0) to produce, as output, a new payroll record, the payroll register, and the current payroll records.

Data Collection and Data Flow Diagrams **135**

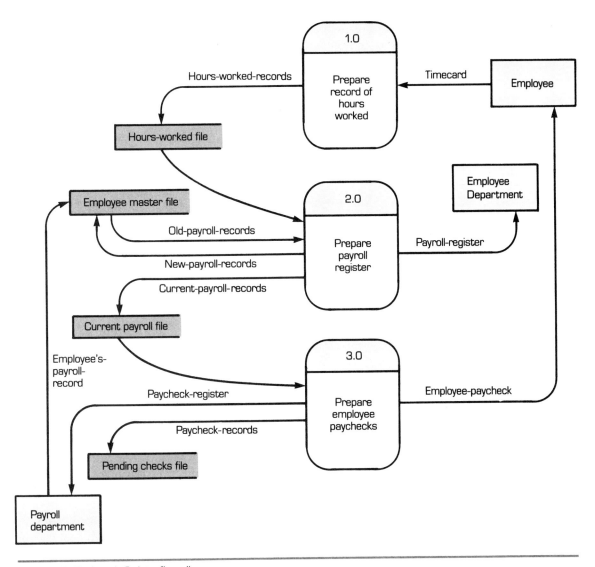

FIGURE 5–2 Level–0 data flow diagram

Data Collection and DFD Decomposition

When the analyst discovers that a DFD does not contain sufficient information to show which part of the system should be studied next, or whom to interview and why, this generally means that the DFD must be *decomposed*. With decomposition, the analyst breaks down the transforms of a system into ever-smaller pieces (e.g., functions are broken down into several subfunctions).

Let's suppose we determine that transform 2.0, prepare payroll register, is too complex. Accordingly, we need to decompose this transform into smaller, more easily understood transforms. **Figure 5–3** illustrates the level-2 DFD, which represents the decomposed view of transform 2.0. Why level-2? Because this number identifies the number of the transform to be described. If we had decided to decompose transform 1.0, we would have constructed a level-1 DFD. Or if we had decided to decompose transform 3.0, we would have constructed a level-3 DFD.

In this case, the level-2 DFD tells the analysis team much more about 1) the procedures required to prepare the payroll register, 2) the way in which the various data flows are interrelated, and 3) the data flows important to subfunctions. To demonstrate, let's consider each of the transforms shown by the level-2 DFD. Transform 2.1 suggests a very simple operation. The hours worked, appearing on hours-worked records, must be summed to arrive at total hours worked. Transforms 2.2 and 2.3 are somewhat more abstract. As indicated, the total-hours-worked figure is combined with the old payroll record to yield current payroll cost. Following this, the current payroll figure is used in calculating year-to-date pay and in producing a new payroll record and a current payroll summary. Transform 2.4 suggests another simple operation. The current payroll summary is needed in order to produce the payroll register—that is, a listing of current payroll costs.

What else does this analysis of a set of DFDs tell us? First, we learn which procedures are required—one procedure specifies how hours are to be summed, one indicates how current pay is to be calculated, one specifies how year-to-date pay is to be calculated, and one states how the payroll register is to be printed. Second, it shows how data flows are interrelated. The old payroll record data enter transform 2.2 to begin the process of calculating employee pay. The new payroll record data leave the system when both current and year-to-date pay have been calculated. Third, the analysis shows which data flows are required to link together subfunctions. Data on total hours worked and current payroll cost do not appear on the level-0 transform, since these flows are internal to transform 2.0. By decomposing the transform, the analysis team can look inside a function, so to speak, to identify data flows that link processing steps internal to a function.

FUNCTIONAL PRIMITIVES

By now some of you will have questions. You might ask, "Can we continue to break down a system into 2.1, 2.1.1, and even lower levels?" The answer is, "Yes, provided we need to decompose further." For example, assume that the transform of calculating current pay is too complex. You might ask, "How are we supposed to calculate current pay?"

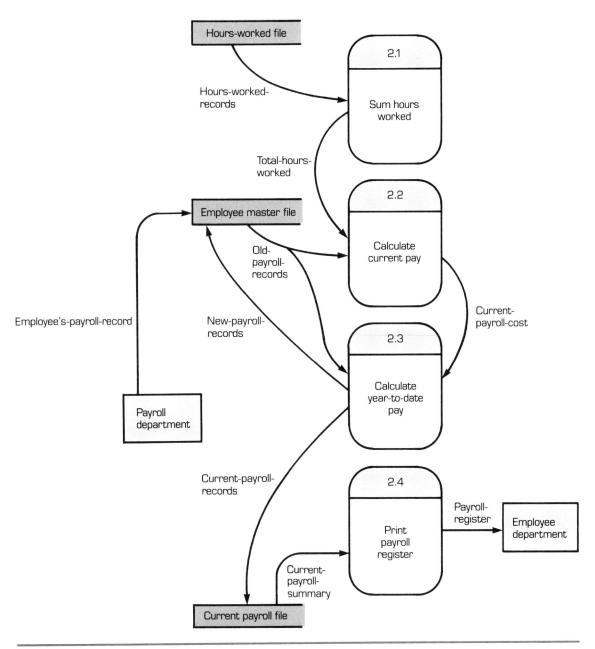

FIGURE 5–3 Level–2 data flow diagram

In outline form, we might decompose the transform further, as follows:

2.2 Calculate current pay
    2.2.1 Calculate regular pay
    2.2.2 Calculate overtime pay
    2.2.3 Calculate taxes
    2.2.4 Calculate other deductions

Following this, we can diagram a level-2.2 DFD, showing these four lower-level transforms and the data flows and data stores required for each transform.

You might also ask, "When do we stop?" In this instance, the answer is more arbitrary. We stop when each transform becomes clear enough to require no further explanation. When a transform reaches this state it is called a *functional primitive*. DeMarco[1] states that a functional primitive is a transform that requires no further decomposition.

Because this decision will differ among analysts, knowing when to stop must remain a subjective judgment. Figure 5–3 provides an example. One analyst might judge transform 2.1, sum hours worked, to be a functional primitive, while another analyst might divide this transform into a lower-level set of transforms, identifying the major functions as: "Obtain an hours-worked record," "Total the regular hours worked," and "Total the overtime hours worked." Fortunately, there is nothing wrong with one analyst stopping at one level and another analyst stopping at another level. This is correct provided an easy transition can be made from the DFD to the software design.

BALANCING

While the extent of a DFD's decomposition can be somewhat arbitrary, balancing cannot. *Balancing* is keeping inputs to and outputs from a transform consistent with inputs to and outputs from the decomposition of that transform. An example will clarify this definition. Level-0 in **Figure 5–4** shows data flows A and B as input to transform 2.0 and data flows C and D as output. When this first-level transform is decomposed into a second-level set of transforms (2.1, 2.2, and 2.3), all higher-level input and output must be accounted for. The level-2 DFD in the figure indicates that balancing exists. Inputs to this DFD consist of data flows A and B; outputs consist of data flows C and D.

To further clarify the concept of balancing, **Figure 5–5** illustrates an out-of-balance situation. The level-2 DFD contains input data flows E and B and output data flows C and F. (You might observe that this level-2 transform is the same as the transform numbered 2.2 on Figure 5–4). In contrast, Figure 5–5's level-2.2 DFD is out of balance. Only a single input, data flow B, is shown. Data flow E is missing.

Figure 5–6 provides another example of balancing, using a more realistic DFD. The level-2 transform, calculate current pay, is shown by the level-2.2 DFD as comprising four functions: 2.2.1, calculate regular

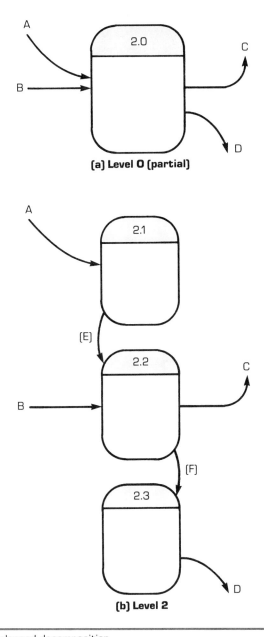

FIGURE 5–4 Balanced decomposition

pay; 2.2.2, calculate overtime pay; 2.2.3, calculate taxes; and 2.2.4, calculate other deductions. Meanwhile, inputs to the set of level-2.2 transforms consist of six different data flows: total hours worked, the hourly rate, the overtime rate, the tax rate, employee deductions, and employee

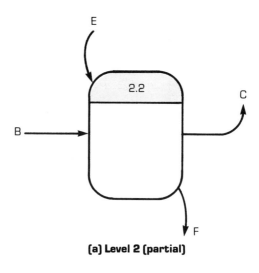

(a) Level 2 (partial)

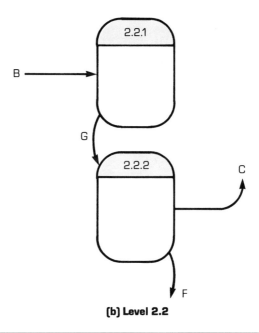

(b) Level 2.2

FIGURE 5–5 An out-of-balance decomposition

payroll history. With this understanding, you might ask, "Is level 2.2 in balance?" Two new rules must be understood before balancing becomes apparent.

First, data flows can be decomposed to clarify their meaning, much like transforms. In Figure 5–6b, the data concerning the old payroll record have been subdivided into four data flows: the hourly rate, the over-

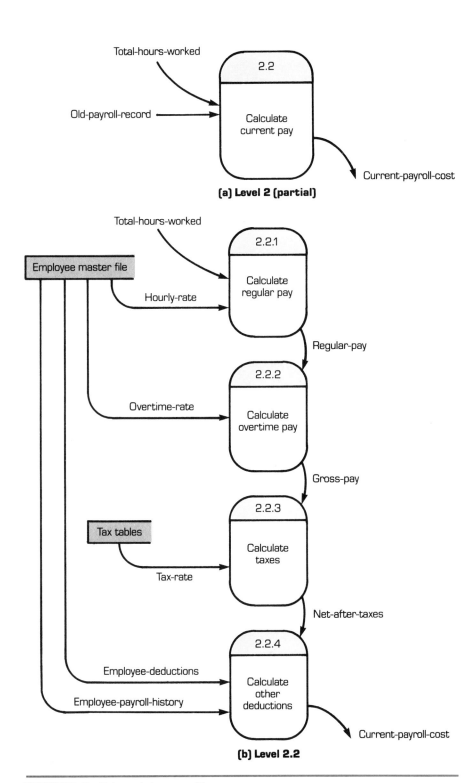

FIGURE 5–6 Balanced decomposition

time rate, employee deductions, and employee payroll history. We can say that these three data flows are equivalent to the old payroll record data flow in that:

old payroll record = hourly rate + overtime rate + employee deductions + employee payroll history

Second, data flows can be local to a particular transform and never appear in higher-level DFDs. In Figure 5–6b, the tax rate data flow is local to the transform 2.2.3, calculate taxes. Because the tax rate is not required by transforms 2.2.1, 2.2.2, or 2.2.4, it need not appear in the level-2 diagram.

With this new information, we can conclude that the new design is in balance. Even so, the concept of balancing is not easy to understand. Therefore, we will continue to emphasize it in subsequent sections. For now, remember that inputs to and outputs from a higher-level transform must remain in balance with the inputs to and outputs from the decomposed, lower-level DFD.

TRACING DATA FLOWS

With a completed and balanced set of DFDs in hand, the analysis team has at its disposal the logical blueprint of a system. By tracing the specified data flows, the team can quickly discover which types of data it should collect and evaluate.

Suppose that you are required to collect data to describe transform 2.1 shown on Figure 5–3. This DFD tells you to determine the contents of two data flows: the hours-worked record and total hours worked. If total hours worked were added to the hours-worked record, you would probably collect two copies of a form: one showing the hours-worked record before hours worked were summed and one showing the updated hours-worked record. This transform also tells you to ascertain how the hours are summed. Do timekeepers sum hours worked, or is a computer used? In either case, the analyst must determine how the hours total is computed to ensure correctness in processing.

PERSONAL INTERVIEWS AND QUESTIONNAIRES

Besides constructing a detailed set of DFDs, a major aspect of systems analysis is the systematic collection of facts relating to a proposed system. When engaged in such fact-finding, the analysis team must consider both primary and secondary sources of data. *Primary sources* are the users of a system, or those who have first-hand knowledge of how a system should operate. *Secondary sources* include information found in manuals, reports, and system documentation. In this section our concern is with gathering information directly from users within an organization through the use of personal interviews and questionnaires. Nonetheless, this is not to imply that the data collection activity is restricted to these two methods. Besides seeking data from people within an organization, the analyst may also seek data from people outside the organization. Through vendor contacts, industry sources, professional groups, and private parties, an analyst often can pick up ideas on how a system should function.

The Personal Interview

Of the many different data collection methods, analysts often have the most difficulty with the personal interview. In such an interview, the analyst must ask the right questions in order to learn how a system operates. Sound easy? Suppose that you are asked to conduct an interview with a person who knows a great deal about the existing system and how a new system might function. Suppose further that you have determined that this person is openly hostile to any new system and will attempt to block any effort to begin work on it. How would you conduct a personal interview with this person?

STRUCTURED VERSUS NONSTRUCTURED INTERVIEWS

Interviews can be either structured or nonstructured. Those asking specific, predetermined questions are called *structured interviews*. To complicate matters somewhat, there are both highly structured and semi-structured interviews. A highly structured interview usually consists of a series of *closed-ended questions*, of which the following are examples.

1. What error rate have you experienced with the existing billing system?
   a. _____ 1 percent or less
   b. _____ 3 percent or less
   c. _____ 5 percent or less
   d. _____ more than 5 percent

2. What error rate can be tolerated?
   a. _____ 1 percent or less
   b. _____ 3 percent or less
   c. _____ 5 percent or less
   d. _____ more than 5 percent

As these examples show, closed-ended questions restrict the user's reply to a predetermined set of response categories.

Interviews are less structured when open-ended questions are included. *Open-ended questions* give the user the opportunity to express both facts and feelings. Examine the following open-ended questions:

1. What types of reports need to be produced by the billing system?
2. How should these reports differ from those prepared in the past?
3. Who will help us design the format and contents of these new reports?

These questions lead to a variety of user responses. The first question, for example, asks users to categorize their reporting needs, while the second question requires users to compare the proposed system with the current reporting system. The third question is more specific. It asks whether users are willing to help design the new reporting requirements.

The less structured an interview becomes, the more it is said to be a *nonstructured interview*. With a nonstructured interview, the analyst might begin with an open-ended question designed to get the user talking. Examples might include:

1. Can you tell me something about how your present billing system works?

2. What types of information do you need to make inventory restocking decisions?

With a nonstructured interview, the analyst attempts to get users to talk freely about their wants, desires, fears, and other personal feelings regarding a new system. The advantage of such an approach is that it often uncovers various little-known or unrecorded facts. For instance, in talking with users an analyst might discover mistakes made in the past. A user might warn the analyst, "Be careful. Our boss has a tendency to implement a new system before it is ready. We seem to always wind up with a disaster."

THE THREE-SESSION INTERVIEW

By now some of you may be asking, "How do analysts choose between structured and nonstructured interviews? Is one type better than another?" These questions are easier to answer if we realize that an analyst usually interviews the primary system users more than once. As a rule of thumb, at least three interview sessions are required: the initial session, the fact-finding session, and the summary session. The *initial session* is a short, nonstructured interview in which the analyst describes the work to be done, clarifies his or her role, and asks the user for his or her ideas. This initial session provides the analyst with facts and opinions necessary for framing questions to be asked during the second session.

This second session, the *fact-finding session*, is longer and more structured than the initial session. The purpose of this session is to determine specific system functions and performance levels desired by the user. For example, a user might be asked to comment on the performance expected from a company's billing system, such as the following question illustrates.

The analyst could ask, "Following the shipment of goods to the customer, how many hours should it take us to prepare and mail the customer's bill?" (By stating four hours, for example, the user states the performance expected from the system). When conducting a fact-finding session, the actual interview is often divided into several question-and-answer categories. One category might be devoted to questions dealing with the processing of information, such as the calculation of summary totals; another might be directed at the flow of data, such as the user's reporting requirements; still another might consider data storage requirements, such as the user's need to dig into files to review old forms.

The third session, the *summary session*, represents a nonstructured reporting session rather than an interview. The purpose of this session is to summarize the results of the fact-finding interview and to determine whether the user agrees with the analyst's findings. This session tests the analyst's interpretive skills. If the analyst has not interpreted the user's remarks correctly, the summary session should bear this out.

The Questionnaire

Unfortunately, the three-session personal interview can often be prohibitively time- and resource-consuming. Suppose that an analyst has to collect facts from 500 people. Using the three-session rule, the analyst would need to schedule 1,500 interviews. Even if three analysts were assigned the task and could average 25 interviews a day, it would take them approximately three months simply to collect data. In most instances, this time requirement is too great.

When personal interviews are determined to be too time-consuming and expensive, analysts rely on questionnaires to survey user opinions and to collect facts. Besides being more economical than the personal interview, questionnaires offer several advantages:

1. They feature standardized instructions for recording responses and ordering questions. The personal interview is rarely uniform from one interview to the next.

2. They allow for greater anonymity in response. Users may state something on a questionnaire that they might not state during a personal interview.

3. They give users added time to think about their response to a question. During a personal interview, users are asked to make a more or less immediate response.

4. They can be administered to hundreds of individuals simultaneously. Compared with the personal interview, the distribution of the questionnaire clearly is much more efficient.

5. They can be analyzed quantitatively. As discussed next, a quantitative (summed) evaluation simplifies the reporting of user attitudes and feelings.

SUMMATED SCALES

Summated scales[2] can be used along with questionnaires. With a summated scale, each response to a question is given a numerical score; the summation of the scores for each question, followed by the calculation of the average score and the variance, indicates the degree to which users agree or disagree. With summated-scale questions, the analyst can make distinctions of degree when collecting data about a system. For example, a team of analysts might want to know the degree to which a particular system function will be accepted. They discover that only 60 percent of the user population actually favors the new function.

Let's look at a more complete example to describe the purpose and design of a summated scale. Banks tend to use either a single-channel or a multiple-channel system to establish a line for their customers awaiting service from tellers. As **Figure 5–7a** shows, a multiple-channel system requires the customer to pick out a teller window in advance and to wait in line until that teller is available. The single-channel system, as shown in **Figure 5–7b,** features a first-in, first-served arrangement. The customer at the head of the line moves to the first available service window.

In deciding which of these two systems is preferable, the analyst might decide to survey customer attitudes. He or she then selects 1,000

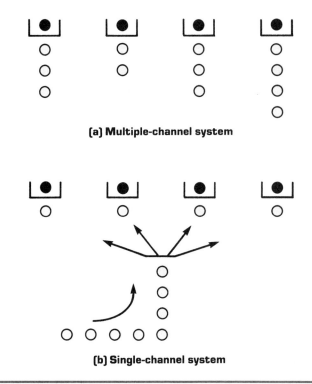

FIGURE 5–7 Examples of multiple- and single-channel systems

customers at random and asks them the following question (among others):

> How would you rate our single-line method of servicing your account? Check the appropriate answer.
>
> | Strongly Approve | Approve | Undecided | Disapprove | Strongly Disapprove |
> |---|---|---|---|---|
> | (1) ___ | (2) ___ | (3) ___ | (4) ___ | (5) ___ |

This particular question follows a *Likert-type scale*, a pattern in which subjects are asked to respond to a numerically weighted question.[3] After the analyst has collected the completed questionnaires, the weight of each response is summed. In this example, the analyst might discover that customers disapprove, however slightly, of the single-channel system, based on a total summed score of 3,400, an average score of 3.40, and a variance of 0.63.

PRETESTING QUESTIONNAIRES

The most difficult aspect of using questionnaires is deciding which questions to ask. Once the content of the questionnaire has been determined, the wording and the format of each question must be finalized. Although that may sound simple enough, consider the following pairs of questions, each of which requires a "Yes," "No," or "Undecided" response:

Pair 1:
Was the instructor fair in the grading of examinations?
Were the examinations fair?

Pair 2:
Are you satisfied with your current wages?
Are the wages paid by the company fair?

Pair 3:
Would you support the proposed point-of-sale (POS) billing system?
Should the company go ahead with the proposed POS billing system?

If these alternative questions were used in questionnaires, with two identical samples, some interesting differences might appear. For example, some subjects would find the examinations fair, but the grading not; some would find the wages paid by the company acceptable, but their personal wage not acceptable; and some would support the POS system, but not advise going ahead with its implementation.

Because of differences in how people interpret questions, an analyst must carefully pretest a questionaire. Pretesting must address several issues: Is each question necessary? Is each question clear? Is any question biased? Does any question generate emotional feelings? Does the subject have the background needed to answer the question? Are the questions in the correct order? With pretesting, the analyst gives the user the opportunity to critique each question and express how each question was interpreted. To clarify the final questionnaire, several forms of the same question are pretested. Pretesting helps the analyst determine which questions communicate the variables that the analyst seeks to measure.

POSTANALYSIS

Postanalysis of the questionnaire generally calls for some form of *discriminative analysis*. Through such a process, the analyst can determine whether subject groups are from one or more populations. The analyst must also test the relative importance of each predictive variable. Although the methods of discriminative analysis fall outside the scope of this book, an example will help explain the value of these techniques. Assume that the analyst wishes to know the relative importance of four system characteristics: ease of use, expense of use, improved reporting features, and processing speed. If the response "Would approve the system" is almost always associated with "Improved reporting features," and the response "Would not approve the system" is almost always associated with "Expense," we might conclude that these characteristics discriminate well in separating strongly supporting from nonsupporting groups (i.e., populations). If, however, ease of use and processing speed scores are about the same for both those who would approve and those who would not approve the system, we can conclude that these characteristics discriminate poorly.

Advantages and Disadvantages

Through personal interviews and questionnaires, the analysis team can collect facts and opinions about both the current and the proposed system. If time and cost were not considerations, most analysts would rely on personal interviews to collect data. The advantages of the personal interview are several: they are flexible, permit users to elaborate on their views, add the personal touch, and allow the analyst to observe reactions to questions as well as record the answers to the questions.

When the population to be sampled is large and when time and cost are important factors, however, a questionnaire becomes a suitable substitute to the personal interview. When prepared, distributed, and analyzed correctly, the questionnaire can be invaluable in determining user reactions to a new system. With the questionnaire, predictive interpretation is possible. In some instances, the analyst can determine whether upper- and lower-level managers, as groups, respond similarly to the same set of questions. In other instances, the analyst can predict whether such factors as the cost, security, performance, and training associated with a proposed system are important to the sample population.

FORMS AND PROCEDURES COLLECTION

A second systematic collection of facts involves collecting forms and documenting procedures. Here, the analyst gathers secondary sources of information—job descriptions, operating procedures, manuals, reports, business documents, and system and program flowcharts. The analyst must study these secondary sources in order to understand how the system works. Moreover, in reviewing these materials, the analyst can begin to depict the *data structure* of a system. Up to this point, little has been said about the way in which data are organized in a system. To some, a data structure is nothing more than a composite of data. To others, the concept of a data structure requires a formal definition. Gane and Sarson[4] maintain that a data structure consists of one or more data elements that usually describe some *object* or *entity*, such as a customer, book, airplane, or product.

At this point, you might ask, "What is a data element and how are data elements organized to give them a structure?" For our purposes, a *data element* is a single unit of information that helps describe an object. A customer's first name, for example, is a data element, as is a customer's telephone number. Both units of information describe a single object (e.g., a customer).

The relationship between a data element and a data structure can be illustrated by a *data description hierarchy* (see **Figure 5–8**). This hierarchy can be interpreted as follows: data flows, data stores, and transforms contain one or more data structures; each data structure contains one or

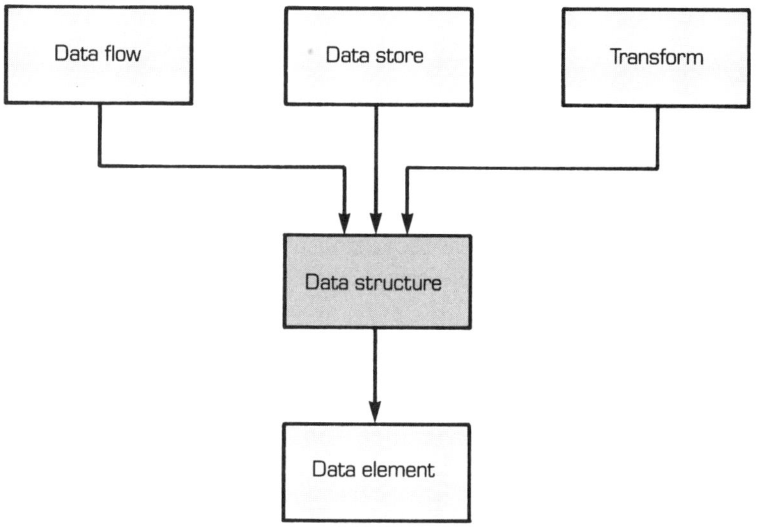

FIGURE 5–8 Data description hierarchy

more data elements. A practical example will help demonstrate the concept of a data description hierarchy.

Suppose that we wish to describe the data flow defined as an old payroll record. We might begin by jotting down the various data structures that make up an old record. These include

1. employee identification,
2. employee name and address,
3. personnel information,
4. wage and salary information,
5. leave and pension history,
6. last current payroll information, and
7. last year-to-date payroll information.

Each of these seven data structures, in turn, can be broken down into a set of data elements. For example, the data structure called personnel information might consist of the following elements:

1. department number,
2. date started employment,
3. sex,
4. date of birth,
5. marital status, and
6. last performance review.

Forms Collection

Because most data entering or leaving a system are recorded on forms (source documents and printed documents), an analyst can learn a considerable amount about a system's data structure by carefully studying these forms. Consider the sample invoice shown on **Figure 5–9.** The invoice form provides space for recording a variety of items, beginning with the customer's sold-to address (1), the ship-to address (2), the invoice number (3), the date the order was shipped (4), and so forth. Through further analysis of these data, the analyst learns that the invoice contains at least four data structures: customer information, customer order information, order filling information, and billing information. **Figure 5–10** shows these four structures and their component parts. This diagram tells the analyst that a customer number, customer sold-to address, customer ship-to address, the terms, and salesperson number all in effect describe a customer. In addition, when a customer places an order, additional information is required: the customer order number, item number, description of each item, price of each item, and quantity of each item ordered. When goods are shipped, order filling information

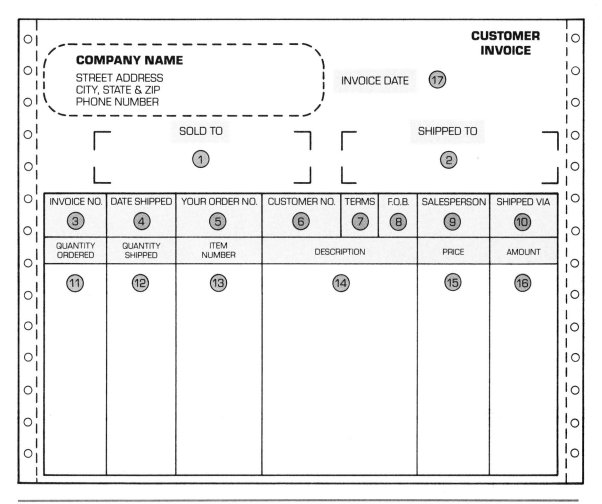

FIGURE 5–9 Customer invoice

is required: the date of the shipment, the quantity of each item shipped, FOB instructions (i.e., with or without an extra shipping charge), and ship-via instructions (e.g., truck, rail, etc.). Finally, billing information is also required: the invoice number, the date of the invoice, and the dollar amount (e.g., the dollar cost to the customer) for shipped items.

Procedures Collection

Along with forms, the analyst must also collect written procedures outlining the details of data flow transformations (e.g., what takes place within a transform shown on a DFD to alter input data). A *written procedure* is nothing more than a written statement describing how work

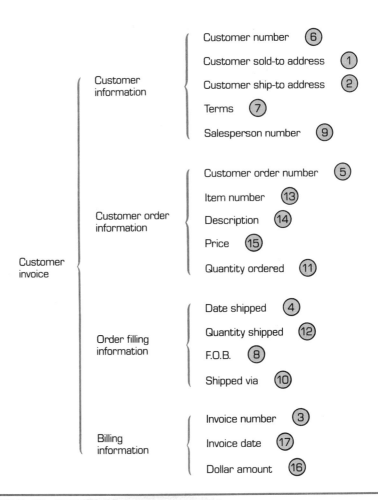

FIGURE 5–10 Customer invoice data structure

should be done or policies carried out. **Figure 5–11** illustrates a standard written procedure that describes the processing of customer payments. To the experienced analyst, this procedure communicates several items, such as the following:

1. Who prepared the procedure and when. This procedure was prepared by the accounts receivable department on January 1.
2. What the purpose of the procedure is. The purpose of this procedure is to explain the steps important to processing customer payments.
3. Which departments have access to this procedure. This procedure is distributed to all accounting departments and to data processing.
4. How often the procedure is utilized. This procedure states that customer payments are processed daily.

```
┌─────────────────────────────────────────────────────────────────────────┐
│                          STANDARD PROCEDURE                             │
├─────────────────────────┬──────────────────────┬────────────────────────┤
│ Department:             │ Issued:              │ Number:                │
│   Accounts receivable   │   January 1          │   A/R 394              │
├─────────────────────────┼──────────────────────┼────────────────────────┤
│ Subject:                │ Effective:           │ Page    of             │
│   Processing of customer payments │ January 1  │   /      /             │
├─────────────────────────┼──────────────────────┼────────────────────────┤
│ Distribution:           │ Frequency:           │ Supersedes:            │
│   Accounting [all] data processing │ Daily     │   A/R 096              │
└─────────────────────────┴──────────────────────┴────────────────────────┘
```

| Abstract | Explains the steps to be taken in processing customer payments |

Position	ACTION
Cash-Management Supervisor	1. Reviews cash-handling procedures to expedite the processing of payments 2. Resolves problems when payments are incorrectly submitted 3. Compares cash-control slip total to computer processed total 4. Compares bank deposit slip total to bank processed total
Accounting Clerk	1. Opens mail sent by customers 2. Reviews payments for accuracy 3. Completes customer payment stub when necessary 4. Prepares cash-control slip 5. Sums total cash payments received 6. Forwards copy of cash-control slip to data entry department 7. Prepares bank deposit 8. Makes bank deposit 9. Forwards copy of cash-control slip and bank deposit slip to cash-management supervisor.

FIGURE 5–11 Structured written procedure

5. Who is responsible for operational tasks and what those tasks are. As indicated, an accounting clerk is responsible for various tasks, including the preparation of two types of documents (a cash-control slip and a bank deposit slip).

6. Who is responsible for supervisory tasks and what those tasks are. As shown, a cash-management supervisor is responsible for reviewing operational tasks, resolving problems, and comparing cash totals.

Armed with the knowledge of who does what and when they do it, the experienced analyst can construct a context diagram showing the main parts of the system described by the procedure and prepare a set of rules showing how company policy is carried out by means of the procedure. **Figure 5–12** illustrates a context diagram that shows the main

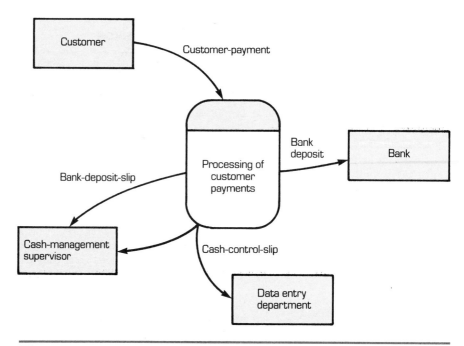

FIGURE 5–12 Context diagram describing the processing of customer orders

parts of the customer payment processing system. As described by the written procedure, the single input to processing is the customer payment. In addition, there are three system outputs: a bank deposit, a bank deposit slip, and a cash-control slip. All three outputs follow the processing of payments.

Procedures collection is relatively straightforward for the analyst, provided DFDs have been carefully designed to isolate the materials to be collected. Even so, problems do arise. The main problems occur when written procedures are missing, incomplete, or inaccurate—or when duplicate forms are discovered.

Let's consider the problem of missing, incomplete, or inaccurate procedures first. Since procedures are needed to describe how work should be done and explain how policies are to be carried out, the analyst must write procedures if they are missing and review existing procedures to verify that they are complete and accurate. Both of these activities are time-consuming. When writing a procedure, the analyst must identify all responsible parties and determine the tasks appropriate to each. When reviewing a procedure, the analyst must examine carefully the work being done to determine whether the steps shown on the procedure are actually being carried out. Suppose that an analyst is reviewing the procedure for processing customer payments (see Figure 5–11). The analyst might ask the accounting clerk, "Why is a copy of the bank deposit slip not forwarded to the cash-management supervisor, as stated on the writ-

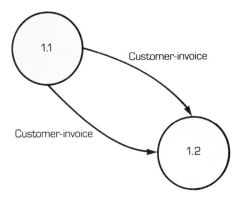

FIGURE 5–13 DFD with duplicate data flows

ten procedure?" The clerk might respond, "Oh, that procedure needs to be revised. The supervisor told us to stop sending the copy, since he received another copy from data entry after they entered data into the computer."

The problem of duplication is equally difficult to resolve. Duplicate forms often indicate the existence of two or more data flows leading from a source or transform when there should be only one (see **Figure 5–13**). Once the analyst discovers a duplicate data flow, he or she might uncover a variety of reasons for the duplication. For instance, the analyst might determine that one form was used for eastern sales and another form was used for western sales. Why? The answer from the staff was, "Because we have always done it this way." Further digging by the analyst might lead to the conclusion that nobody really knew the answer.

As another example, an analyst might determine that one customer invoice contained space for recording the quantity shipped, while another customer invoice contained space for recording the quantity ordered and the quantity shipped. Why this difference? The analyst would need to question management to determine the importance of recording the different types of quantity information.

The discovery of duplicate forms usually requires some form of policy decision. Working with user management, the analyst must strive for the design of a single, standard form. The form thus becomes more than a piece of paper. It specifies the data structure of a data flow and shows how the pieces of a system are knitted together.

Flowchart Collection

When the current system is a computer-based one, the analyst must collect system flowcharts along with forms. As stated in chapter 3, system flowcharts are used to indicate the main steps in computer processing and to illustrate input to and output from processing. Consequently, system flowcharts provide three important types of data collection information:

1. Forms to be collected for each computer program—these include source documents, along with hard-copy output registers, reports, summaries, and formal business documents.
2. Displays to be studied for each interactive computer program—these displays show which data must be keyed to processing, how keyed data are verified, and so forth.
3. Jobs steps for each computer program—job steps help the analyst understand how the various pieces of a system are organized; they also help the analyst comprehend the structure of the computer code written for each computer program.

To demonstrate the value of system flowcharts, let's review the processing of the payroll register, as shown earlier on Figures 5–2 and 5–3. **Figure 5–14** illustrates the system flowchart for the transform, prepare payroll register. As indicated, input to the print payroll register program includes records stored on an hours-worked file and an employee master file. Output to the program includes the printed payroll register, a hard-copy document, and the current payroll file. Why would the analyst collect system flowcharts if DFDs and structure charts were available? Observe that a system flowchart provides clues of how a computer program, with its associated internal structure, is interrelated with computer files and their corresponding data structures. On examination of the figure,

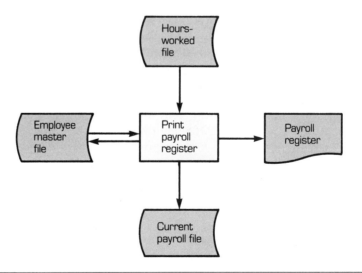

FIGURE 5–14 System flowchart: transform 2.0

the analyst can determine that he or she needs not only to understand the internal structure of print payroll register but also to determine the contents and structure of three physical files and one hard-copy output report.

The Mansfield, Inc., Case Study

Roger Bates started the data collection effort by asking the analysis team to prepare lower-level DFDs of the existing billing system. "The processing of customer invoices is more complex than it looks," Roger commented. "We need to identify which steps can be handled by the computer and which steps must be done by hand. We must define each data store, each important transform, and all data flows."

Existing Computer System

Ken Westerling admits that he made a mistake when the Mansfield computer was installed. "I bought the vendor's billing package, believing that it would be able to meet our needs," he told the analysis team. "Well, the product might work for direct mail companies, but not for us. Tying the customer account to the customer zip code, for example, is not acceptable. For our business, such a scheme is impossible to maintain. Moreover, the update procedures included in the package were never integrated. If we change the customer account number in one part of the system, it may or may not change the same account number throughout the system. Finally, attempting to match all accounts at month end is unworkable. We need some type of cycle billing, where we can bill several times during the month. Given these failings, I think that we should scrap the existing system and replace it with a new one."

With the knowledge of the problems with Mansfield's existing billing system and Ken's approval to go ahead, Roger moved quickly. "First, we need to document the logical characteristics of our existing system," Roger stated. "After this, we can decide how to go ahead and build a new system."

Level-1 DFD

Figure 5–15 shows the level-1 DFD prepared by the analysis team for the logical description of the existing billing system. "Initially, we create invoice records and store them in an invoice transaction file," remarked Carolyn Liddy. "This represents the interactive part of processing. When we have entered all invoices for the day (or for a half-day if we decide to process invoices twice a day), we close the invoice transaction file and move to the next step, which as shown is printing the invoice register."

Carolyn continued, "The invoice register is a listing of the invoices processed during the day. After the listing is printed, it must be approved by a member of John Havensek's staff. John, maybe you can fill us in on what happens next."

"We visually review the invoice register to determine whether the number

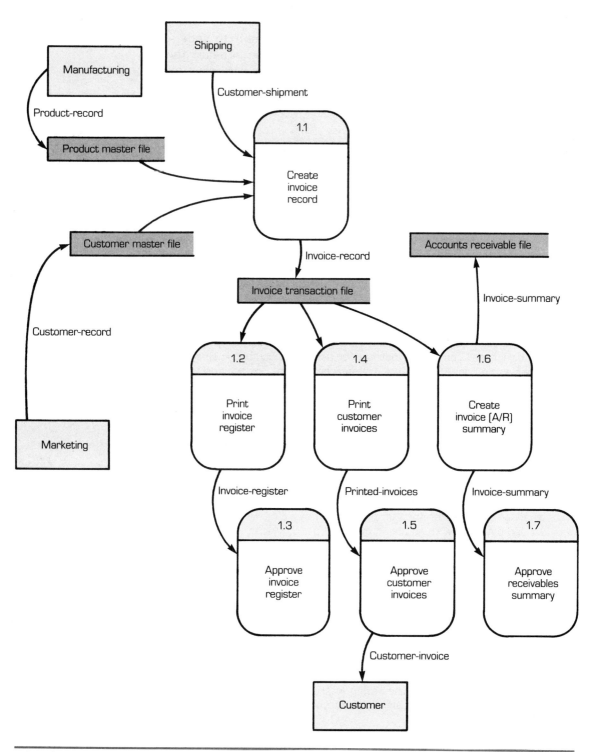

FIGURE 5–15 Level-1 DFD showing the transforms internal to preparing a customer invoice

Customer number
Record code
Invoice (or credit) number
Date of transaction
Salesperson number
Total invoice charge (or credit)

FIGURE 5–16 Data elements required for each invoice summary record

of invoices to be printed matches the number of orders filled and shipped," John explained. "We also sample shipping papers to determine whether name and address information is correct. During slow periods, we have few problems with this review. However, when processing volumes are at peak levels, we discover all kinds of problems. We sometimes find that the customer address is wrong, although we have also found cases where we are billing the wrong customer."

Following a lengthy group discussion of these and other problems, Carolyn continued with the description of the existing system. "After the invoice register is approved, we go ahead and begin to print customer invoices," she stated. "Next we sample these invoices to determine whether invoice registration and printer alignment are acceptable."

"The last two steps of the billing system are the easiest to explain," she concluded. "After customer invoices are approved, we create the invoice summary [see **Figure 5–16**]. The invoice summary consists of only six data elements, beginning with the customer number and ending with the total invoice charge. Compare the contents of this file with the larger invoice transaction file [see Figure 5–10]. As you can see, all product information is stripped from the summary file.

"As shown by the DFD, the last step is to approve the receivables summary," Carolyn added. "A member of John's staff does this."

System Flowcharts

Carolyn and Neil Mann undertook one additional task in describing the current billing system: they prepared system flowcharts for each processing step. Neil described the process associated with entering customer orders and updating the product master file.

"**Figure 5–17** shows how customer orders are processed by the computer," he explained. As illustrated, the source document is the customer shipping slip. We enter customer shipping information (customer number, order number, quantity ordered, quantity shipped, and so forth) by keyboard into processing and display the results. Processing by the computer consists of reading data stored on the product and the customer master files, combining these data with keyed data, computing the invoice charge, and writing the completed invoice record to the invoice transaction file.

"**Figure 5–18**," continued Neil, "describes how the product master file is updated. Observe that this update process is not shown on the DFD. We update the product master file before we enter customer shipping information. To begin processing we must indicate whether we plan to add, modify, or delete

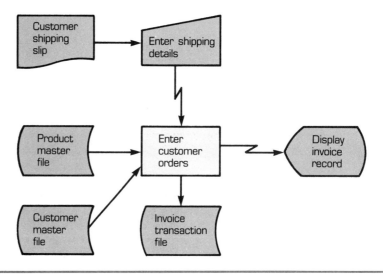

FIGURE 5-17 System flowchart showing how an invoice record is created and stored

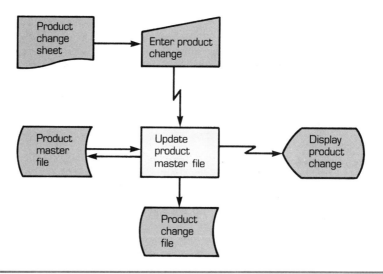

FIGURE 5-18 System flowchart showing the update of the product master file

a product record. Let's suppose we decide to modify a record. To begin, the program reads the original product record into memory, where its contents are changed. Once changed, the revised record is written back to the product master file. The last step is to record the change on a change file. This file stores the original and the revised version of the product record."

SUMMARY

Data collection consists of tracing data flows and conducting interviews, while data analysis involves making sense of the collected data. Tracing data flows begins with the preparation of DFDs. DFDs provide the analysis team with a clear idea of who should be interviewed.

In constructing DFDs, the analysis team decomposes higher-level transforms until each transform is sufficiently clear to require no further explanation. These lowest-level transforms are called functional primitives.

Balancing is necessary in the decomposition of a higher-level transform. Balancing ensures that input to and output from a transform are consistent with input to and output from the decomposition of that transform.

Usually, there is no clear time separation between constructing DFDs and conducting personal interviews, distributing questionnaires, and collecting documentary evidence about a system. In practice, these different data collection activities are often done concurrently. In this way, the analyst can determine how far to decompose a transform. Decomposition continues until the analyst can fully explain the steps that take place within a transform.

Personal interviews and questionnaires provide primary information about a system. The personal interview can be either structured or nonstructured, consisting of closed-ended or open-ended questions. The three-session interview begins with open-ended questions, moves to a more structured interview (the fact-finding session), and ends with a nonstructured interview (the summary session).

Questionnaires typically are used when the user population is large. Summated scales can be devised with the use of questionnaires. These scales help the analyst determine the degree to which users approve or disapprove of particular system features.

The analyst is advised to conduct both pretesting and postanalysis of questionnaires. Both help ensure that question responses reveal what the analysis team set out to discover about a particular subject area.

The systematic collection of secondary information about a system is accomplished, in part, by collecting forms, procedures, and flowcharts. These secondary sources tell much about the data structure and the data description hierarchy of a system.

Forms collection, the collection of source and computer-printed documents, is undertaken to determine which specific data elements are either entered or printed as a result of processing.

Procedures collection, the collection of written documents describing the work that should be done and the policies that should be carried out, is required for the analyst to understand the behavior that takes place within a work area.

Flowchart collection, especially the collection of system flowcharts, is recommended when the analyst is studying an existing computer-designed system. System flowcharts help the analyst identify forms to be collected, along with displays and job steps to be studied.

REVIEW QUESTIONS

5–1. How does data analysis differ from data collection?

5–2. Why are context diagrams and level-0 DFDs important to the work of data collection and analysis?

5–3. What is a functional primitive?

5–4. What is the difference between primary and secondary sources of data?

5–5. What is the difference between a structured and a nonstructured interview?

5–6. Name the three sessions in the three-session interview.

5–7. What advantages does a questionnaire offer?

5–8. Why is discriminative analysis important in the use of questionnaires?

5–9. What is a data element? What is the relationship between a data element and a data structure?

5–10. Why are written procedures prepared by an organization?

5–11. Name the three types of data collection information that are supplied by system flowcharts.

EXERCISES

5–1. Suppose that you are required to prepare a term paper. Before starting to collect data, you decide to draw a context diagram showing the various parts of the term paper system. (The system shown below describes the sources and sinks of such a system.) First, complete this system by defining each data flow. Second, based on your complete system, which types of data must be collected?

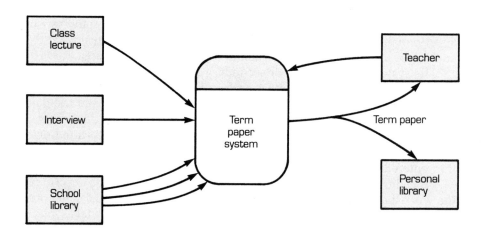

5–2. Write a question and devise a summated scale to determine whether users will be able to tolerate a billing system with a 3 percent error rate.

5-3. Draw two diagrams to illustrate how the customer master file for the billing system is to be updated (see Figure 5-15). Similarly to the update of the product master file (see Figure 5-17), you can assume that all changes to the master file are to be written to a change file. First, draw a DFD to show the steps in this process. Second, draw a system flowchart to show the steps in this process. Compare these two diagrams. What are the advantages of using a DFD? What are the advantages of using a system flowchart?

5-4. Suppose that you obtain a procedure describing the actions taken by a timekeeper in processing employee timecards. The procedure reads as follows:

1. Receives weekly time cards from employees.
2. Reviews time cards for accuracy.
3. Totals regular hours and overtime hours.
4. Prepares register of departmental hours worked.
5. Prepares register of project hours worked.
6. Forwards departmental hours-worked register to departments for supervisory approval.
7. Forwards project-hours-worked register to project planning and control.
8. Forwards departmental-hours-worked register to payroll.

Prepare a context diagram to show the sources and sinks and the data flows described by this procedure. What information is needed before you can complete the level-0 DFD?

WORLD INTERIORS, INC.—CASE STUDY 5
UNDERSTANDING THE CURRENT SYSTEM

Introduction

After reviewing John Welby's analysis of WI's problems, Rena Logen concluded that the key to cleaning up many of WI's administrative problems was a much improved computer system. She came to this conclusion despite the fact that her experience with computers and computer personnel had been disappointing. "We were sold a bill of goods," she would go on to say. "We should have never purchased a microprocessor to begin with. The off-brand machine was a silly notion. Next time we'll stick to well-known brand-name equipment and thoroughly tested software. And we'll try to select equipment that can be more easily maintained."

John was relieved when he heard that Rena would support a new computer system. "Before moving ahead too fast, however, I'd better get a more detailed view of how WI's processing currently works," he thought to himself. "There is no point in developing an entirely new system if parts of the existing system can be saved. What I need to do is start with direct sales processing, study their automated procedures, and then move on to retail store procedures. After I'm finished, I should be able to propose the design of a more fully operational system."

John began by reviewing his notes on WI's current system.

Notes from Interview with Krintine Rinehart

1. Customers are required to send two items with their order: a completed customer order form and a personal check.
2. Once an order is received, it is cleared. Clearing the order consists of making sure that all information shown is correct and that payment is in full.
3. Cleared orders are placed in an accepted orders file.
4. Cleared orders serve as input to invoice processing.
5. Three copies of the invoice are printed as output of invoice processing, with distribution as follows: original copy to manufacturer, one copy to customer, and one copy to customer invoice file.

Notes from Interview with Ralf Elders

1. Inputs to pay manufacturer's charges consist of the customer invoice and the factory (manufacturer's) invoice.
2. Charges from products shipped are matched and the factory invoice is either approved (for payment) or disapproved.
3. Approved factory invoices are stored in the accounts payable file.
4. Input to the actual payment of a factory's charge is limited to approved factory invoices.
5. Outputs from the payment of an approved factory invoice consist of a vendor-voucher check and a copy of the check voucher.
6. All outstanding checks are kept in an outstanding checks file.

Finding his notes to be incomplete in spots, John decided to collect additional information about the current system from Krintine.

Order Processing

Krintine Rinehart explained the computer aspect of order processing as follows: "Every day we enter customer orders into the computer and write these to a daily order file. To begin processing, we enter the date and the number of the first invoice to be printed. After this, we enter customer order details, one order at a time. For each order [see **Figure 5—19**], we are required to key in the customer's first and last name; street, city, and state address; postal code; and phone number. Next, we move to the bottom of the order in order to key in the customer order number (which we assign), the date of the order, the terms code (cash, charge, and so forth), the supplier number, the ship-to code, the ship-via code, and any special instructions. Finally, we are required to key in for each item ordered the item number, item name, the quantity desired, the finish desired (e.g., metallic, nonmetallic, etc.), and the unit price. After the entry of the last item, we enter the sales subtotal, the sales tax, and the customer payment."

"Once we are finished with an order," Krintine continued, "the computer automatically writes to two files, one called the customer order history file and the other, the daily order file. The customer order history file stores orders from customers who have purchased items from us in the past. Besides this information, the customer order history file contains two fields used later on: a problem-flag field and a date-filled field. We turn the problem flag on whenever we have a problem with an order. We enter the date the order was filled when we receive confirmation from the manufacturer. The second file, the daily order file, contains the information keyed to processing.

"Once we have entered all new orders for the day, we tell the computer to process the orders stored on the daily order file. The computer retrieves information stored on the daily order file and starts two updates: of the mailing labels file and of the supplier product file. Initially, the computer responds with a display telling us that the first update is in process. If a customer is new, the update consists of adding the customer's key, name and address, postal code, and date of the latest order to the mailing labels file. If the customer is a repeat customer, the computer simply changes the date of the latest order.

"After the first update is completed, the computer responds again. It tells us that a second update is in process. This update is the modification of the supplier-product file; it is also used to add the supplier's name and address to a daily order record. Using the supplier number, the computer can extract a product record for a supplier

WORLD INTERIORS, INC.
1919 FAIRWAYS DRIVE
WILSONVILLE, OR 97000

CUSTOMER ORDER

Name: Esther Peters
Phone: (312) 555-3300
Address: 1986 Wilson St.
City: Brooklyn
State: N.Y.
Zip: 11200

VISA/M.C.#:
Expiration Date:
Customer Signature:

Item Number	Name of the Item	Quantity Desired	Finish Desired	Price Each	Total
H37 F352	Vertical Blind	1	100	74.70	74.70
H37 V352	Valance	1	100	8.10	8.10
				Sales Subtotal	82.80
				Sales Tax (In-State Residents only)	
				Sales Total	82.80

FOR OFFICE USE ONLY

Date:
Customer Order Number: 21050
Terms: 1
Supplier Number: H37
Ship To: C
Ship Via: UPS
Special Instructions:

FIGURE 5–19 Customer order form

WORLD INTERIORS, INC.
1919 Fairways Drive
Wilsonville, OR 97000

CUSTOMER INVOICE

INVOICE	DATE
22345	5/04/XX

ORDER TO:
B & H Coverings
45th and Windsor
Minneapolis, MN 55045

CUSTOMER:
Esther Peters
1986 Wilson Street
Brooklyn, N.Y. 11200
(312) 555-3300

Customer Order No.	Customer Order Date	Supplier Number	Terms	Ship To	Ship Via
21050	5/03/XX	51287	Cash	Customer	UPS

ITEM NUMBER	DESCRIPTION	QUANTITY	FINISH	PRICE	EXT. PRICE
H37 F352	Vertical Blind	1	100	74.70	74.70
H37 V352	Valance	1	100	8.10	8.10

Special Instructions: **None**

Sales Subtotal	82.80
Sales Tax (In-State Residents only)	—
Sales Total	82.80

FIGURE 5–20 Customer invoice form

and change the total number of orders taken and the dollar value of the orders taken for this product. Once updated, the computer writes the revised record to the file. By the way, this supplier product file contains very little information. It contains the supplier number, the name and address of the supplier, and the product numbers and name of each product supplied by the manufacturer.

"After the two updates are finished, the computer waits until we instruct it to begin printing. The first print program produces a printed listing of the orders for the day, by name of supplier. We call this listing the daily order register. A sort precedes this print run because the required sequence typically is different from the sequence followed when we entered customer order details into processing. If we approve the daily order register (that is, we find no mistakes), we insert special forms and tell the computer to begin printing customer invoices [see **Figure 5–20**]. After all invoices are printed and approved, we tell the computer to print customer and manufacturer labels [see **Figure 5–21**]. Labels are attached to the envelopes that hold the copy of the invoice we return to the customer and the invoice sent to the manufacturer."

```
Esther Peters
1986 Wilson Street
Brooklyn, NY 11200
```

FIGURE 5–21 Mailing label

CASE ASSIGNMENT 1

Prepare a level-1 DFD showing how customer mail orders are processed. Review John's notes and the level-0 DFD prepared earlier (see chapter 3) before beginning this assignment.

CASE ASSIGNMENT 2

Prepare a level-2 DFD showing how the manufacturer's charges are paid. Review John's notes and the level-0 DFD prepared earlier (see chapter 3) before beginning this assignment.

CASE ASSIGNMENT 3

Decompose the level 1.2 transform to better explain the process followed in preparing customer invoices. The input to level 1.2 is a cleared customer order; the output consists of three copies of the customer invoice.

CASE ASSIGNMENT 4

Prepare a system flowchart to describe how customer orders entered into processing are processed by transform 1.2.1. Assume that the source document is a cleared customer order; the main output is the daily order file. Show the name of this program and the names of all files, and label the display screen.

CASE ASSIGNMENT 5

Prepare a system flowchart to show how orders stored on the daily order file are processed by transforms 1.2.2 and 1.2.3. The main input to this program is the daily order file, while the main output is the expanded daily order file. Name the program (or each job step), label all files, and label each display screen, as required.

REFERENCES

1. T. DeMarco, *Structured Analysis and System Specification* (New York: Yourdon Press, 1978).
2. See C. Selltiz, et al., *Research Methods in Social Relations* (New York: Holt, Rinehart, Winston, 1964).
3. See P. Green and D. Tull, *Research for Marketing Decisions* (Englewood Cliffs, N. J.: Prentice-Hall, 1966).
4. C. Gane and T. Sarson, *Structured Systems Analysis: Tools and Techniques* (Englewood Cliffs, N. J.: Prentice-Hall, 1979).

6
Data Organization and Documentation

INTRODUCTION

THE process of data collection and analysis should lead to a completed set of data flow diagrams (DFDs) and written procedures, a wide variety of reports and forms, interview notes and questionaires, plus system flowcharts. What the analyst must do next is devise a method of data organization that will define how a system is logically put together.

You may ask, "Why do I have to define the design? Aren't data flow diagrams self-explanatory, and can't I simply show them to others?" These questions can be answered in two ways. First, in the analyst's mind, DFDs are often self-explanatory; however, in the eyes of others, they generally are not. (Nevertheless, DFDs are easy to explain to others.) Moreover, DFDs do not define the terms used in their design. For example, the word *invoice* might mean a customer invoice (a bill prepared and sent by a firm) or a vendor invoice (a bill to be paid by a firm). Thus, the analyst must define precisely how a system is put together.

This chapter examines a specific set of methods for organizing and documenting the data illustrated on DFDs. When you complete this chapter, you should be able to

- describe the purpose of a data dictionary and show how terms are added to such a dictionary;
- describe the contents of a data store;
- draw a data structure diagram;
- construct a Warnier-Orr diagram; and
- write a transform description using structured English.

Once you are familiar with these methods of organizing and documenting collected data, you will be able to develop a *data specification*. Why is such a specification required? Primarily, it is used to provide the rules and conventions for describing each data flow, data store, and data transform required by a system, and, as important, to show how the various pieces of a system work together.

DATA DICTIONARY

Let's suppose that you have been assigned the task of creating a data dictionary. From your knowledge of dictionaries and of the brief overview of a data dictionary in chapter 3, you might assume that a data dictionary would consist of an alphabetized list of words, with each word followed by its definition. This assumption is partially correct. In structured analysis, each data element is listed and defined. A data dictionary contains more than data elements, however. Besides listing and defining all data elements, the data dictionary includes the definition of each data structure, using a prescribed convention. Employee pay rate, for example, could be defined as:

```
Employee-pay-rate = Hourly-rate + overtime-rate +
                    employee-deductions
```

Stated somewhat differently, employee pay rate is equivalent to the combination of three data structures: hourly rate, overtime rate, and employee deductions.

Contents

A data dictionary contains the definitions of all data flows, the contents of all data stores (files), and, at a minimum, the keywords used to describe each transform. Up to now we have used such phrases as total hours worked and overtime rate to describe data flows, employee master file and tax table to describe data stores, and calculate regular pay and calculate overtime pay to describe transforms. A data dictionary must clarify what these words mean. Moreover, it must be written in a special way. All terms must be self-defining and, as needed, partitioned into component parts. A term is *self-defining* when it can be understood from its name alone. A term is *partitioned* much like DFDs are partitioned, by subdividing a term into its component parts.

As an illustration of self-defining and partitioning, let's examine the terms used in defining gross pay, the single output produced by transform 2.2.2, calculate overtime pay (see **Figure 6–1**). Suppose we know that gross pay is made up of regular pay and overtime pay. We can write:

```
Gross-pay = Regular-pay + overtime-pay
```

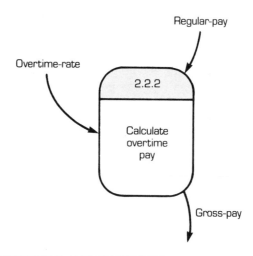

FIGURE 6–1 DFD showing the calculation of gross pay

However, both regular pay and overtime pay require further partitioning, since neither is self-defining. We can write:

```
Regular-pay = Regular-hours-worked × hourly-pay-rate
```

and

```
Overtime-pay = Overtime-hours-worked × overtime-rate
```

Even now, not all of the terms are *self-defining*. We can write:

```
Overtime-hours-worked = Total-hours-worked - 40
```

and

```
Regular-hours-worked = Total-hours-worked - overtime-
                       hours-worked
```

At this point, we decide to stop.

This example illustrates the concept of *top-down* partitioning of terms and expressions. Through partitioning, a complex concept such as gross pay is subdivided into more easily understood component terms. Such a process continues, with one subdivision after another, until all lower-level terms are self-defining. In our example, the data flow called total-hours-worked is considered to be self-defining.

Conventions

Several defining conventions are required to explain the words and expressions placed in the data dictionary. Because most dictionary entries are expressed as formulas, standard relational operators can be used. Thus,

- = means is equivalent to
- + means add
- − means subtract
- × means multiply by
- ÷ means divide by

These relational operators handle most sequential terms quite well and allow an analyst to *concatenate*, or link together, two or more data items. However, these operators do not allow an analyst to depict two important types of logical constructs: decision constructs and repetition constructs.

Figure 6–2 illustrates the three logical constructs of structured English. They can be summarized as follows:

1. The *sequential construct* consists of a series of concatenated expressions.

2. The *decision construct* contains two or more expressions; however, only one applies in any given case.

3. The *repetition construct* involves the repeating of an expression, generally within some established limit.

An advantage of these three structures is that they are easy to read and use. As important, they can be used to define any data flow. Each of these constructs will be discussed later in this chapter.

Table 6–1 compares the DeMarco[1] and Gane and Sarson[2] conventions for showing decision and repetition constructs. While both sets of conventions are adequate, they cannot be mixed. Forced to make a choice, our selection would be the DeMarco set of conventions; however, either set works in practice.

The following examples help demonstrate how sequential, decision, and repetition constructs are used to define data flows, using these conventions.

EXAMPLE 1

Suppose that the color of a computer terminal screen can be set as either black on white or white on black. We can define terminal color as follows:

```
Terminal-color = [Black-on-white/white-on-black]
```

This definition represents a decision construct.

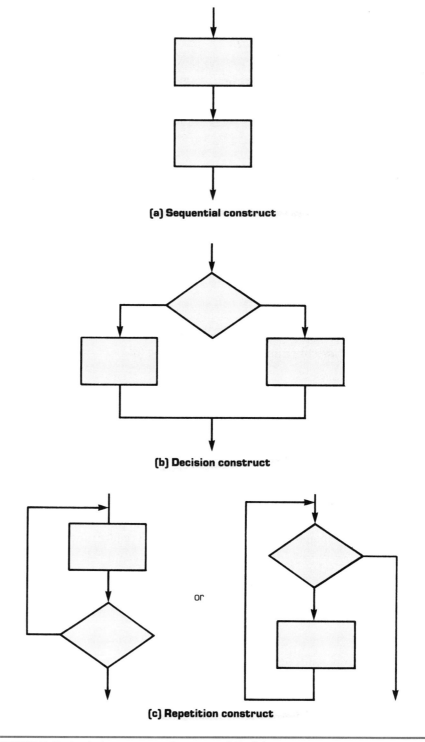

FIGURE 6–2 Logical constructs of structured English

TABLE 6–1 DeMarco and Gane and Sarson data dictionary conventions

Type of Construct	DeMarco Conventions	Gane and Sarson Conventions
Decision	[either/or]	{either/or}
Repetition	m{iterations of}n	iternations of* (m to n)

EXAMPLE 2

Suppose that we wish to define the number of line items to place on an invoice form (such as the invoice shown on Figure 5–9). We know that a maximum of ten line items can be printed. We can define invoice line item as follows:

```
Invoice-line-item = 1 {quantity-ordered + quantity-
                      shipped + item-number +
                      item-description + item-
                      price + item-extended-cost} 10
```

This definition indicates that a minimum of one and a maximum of ten invoice line items can appear on a single customer invoice.

Table 6–2 illustrates the complete set of DeMarco conventions. Besides the five relational operators, the conventions for a decision construct (either/or) and a repetition construct (iterations of) are shown. Two somewhat unique operators are also illustrated. The *optional operator* specifies when a term may or may not appear in a definition. For example, an employee's middle name might be shown as an optional element in a data dictionary as follows:

```
Employee-name = First-name + (middle-name) + last-name
```

This indicates that some employee names will consist of first, middle, and last names, while other employee names will be limited to first and

TABLE 6–2 Complete set of DeMarco conventions

=	is equivalent to	[ ]	either /or
+	add	{ }	iterations of
−	subtract	( )	optional
×	multiply	* *	comment
÷	divide		

last names only. Likewise, the *comment operator* is used to append notes to expressions within a dictionary. We might write:

```
Overtime-pay = *Dollars paid for hours worked in ex-
               cess of a set limit*
             = Overtime-hours-worked × overtime-rate
```

Organization

While conventions show how words and expressions are defined for placement in a data dictionary, they do not show how the data dictionary is to be organized or maintained. One way to organize the data dictionary is to list all expressions in alphabetic order, much like the order found in a standard dictionary. The advantage of this method is that it helps the analyst avoid *redundancy*, of which there are two types: using more words than necessary and duplicating words. For example, in creating an alphabetized listing, an analyst might discover the following redundancy:

```
Regular-pay = regular-hours-worked × hourly-pay-rate
```

and

```
Regular-hours-pay = Regular-hours-worked × hourly-
                    pay-rate
```

Fortunately, this redundant situation is easily corrected.

A major disadvantage of an alphabetized listing is that it does not show how words and expressions are nested. **Figure 6-3** illustrates a nested listing. This type provides information about the *composition* of a data structure. It shows exactly what the components are and how they are interrelated. Even then, a major disadvantage with a nested listing is that it makes it difficult for the analyst to discover redundant situations.

```
Gross-pay = Regular-pay + overtime-pay
   Regular-pay = Regular-hours-worked + hourly-pay-rate
      Regular-hours-worked = Total-hours-worked - overtime-hours-
                             worked
```

and

```
Gross-pay = Regular-pay + overtime-pay
   Overtime-pay = Overtime-hours-worked + overtime-rate
      Overtime-hours-worked = Total-hours-worked - 40
```

FIGURE 6-3 Nested data dictionary listing

Data Element Definition

Name:	Regular-pay
Description:	*Dollars paid to an employee for regular hours worked*
Composition:	Regular-pay = Regular-hours-worked + hourly-pay-rate where Regular-hours-worked = Total-hours-worked − overtime-hours-worked
Aliases:	Standard-pay
DFD Reference:	2.2.1 and 2.2.2
DD Reference:	See gross-pay
DS Reference:	Current-payroll-file, employee-master-file
Characteristics:	Maximum characters: 6 Type: F Minimum characters: 0 Format: XXXX.XX

FIGURE 6-4 Data element listing in the data dictionary

Which type of listing is favored, given that each is advantageous for a different reason? The answer is that both listings are generally required. Besides identifying redundant listings, the alphabetic listing helps an analysis team to quickly determine whether a term is used and, if it is, what it means. If the analyst needs to know more about the composition of a term and its use in a design, however, a nested listing is advised.

Figure 6-4 illustrates how a single data element or data structure might be represented on a page in a data dictionary. The contents of this page can be summarized as follows:

1. *Name* identifies the data element or structure.

2. *Description* provides a concise definition of the data element or structure.

3. *Composition* shows the component parts of the term and, as necessary, a breakdown of those parts.

4. *Aliases* are the names of other terms that have the same meaning. As illustrated, regular pay can also be represented as standard pay.

5. *DFD reference* indicates where the term is shown on a DFD or found in a transform description. Two DFD numbers show connections between transforms. In our example, two numbers tell the analyst that regular pay flows from transform 2.2.1 to 2.2.2.

6. *Data dictionary (DD) reference* shows which other terms in the dictionary contain the data element name.

7. *Data store (DS) reference* indicates whether the term is contained within a data store. As illustrated, regular pay is used in describing the employee master file and the current payroll file.

8. *Characteristics* provides information about the format of the data element or structure. As indicated, the maximum number of characters in regular pay is six, the minimum is zero, the type is F (floating point), and the format is XXXX.XX.

Reasons for Creating the Data Dictionary

For the experienced systems analyst, there are several good reasons for insisting that a data dictionary be created. First, a data dictionary provides all people working on a development project with a standard terminology. It allows people to decide what they mean by each term. For example, suppose a manager asks, "Does the definition of gross pay cover all overtime pay?" Unless an analyst had defined gross pay in advance, he or she would be pressed to give the manager an answer. However, with a dictionary, the analyst can respond, "Yes, the definition of gross pay does account for all overtime pay. Gross pay is defined as regular pay plus overtime pay."

Second, the data dictionary provides the analyst with several types of information. These include:

1. *ordered listings,* including all data structures and elements as well as nested listings showing the composite of each complex data structure or element; and

2. *cross-reference listings,* including which DFD, data store, or transform uses the data structure or element as well as which chains of terms are contained in the data dictionary.

Cross-reference listings can be as important as ordered listings. They allow the analyst to move freely from a set of terms to a set of DFDs.

Third, the concept of a data dictionary provides the systems team with a method by which to document a system as it is logically defined. Continued on a daily basis, the task of documentation becomes less burdensome. Indeed, system documentation becomes as important as the detailing of system components and interrelations. The conventions required to explain words and expressions force the analyst to select terms that can be defined.

Currently, commercial software packages support the creation and maintenance of a data dictionary. These packages simplify the adding of new terms to the dictionary and the modification of previously stored terms. They feature the capability to produce a variety of ordered listings and support several types of cross-referencing. Even with the help of software, however, there is no short cut to creating and maintaining the dictionary. It takes considerable time to define each term and expression—a task generally required of a systems analyst. In addition, it takes considerable time to enter the data dictionary details into the computer. For large projects, this task is generally assigned to a *data manager*, the individual responsible for organizing and managing the data to be placed in a data dictionary.

DATA STORE DESCRIPTIONS

Besides defining all data flows and entering them in a data dictionary, the analyst must describe the contents of each data store. The data store, as defined in chapter 3, serves as a repository of data in a system. As shown on DFDs, data can be placed in a data store as an output of a transform, or it can follow from a source.

Data store descriptions are prepared similarly to data flow descriptions, in that the same rules apply: each complex term must be partitioned until all lower-level terms become self-defining. For example, suppose that the Mansfield analysis team decided that forty-five different data elements were required to describe each of the organization's employees. As shown in **Figure 6–5,** the listing of these elements forms a payroll record, while a collection of these records makes up the employee master file (see Figures 3–4, 3–5, 5–2, and 5–3). Initially, the team members decided to identify each employee with a unique employee number. Following this, they grouped the remaining elements into six major categories. In this way, they could define the employee master file as follows:

```
Employee-master-file = {employee-record}
    Employee-record = Employee-number +
                      employee-name-and-address +
                      personnel-information +
                      wage-and-salary-information +
                      leave-and-pension-information +
                      current-payroll-information +
                      YTD-payroll-information
```

Next, the team went on to define the employee number and data elements within each of these six categories. Some selected examples are shown below.

```
Employee-number = 5 digits
Employee-home-address = ([PO box/apartment-number]) +
                        street-address + city-address
                        + postal-code
Marital-status = [married/single]
Accumulated-vacation= 1{digit}2
Exempt-status = [yes/no]
Current-gross-pay = 1{digit}6 *9999.99*
```

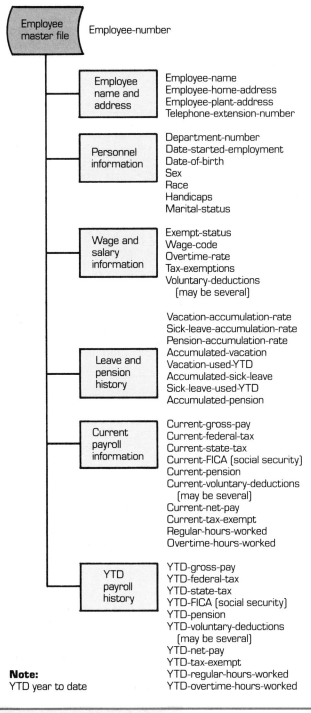

FIGURE 6-5 Employee master data store

These definitions can be interpreted as follows:

1. Each employee number consists of five digits.//
2. Each employee address consists of an optional post office box or apartment number, plus a street address, city address, and postal code.
3. Marital status is either married or single.
4. Accumulated vacation is either one or two digits and cannot exceed ninety-nine days.
5. Exempt status is either yes, meaning exempt, or no.
6. Current gross pay ranges from one to six digits, with an upper limit of 9999.99.

As this example demonstrates, the process of definition for a data store appears to be identical to the process of definition for a data flow. Actually, there is an important difference, as we will examine shortly. First, though, we need to define a data store in more detail.

By definition, a data store describes an *object* or entity and is made up of *records*. A record, in turn, is described by its *attributes*, is identified by one or more *key attributes*, and may contain one or more *pointer attributes*. Because these terms are somewhat different than those used earlier, we will define each briefly.

An object is a general class of items about which information is collected. We might collect information about customers, books, taxpayers, or bills to be mailed to customers. Each of these represents an object.

A record describes a single, logical item within the set of items that make up the object. For example, a customer is a single item within a set of customers; a customer invoice is a single item within a set of customer invoices. Thus, we can define each of these records as follows:

{customer} and
{customer-invoice}.

An attribute is a property of an object. A customer number is an attribute of a customer. Likewise, the customer sold-to address, customer order number, and date shipped are attributes. Examine Figure 6-5 once again. Each data element in this data structure represents an attribute of the object employee; each set of data elements describes an employee record.

A key attribute is a primary key or identifier. It is required to identify a single record within a set of records. The federal government, for example, requires all taxpayers to use a social security number. This number serves as a key attribute. It is unique to the taxpayer.

A pointer attribute is an attribute embedded in a record that points to another record in the same object or in another object. The department number embedded in an employee's record (see Figure 6-5) could be used to locate another department record. The department number, in this instance, is a pointer attribute.

With these new definitions in mind, let's examine the data to be keyed in our payroll processing example. As shown in **Figure 6–6,** data to be keyed from an employee's timecard are limited to eight attributes beginning with the employee number. (This assumes that before employee data are entered, the payroll number has been entered and is to be appended to each record.) The hours-worked file that follows this key entry process (see Figure 5–2), can be defined as follows:

```
Hours-worked-file   = {hours-worked-record}
Hours-worked-record = Employee-number + department-
                      number + payroll-number + rec-
                      ord-code + regular-hours-
                      worked + overtime-hours-worked
                      + vacation-hours + sick-leave-
                      hours + holiday-hours
```

The key attribute in this example is the employee number. Because of its importance, each key attribute is underlined.

Data Structure Diagrams

At times the definition of an object fails to convey logical database requirements. By definition, a *database* is a collection of interrelated data, with minimum redundancy, that serves one or more computer applications. The analyst must select a method by which to depict these interrelationships.

One method of showing the logical relationships within a database structure features *data structure diagrams*, or DSDs. **Figure 6–7** illus-

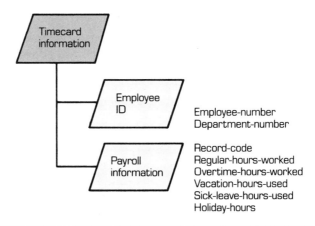

FIGURE 6–6 Payroll data keyed to processing

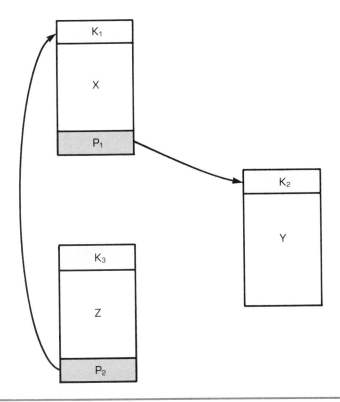

FIGURE 6-7 A data structure diagram (DSD)

trates a somewhat abstract DSD. The notation shown indicates the following:

1. X, Y, and Z represent three different objects.
2. K_1, K_2, and K_3 are key attributes.
3. P_1 and P_2 are pointer attributes.

This DSD frames the following logical database interrelations:

1. Given K_1, we can access a record stored on object X and, as indicated by P_1, a record stored on object Y.
2. Given K_2, we can access a record stored on object Y.
3. Given K_3, we can access a record stored on object Z and, as indicated by P_2, a record stored on object X.

LOGICAL EXTERNAL POINTERS

The lines drawn between P_1 and K_2, and P_2 and K_1 on Figure 6-7 are called logical external pointers. A *logical external pointer* identifies how access to information between objects is accomplished. Let's consider a more practical example to clarify this point. Suppose we wish to expand the payroll system at Mansfield by key entering the number of hours employees spend working on projects, followed by posting these hours to a project master file. The DSD shown on **Figure 6–8** illustrates the resulting interrelationships between stored data. Each employee record now contains a department number pointer attribute and a project number pointer attribute. Moreover, a single arrowhead is used to illustrate a one-to-one relationship, while a double arrowhead is used to indicate a one-to-many relationship. Thus, we can interpret this DFD as follows:

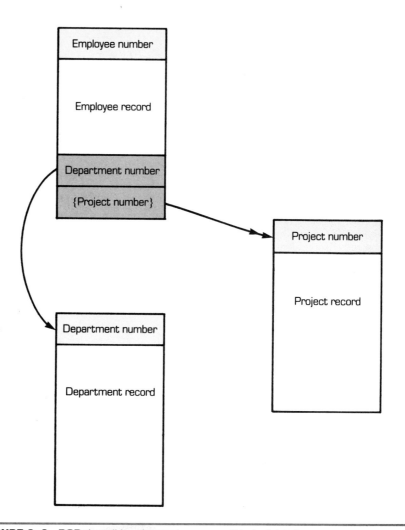

FIGURE 6–8 DSD describing the employee payroll logical database

1. Given one employee number (a key attribute), we can obtain the employee's record, the employee's departmental record, and a project record for each project worked on by the employee.
2. Given a department number (a key attribute), we can obtain a department record.
3. Given a project number (a key attribute), we can obtain a project record.

Thus, logical external pointers clarify how links between different data structures are set and maintained. It is not necessary to store all data relating to an employee (employee information, department information, and project information) in a single data store. Instead, it becomes possible to link multiple structures, using pointer attributes as keys.

LOGICAL INTERNAL POINTERS

Besides logical external pointers, the analyst must be able to depict how access to similar data within the same object is possible. A *logical internal pointer* is used to identify object data interrelationships within a record. **Figure 6-9** provides two examples of logical internal pointers. In Figure 6-9a, the department number is shown as both an internal and an external pointer. As an internal pointer, the number is used to link together all employees who work in the same department. In Figure 6-9b, the employee number is shown as embedded within each project record. This number makes it possible to link together all projects worked on by a single employee. Moreover, the employee number is also defined as an external pointer.

LINKED LIST

A linked list can be created with internal logical pointers. By definition, a *linked list* is a list in which each record points to its logically related successor, except for the last record, which contains an end-of-list indicator. **Figure 6-10** illustrates the simplest type of linked list, a *one-way linked list*. As illustrated, the link indicates whether another project has the same employee number. In this example, employee 1724 is shown as working on projects A112 and A183. An asterisk signifies that the end of a list has been reached. Likewise, tracing through this example shows that employee 3142 has reported time for two projects, employee 2101 has reported time for three projects, and employee 0981 has reported time for a single project.

INVERTED FILES

One reason to embed pointers as links within a record is to allow the creation of inverted files. An *inverted file* arranges records by a pointer attribute rather than by a key attribute. Study **Figure 6-11** for a moment. This figure illustrates how the project file can be transformed into an

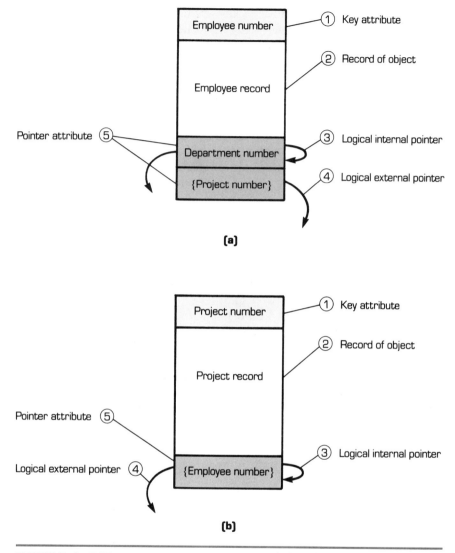

FIGURE 6–9 DSD showing logical internal and external pointers for an employee record (a) and a project record (b)

inverted file. Could several inverted files be created from a single primary file? Yes. Several inverted files are created from a single file, such as a project file when such a file contains several pointer attributes.

Why are linked lists and inverted files important? Generally, to save time and expense. Consider the following. Although employee records are arranged in alphabetical order within the employee data store, employee paychecks are distributed by department. Since the records are

Project Number	Project Description	Employee Number	Link
A112	....	1724	A183
A113	....	3142	B664
A161	....	2101	AB16
A183	....	1724	*
AB12	....	2101	B132
AB16	....	0981	*
B132	....	2101	*
B664	....	3142	*

A183 → Pointer to key of next record on list

* → End of list

FIGURE 6–10 One-way linked list

not in the desired order, one alternative is to sort the entire file, rearranging the employee records in departmental order, printing the paychecks in that order, and then resorting the entire file in alphabetic order. As expected, these sorts can take considerable time. Another, less costly alternative is to embed a department pointer attribute within a

Employee Number	Project Number	Project Description
1724	A112	....
1724	A183	....
3142	A113	....
3142	B664	....
2101	A161	....
2101	AB16	....
2101	B132	....
0981	AB12	....

FIGURE 6–11 Inverted project file showing project records arranged by employee number (secondary key)

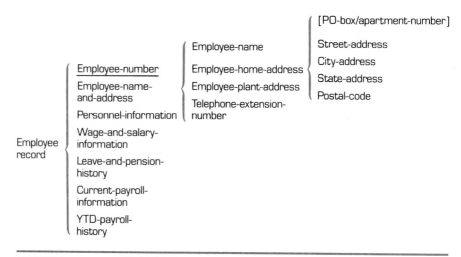

FIGURE 6–12 Warnier-Orr diagram showing decomposition of the employee record

record. Such a pointer permits records to be printed directly in a desired sequence. The disadvantage of this approach is that all embedded pointers must be maintained. When records are continually being added or deleted from a file, the maintenance of pointers can become complicated and troublesome.

WARNIER-ORR DIAGRAMS

Some analysts prefer to draw Warnier-Orr diagrams instead of DSDs in describing the logical contents of a data store. Introduced by Warnier[3] and later modified by Orr,[4] Warnier-Orr diagrams are used to decompose the contents of a data store, much like DFDs are used to decompose the functions of a system.

Figure 6–12 illustrates how a Warnier-Orr diagram would be drawn for the employee master file. As before, the employee master file is defined as a set of employee records, thus leading to the equation:

```
Employee-master-file = {employee-records}
```

However, the Warnier-Orr diagram tells us how to define an employee record. We can write (as before):

```
Employee-record = Employee number + employee-name-
                  and-address + personnel-information
                  + wage-and-salary-information +
                  leave-and-pension history +
                  current-payroll-information + YTD-
                  payroll-history
```

Let's suppose next that we wish to know which attributes make up the category of employee name and address. The Warnier-Orr diagram informs us that

```
Employee-name-and-address = Employee-name + employee-
                            home-address + employee-
                            plant-address +
                            telephone-extension-number
```

If we wanted to know which attributes make up employee home address, further decomposition would be necessary. The Warnier-Orr diagram tells us that

```
Employee-home-address = ([PO box/apartment number]) +
                        street-address + city-address
                        + state-address + postal-
                        code,
```

where the post office box or the apartment number is optional.

The Warnier-Orr diagram continues to be pushed to the right until all right-hand attributes can be defined by their *field length*, or physical storage requirements. City address might be defined as

```
City-address = 1{character}12
```

or as requiring from one to twelve characters. Telephone extension might be defined as

```
Telephone-extension = 4 digits
```

In this instance, an extension always consists of a four-digit number.

As this example suggests, there are several good reasons for using Warnier-Orr diagrams instead of DSDs to describe the contents of data stores. First, Warnier-Orr diagrams are easy to construct, read, and interpret. Second, they permit larger, complex terms to be decomposed into constituent parts. Third, they can be used to analyze both actions and things. In later chapters, Warnier-Orr diagrams will be used to define and order the actions within a computer program. Fourth, they simplify the definition and order of terms to be entered into a data dictionary.

The main disadvantage of Warnier-Orr diagrams is also the main advantage of DSDs. Warnier-Orr diagrams do not show either external or internal logical pointers. As a consequence, many analysts use both types of diagrams in describing the contents of data stores. Warnier-Orr diagrams are used to decompose terms until a self-defining set of terms is realized; DSDs are used to show interrelationships within and between sets of data.

TRANSFORM DESCRIPTIONS

The last type of data to be organized by the analyst is material dealing with transforms. As discussed briefly in chapter 3, the analyst relies on structured English to explain which actions are to take place within a transform and what policies govern these actions. Suppose we wish to hold customer orders if customers have not paid their previous bills. We could write:

```
      Hold the customer's order
          until the customer's account is paid.
```

This statement describes the action to take ("hold the customer's order"); however, it does little to clarify the policy that governs this action. Accordingly, we might rewrite this transform as follows:

```
IF a customer-order exceeds $400 and
    IF the customer-account-balance is 60 days past due
        THEN hold the customer-order
        ELSE process the customer-order
```

In this instance, the actions to take are clear: either hold or process the order. More important, the policy that governs these actions is clarified: hold the order only if the new order exceeds $400 *and* the customer's account balance with the firm is sixty days past due.

Rules of Structured English

Let's briefly review some rules of structured English before providing further examples of transform descriptions. As suggested earlier, structured English consists of *imperative statements* that either tell us how to do something or command us to do something. The following is a statement that tells us how to do something:

```
Move the next payroll-transaction to the end of the
    payroll-transaction-file.
```

An example of a command that tells us what to do is:

```
      Close the payroll-transaction-file.
```

The rules associated with the writing of such imperative sentences may appear obvious. They include:

1. Make each statement concise and to the point.
2. Use an action verb such as *move, close,* or *open* to describe what is to be done.

3. State the object of the verb explicitly. We might have written "Close the file," instead of "Close the payroll transaction file." By naming the file, we make the object of the verb explicit.
4. Document all keywords and terms and place these definitions in the data dictionary.
5. Minimize the use of adjectives.
6. Avoid the use of relative words, such as *improve, faster, higher, increase,* or *good.*

Structured English Conventions

Similar to designing the data dictionary, writing transform descriptions requires following specific conventions. With structured English, a standard notation is needed to identify each of the three logical constructs that frame the language. All transform descriptions can be written using combinations of sequential, decision, and repetition constructs.

Figure 6–13 shows a set of structured English conventions that follow the process flow notation shown earlier (see Figure 6–2). Let's consider several examples to show how these conventions can be applied.

Type of Construct	Structured English Is Written as
Sequential	Simple imperative sentence.

Decision	IF THEN
	⋮ ELSE
	END-IF
	
	or as
	
	SELECT the following that applies:
	    CASE 1
	    CASE 2
	      ⋮
	    CASE *n*

Repetition	REPEAT-UNTIL
	    ⋮ (the condition is false)
	END-REPEAT-UNTIL
	
	or as
	
	DO-WHILE
	    ⋮ (the condition is true)
	END-DO-WHILE

FIGURE 6–13 Structured English conventions

SIMPLE DECISION STATEMENTS

Suppose that we know the following about overtime hours worked. Past employee history records indicate that 95 percent of all weekly overtime hours reported are less than twenty-five per employee. For this reason, the Cardinal Corporation decides to require supervisors to review any weekly timecard in which overtime hours are reported in excess of twenty-four hours.

Structured English reduces this narrative to the following:

```
IF overtime-hours are greater than 24
    THEN ask the supervisor to review the employee-
      time-card
    ELSE continue on
END-IF
```

NESTED DECISION STATEMENTS

Nested decision statements are handled much like simple decision statements. We will complete the invoice example shown earlier:

```
IF a customer-order exceeds $400
    (THEN) and IF the customer-account-balance is 60
      days past due
                    THEN hold the customer-order
                    ELSE process the customer-order
                END-IF
    ELSE process the customer-order
END-IF
```

This description is nested because one set of IF-THEN-ELSE statements fits inside (e.g., becomes nested in) a second set of IF-THEN-ELSE statements.

CASE STATEMENTS

Case statements can also be used to describe a transform. Consider the following:

```
SELECT the dollar amount that applies:
    CASE 1  IF the planned expenditure > $10,000
              THEN send expense-justification-report
                to corporate headquarters.
    CASE 2  IF the planned expenditure is in range
              $1,000 to $10,000
              THEN send expense-justifiation-report
                to plant manager.
    CASE 3  IF the planned expenditure is < $1,000
              THEN send expense-justification-report
                to area supervisor.
```

Case statements allow for the existence of several possibilities. The ELSE condition is not needed because the reader understands that if a case does not apply, the process will jump to the next case to determine its applicability.

REPETITION STATEMENTS

Writing repetition statements can take two forms. As shown in **Figure 6–14a,** the REPEAT-UNTIL construct works on a process until it meets some condition—that is, until the condition becomes false. Consider the following:

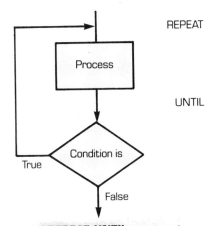

(a) REPEAT-UNTIL construct

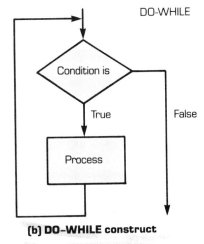

(b) DO-WHILE construct

FIGURE 6–14 Repetition constructs

```
REPEAT-UNTIL all invoice-line-items are extended.
    Multiply quantity-shipped by item-cost to arrive
      at item-extended cost
END-REPEAT-UNTIL
```

This description is consistent with the earlier definition of an invoice line item and is thus consistent with the definitions recorded in the data dictionary. In addition, the REPEAT-UNTIL construct is quite acceptable in this instance, since every customer invoice will always have a minimum of one invoice line item.

The DO-WHILE construct is similar to the REPEAT-UNTIL construct except that a condition is tested before a process is executed (see **Figure 6–14b**). As illustrated, the process is not executed if the condition is false to begin with. Consider the following:

```
DO-WHILE there are overtime-pay-records.
    Read a record.
    Add overtime-hours-worked to overtime-hours-total.
    Add overtime-pay to overtime-pay-total.
END-DO-WHILE
```

This structured English example also contains terms described earlier and placed in the data dictionary. The DO-WHILE construct is necessary because instances may arise in which there are no overtime pay records. When this occurs, the process of adding overtime hours and dollars would be bypassed.

As these examples demonstrate, transform descriptions provide the analyst with a documented record of the logical processing steps in a system. Once these descriptions are completed and combined with the data dictionary and data structure diagrams, the analyst has a comprehensive knowledge of

1. the specific meaning of terms used in constructing DFDs;

2. the specific contents of data stores and the interrelations that exist between and within data stores; and

3. the logical contents of each transform and the specific meaning of keywords used in each transform description.

Collectively, then, the analyst can document in an organized manner the way in which a system is logically put together.

Experienced systems analysts will spend considerable time attempting to clarify the policy that governs the transforming of data. In many cases, analysts will help clarify company policy by coordinating it among departments, plants, or even among employees in the same department. As an example, one employee might hold a new order if a customer's account balance were sixty days past due, while another employee might hold the order only if the balance were ninety days past due. Discovering these differences is part of the problem-finding side of systems work. Once a problem is found, it usually can be solved.

Preparing transform descriptions also helps the analyst discover errors in logic. For example, an analyst might find overlapping conditions, such as the following:

```
SELECT the weight that applies:
    CASE 1  IF the weight is in range 5 to 10
               THEN rate is weight × 1.5.
    CASE 2  IF the weight is in range 3 to 6
               THEN rate is weight × 1.3.
```

Similarly, an analyst might discover steps missing in a process, improperly nested decision steps, cases of repeat-until in which the condition is false to begin with, and so forth.

Finally, structured English is easy to adopt as a kind of shorthand, which is an advantage as well as a disadvantage in describing transforms. The problem is that the descriptions written may be overly cryptic and unique to the analyst. Structured English is especially troublesome if the analyst has a tendency to abbreviate heavily. In addition, attempting to show structured English to a user group may create more problems than it solves. The exception is when users and the analysis team work together to specify logical steps in a process, such as the method to be used in calculating a finance charge. In this instance, the availability of transform descriptions written in structured English is invaluable.

The Mansfield, Inc., Case Study

Roger Bates gathered his small staff and told them what he wanted next. "I want you to define all data flows appearing on the DFDs you have prepared, to construct DSDs of all data stores and define the contents of each data store, and to write structured English descriptions of all transforms." He then added, "I think you will find that this work will require a fair amount of time to complete. However, it will be time well spent once we begin work on the final specifications for the new system."

Data Dictionary

Carolyn Liddy found that defining all data flows was in fact easier than she had imagined. She explained how she defined a customer invoice. (See Figure 5–10 for a description of the logical contents on a customer invoice; see Figure 5–15 to review the logical description of the system.)

She began: "We define each customer invoice as follows:

```
Customer invoice = Customer-information + customer-
                   order-information + order-filling-
                   information + billing-information
```

"Next, we test each term on the right-hand side of this equation to determine whether it is self-defining. Consider customer information, for example. Since this term is not self-defining, we break it down as follows:

```
Customer-information = Customer-number + customer-
                       sold-to-address + customer-
                       ship-to-address + terms-code +
                       salesperson-number
```

"In this instance, some terms are self-defining, such as customer number and salesperson number, while others, such as terms code and customer ship-to address are not self-defining. For terms that are not self-defining, a further breakdown is required. As another example, we defined terms code as follows:

```
Terms-code = 1 = [2 percent 10/ net 30 on receipt]
```

"When the terms code is equal to 1 we specify a 2 percent cash discount if the customer pays within ten days, or ask that the customer pay in full on receipt of the invoice."

Data Structure Diagrams (DSDs)

"The DSDs were more difficult to prepare than the listings to be placed in the data dictionary," Neil Mann remarked. "Initially, we looked at what the existing system provided and compared this with what we expected the new system to be able to do. **Figure 6–15** illustrates that the existing system allowed us to

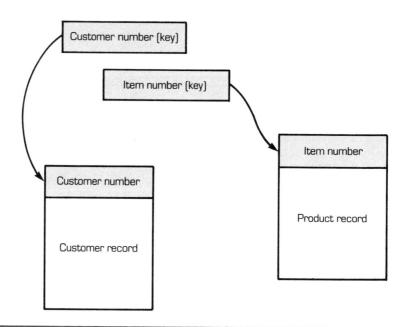

FIGURE 6–15 DSD showing existing logical relationship

do very little. Given a customer key, we could obtain a customer record, and given an item number, we could obtain a product record. However, the existing system did not permit us to begin with a customer order number and find the customer record, product records, or the salesperson record."

Figure 6–16 illustrates the DSD that Neil first proposed for the new system. "What we want to do," he said, "is to keep online a record of all customer shipments and to preserve ties between this record and corresponding customer, product, and salesperson records, so that the following holds:

1. Given a single customer order number, we can obtain the customer shipment record, the customer's record, and a product record for each item ordered.
2. Given a customer number, we can obtain a customer record.
3. Given an item number, we can obtain a product record.

"It is also possible to make our new system even more elaborate," he added. "As shown in **Figure 6–17,** let's suppose we wish to tie salesperson records to our design. This can be done by adding the salesperson key to each customer

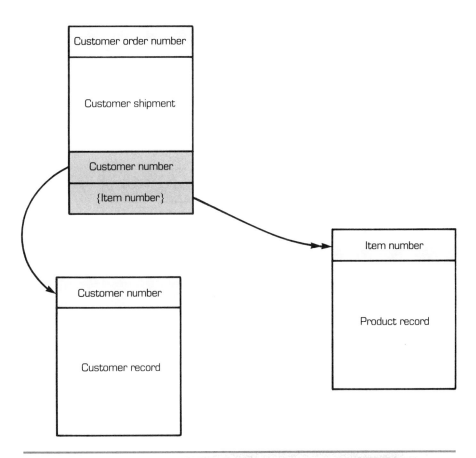

FIGURE 6–16 DSD describing the proposed customer invoice logical database

196   Chapter 6: Data Organization and Documentation

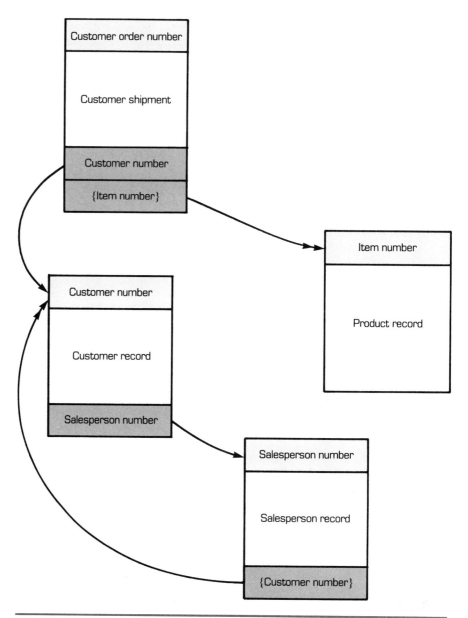

FIGURE 6–17 Revised DSD showing the proposed relationship between the customer invoice and the salesperson record

record and all customer numbers assigned to the salesperson to the salesperson's record. We can then say:

1. By knowing the customer order number, we can determine the customer number, the customer record, the salesperson number, and the salesperson's record.
2. By knowing the customer number, we can identify the customer record,

the salesperson assigned to the customer's account, and the salesperson's record; we can also find out all customers assigned to that salesperson's account."

Structured English

Neil discovered that writing structured English also took considerable time and effort, not so much in the actual writing of each transform description, but in thinking about how to clarify the processing steps. Consider the following description prepared by Neil in describing the transform for creating an invoice record.

```
1.  Enter customer shipment data as follows:

    Enter customer-order-number
    Enter customer-number
    Get customer-record from customer-master-file
    IF no record exists
        THEN display no-customer-record found
        ELSE display customer-record
    END-IF
    DO-WHILE there are item numbers
        Enter item-number
        Get product-record from product-master-file
        IF no record exists
            THEN display no-product-record found
            ELSE display product-description and
              product-price
        END-IF
        Enter quantity-ordered
        Enter quantity-shipped
        Set new-dollar-total
    END-DO-WHILE
    Enter date-shipped
    Enter FOB-instructions
    Enter shipped-via-instructions

2.  Then process invoice as follows:

    Assign invoice-number
    Assign invoice-date
    Write invoice to invoice-transaction-file
```

SUMMARY

Systems analysts must develop data specifications to accompany their DFDs. The specification includes a data dictionary, a description of each data store, and a description of each transform.

The data dictionary provides the analysis team with a complete set of terms associated with a system project. It clearly defines all terms used in constructing a DFD, including the meaning of each data element within each data flow, the meaning of each data element within each data store, and the meaning of key terms within each transform.

There are several good reasons for supporting the creation of a data dictionary: it provides the systems team with a standard terminology, it orders listings of all terms, and it establishes a method of documenting a system as it is logically defined.

Data store descriptions include the definition of all terms placed in the data store, plus data structure descriptions. A data structure diagram (DSD) is helpful in showing the logical relationships within a database structure. Logical external and internal pointers are clarified with DSDs. Warnier-Orr diagrams are useful in describing the decomposition of terms used in describing each data store. These diagrams simplify the definition of data store terms to be placed in the data dictionary.

Transform descriptions provide a documented record of the processing steps in a system. Using structured English, the analysis team can document the actions that take place within a transform and the policies that govern these actions.

The make-up of structured English is restricted to simple decision statements, nested decision statements, case statements, and repetitive statements. Collectively, these statements provide the analysis team with a highly structured method of preparing a description of how transforms change the data flows within a system.

REVIEW QUESTIONS

6–1. Why is a data dictionary required for describing a system?

6–2. When is a term placed in the data dictionary said to be self-defining?

6–3. Name the three logical constructs of structured English.

6–4. In constructing a data dictionary, what is the main advantage of a nested listing? What is the main disadvantage of a nested listing?

6–5. What types of references are important for describing each data element contained in the data dictionary? Why is each type important?

6–6. What is a record and how does it differ from an object?

6–7. What is the difference between a key attribute and a pointer attribute?

6–8. How does a logical external pointer differ from a logical internal pointer?

6–9. What advantages can be associated with the use of Warnier-Orr diagrams in describing the contents of data stores?

6–10. List the rules to be followed in writing structured English?

6–11. What are transform descriptions and why are they prepared?

6–12. What is the main advantage and the corresponding main disadvantage associated with the use of structured English?

6–1. How would you record the following in a data dictionary:
1. a person's sex?
2. a person's age?
3. a person's birthday?
4. a person's telephone number?
5. a person's grade in a course, where letter grades of A, B, C, and N are possible?

6–2. Suppose an analyst discovers the following when studying a copy of the data dictionary.

```
Item-extended-cost = Item-price × quantity-shipped
```

and

```
Extended-cost = Item-price × quantity-shipped
```

What does this situation suggest? What, if anything, should be done to correct it?

6–3. Suppose we wish to describe the data dictionary term for the number of regular hours worked during a week. We might use several reporting formats, including Regular-hours-worked, Regularhoursworked, and Regular hours worked. Which format would you recommend? Explain your response.

6–4. Examine the data structure diagram shown below and define the logical database interrelations.

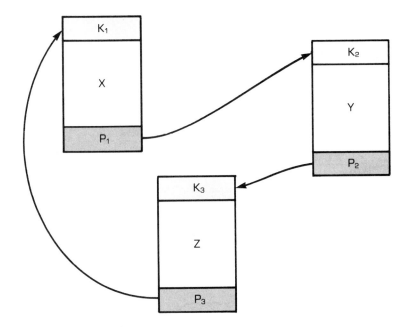

6-5. Examine the two data structures shown below and define the logical database interrelations for each, when access begins at K_1.

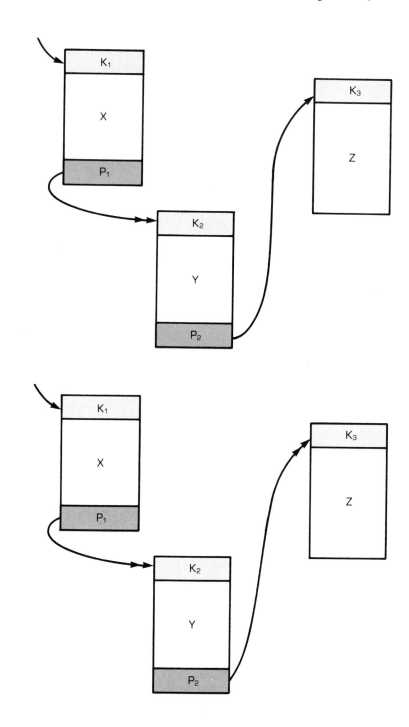

6–6. Design a one-way linked list for the set of records shown below.

Customer Number	Postal Code
10	95401
12	95403
16	95405
18	95403
19	95401
30	95403
32	95403
36	95405

6–7. Design an inverted file from the set of records shown in exercise 6–6.

6–8. Suppose that more than one salesperson can be assigned to each Mansfield account. Modify the DSD shown in Figure 6–17 to show this new relationship. Describe the new relationship between customer and salesperson records.

6–9. Suppose that postal code and customer number both serve as internal logical pointers in processing Mansfield's customer shipments records. Modify the customer shipment data structure shown in Figure 6–17 to show this new relationship.

6–10. Design an inverted file from the data store shown below.

Customer Order Number	Customer Order Details	Salesperson Number	Salesperson Pointer
1011	....	100	1022
1014	....	200	1016
1016	....	200	1101
1022	....	100	1104
1099	....	300	1107
1101	....	200	*
1104	....	100	1112
1107	....	300	1115
1112	....	100	*
1115	....	300	*

6–11. Fred plans to attend school full-time if he can obtain a student loan and part-time if he cannot obtain a loan but can get a part-time job. If

Fred cannot obtain a student loan or find a part-time job, he will keep the job he has and attend school next year.

Describe this situation using structured English notation.

6-12. We want to find out how well a weight-loss class is doing. We plan to obtain each student's weight and to compare it with his or her weight when the class started. For those people who have not lost sufficient weight, say two pounds or less, a personal consultation with the class instructor will be arranged. Moreover, we will relax the fitness schedule for those people who have lost ten pounds or more. Finally, for those in between, we will say, "Keep going. Your fitness schedule seems about right for this point in our program."

Describe this situation using structured English notation.

WORLD INTERIORS, INC.—CASE STUDY 6
DOCUMENTING THE CURRENT SYSTEM

Introduction

"I think I understand the current system," said John Welby. "After all, Rena reviewed the DFDs I prepared and indicated that they were accurate. Even so, I do need to define all data flows and the contents of all data stores. And I had better define the relationships between data stores. Perhaps the key to the new design will be the way in which data are organized. A major problem with the current system is that a great deal of data is collected, but never used."

As part of the data collection process, John documented the contents of the customer order history file, the mailing labels file, the supplier product file, and the daily order file. John provides us with a brief summary of each file.

The Customer Order History File

"The customer order history file stores records from customers who have purchased from us in the past," John explained. "Each record in the file contains a customer key, supplier number, customer invoice number, customer order number, date of the order, terms (code), special instructions, the sales total, a problem-flag field, and a date-filled field. The customer key consists of the customer's five-digit postal code plus the first three characters of the customer's last name. The problem flag is set in the 'off' position when an order is added to the file and is set to 'on' whenever we have a problem with the order. The date the order is filled (the date-filled field) is updated when we receive confirmation from the manufacturer that goods were sent to the customer."

The Mailing Label File

"The mailing label file is limited to information about each customer," John continued. "We store the customer key, customer name (first and last), customer address (street, city, state, and postal code), and date of last order (the date of the latest order from the customer)."

The Supplier Product File

"The supplier product file is similar in design to the mailing label file. We store the supplier number (a five-digit number), the supplier name and address (street, city, state, and postal code), each item number sold by the manufacturer, and each item name. Then we accumulate the number of orders taken (for each item), and the dollar value of orders taken (for each item)."

The Daily Order File

John went on: "The daily order file is the largest of all files. As shown in **Figure 6–18,** this file contains three types of information: mailing information, supplier product information, and current order information. Mailing information is identical to the information stored on the mailing label file with the exception that the date of last order is not stored on this file. Supplier product informa-

Daily order file
{
Mailing-information
Supplier-product-information
Current-order-information
}

FIGURE 6–18 Types of information stored on the daily order file

tion is identical to the information stored on the supplier product file with two exceptions: neither the number of orders nor the dollar value of orders taken is stored on the file. Current order information provides the details specific to the actual customer order (see chapter 5), including the information supplied by the office or set during processing and appended to each order by the computer. We store the telephone number when we can (this number is optional). Moreover, we add a customer order number (a five-digit number), a terms code (one through four), a ship-to code (A through D), a ship-via code (UPS or PEXP), and any special instructions (up to sixty characters). For each item ordered by the customer we add the item number, the quantity desired (up to three digits), the finish desired (a three-digit code), and the unit price (up to five digits). Next, we store a variety of totals—the sales subtotal (up to six digits), sales tax (up to four digits), and customer payment (up to six digits). Finally, an invoice number and the date of the invoice are appended to each daily order."

CASE ASSIGNMENT 1

List the contents of the customer order file. Why are the customer key, supplier number, customer invoice number, and customer order number placed in this file? What, if any, design flaws are apparent?

CASE ASSIGNMENT 2

List the contents of the supplier product file. (You will need this information later when working on chapter 11 assignments.) What, if any, design flaws are apparent?

CASE ASSIGNMENT 3

The current order information shown on Figure 6–18 is too difficult to understand. Subdivide the current order set into three subsets: customer-supplied information, WI-supplied information, and computer-supplied information. Draw a revised Warnier-Orr diagram to illustrate the structure of current order information, once subdivided into subsets. Fifteen items should appear on the far right-hand side of this diagram.

CASE ASSIGNMENT 4

Define the following terms for placement in a data dictionary:

- Current-order-information
- WI-supplied-information
- Telephone-number
- Quantity-desired
- Finish-desired
- Ship-via-code

CASE ASSIGNMENT 5

Draw the DSD to show the logical relationship between the daily order file and the supplier product file.

CASE ASSIGNMENT 6

Use structured English to describe how the mailing labels file is updated. (Use the level-1 DFD prepared for chapter 5 to help you write this description.)

REFERENCES

1. T. DeMarco, *Structured Analysis and System Specification* (New York: Yourdon Press, 1978).
2. C. Gane and T. Sarson, *Structured Systems Analysis: Tools and Techniques* (Englewood Cliffs, N.J.: Prentice-Hall, 1979).
3. J. Warnier, *Logical Construction of Programs* (New York: Van Nostrand Reinhold, 1976).
4. K. Orr, *Structured Systems Development* (New York: Yourdon Press, 1977).

7
Feasibility Analysis

INTRODUCTION

ONCE the analysis team understands the characteristics of an existing system, with all its good points, flaws, and idiosyncrasies, new questions arise: "Should a different system be designed?" "If so, what should be its characteristics?" The first question calls for another type of feasibility analysis—the proposed system feasibility. This type is different from determining the feasibility of beginning the project.

Let's review the purpose of project feasibility before discussing proposed system feasibility. With project feasibility the question to be resolved was whether or not to begin a project, based on a preliminary project analysis (see chapter 4). In preparing the project specification, the analyst had to document project requirements—in order to determine whether the project was feasible from an economic, technical, social, management, and legal standpoint, and to acknowledge the implications of the fully implemented project. The difficulty with this specification is that the analyst's information is limited and often based on individual opinions (e.g., the budget system is ineffective) or on cost projections (e.g., the project should return $24,000 yearly for an expected return on investment of 160 percent).

With system feasibility, a different set of issues comes into play. Some of the questions to be raised include:

1. Will user requirements be satisfied by the new design?

2. What are possible design alternatives?

3. What message forms and displays are needed? How will these be approved by users?

4. What resources will be required in building the new system? In operating the system? What resources will be produced by the new system?
5. What will be the impact of the new system on user groups? On management groups?

In other words, system feasibility is design related as opposed to project related. The question of system feasibility lies not so much with whether or not to continue a project, although such a question can be raised, but with whether or not it is possible to evolve a suitable *logical design specification*—namely, one that can incorporate user-group design and acceptance requirements.

By now some of you are probably thinking, "Are you stating that after I have done the work of finalizing project specifications, followed by collecting and analyzing data and organizing and documenting data, that only then do I think about the design of a new system? Doesn't an analyst have a good idea early on of whether it is feasible to design a system that will be acceptable?"

Obviously, the analyst does consider system feasibility before data collection begins. Otherwise it would be impossible to prepare a meaningful project specification. Likewise, during the process of collecting and organizing data the analyst begins to derive new system ideas and designs. Determining system feasibility might thus be characterized as a process of continuous evaluation rather than something that occurs after the fact. Nonetheless, a more formal feasibility study is generally required. The purpose of such a study is fourfold:

1. to specify how users expect the new system to perform,
2. to develop the new design and possible design alternatives,
3. to decide on the most feasible design alternative, and
4. to determine the technical and social impact of the new design.

When you complete this chapter, you should be able to

- prepare a user's performance definition to specify in detail how users expect a new system to perform;
- describe why design alternatives are required and how they are developed;
- perform an evaluation of design alternatives to determine which is the most feasible; and
- describe the technical and social impact of a new design.

USER'S PERFORMANCE DEFINITION

The analysis team must complete several critical activities before it can determine the most feasible design alternative. We have seen the following:

1. System problems must be identified and evaluated.
2. An initial investigation must be undertaken in order to understand the current system.
3. People involved with the existing and the proposed system must be interviewed and questioned to determine their requirements and expectations.
4. The existing system is often modeled and documented.
5. Forms and data structures pertinent to the new system must be collected and described.

Once these steps are completed, both the user and the analysis team should know which system problems need to be solved and understand the details of the existing system (or problem environment) sufficiently to grasp what is to be accomplished by the new system.

The next stage in analysis might appear obvious: to define in detail how the new system is expected to perform from the user's perspective. This stage is often called the *user's performance definition*. The steps leading to such a definition are

1. statement of new system goals,
2. determination of the appropriate design strategy,
3. identification and ranking of specific design objectives, and
4. identification of design constraints.

Once the user performance definition is complete—when the analysis team and end users agree on all of these steps—the analysis team is better able to evolve the most appropriate design.

Statement of New System Goals

User groups often have difficulty stating new system goals even after considerable time has been spent studying a new system. Why?

When it comes to computer-related projects, many people have difficulty separating the technology of computing from its application. As an example, a company might list its new system goals as follows:

1. improve computing facilities to modernize the way in which the company conducts its business and
2. design an online computer system to provide managers with immediate access to information useful for decision making.

The problem with goals such as these is that they assume that a change in technology will solve organizational problems. Study the first goal. Rewritten, it states:

$$\begin{bmatrix} \text{Improving computer} \\ \text{facilities} \end{bmatrix} \text{ will modernize } \begin{bmatrix} \text{the way in which} \\ \text{the company} \\ \text{conducts its business.} \end{bmatrix}$$

Next, rewrite the second goal, restating it somewhat.

$$\begin{bmatrix} \text{Designing an online} \\ \text{computer system} \end{bmatrix} \text{ will provide } \begin{bmatrix} \text{information useful to} \\ \text{decision making.} \end{bmatrix}$$

From your understanding of functional analysis, both of these goals should appear suspect.

There are better ways to write goals. Consider the following:

1. Decentralize company billing procedures to speed the processing of customer bills and reduce outstanding accounts receivable.

2. Tie together budgeting with accounts payable to improve the planning and control of company expenditures.

3. Develop a sales history database to provide immediate access to sales analysis information.

In these instances, a different assumption is made: a business will be improved by changing the way in which it processes, stores, or transmits information.

As a test, let's rewrite the first goal as follows:

$$\begin{bmatrix} \text{Changing the way in} \\ \text{which a company} \\ \text{processes billing} \\ \text{information} \end{bmatrix} \text{ will improve } \begin{bmatrix} \text{the sending of bills to} \\ \text{customers and the size} \\ \text{of outstanding accounts} \\ \text{receivable.} \end{bmatrix}$$

This goal is more consistent with functional analysis.

Let's also rewrite the third goal.

$$\begin{bmatrix} \text{Changing the way in} \\ \text{which a company stores} \\ \text{sales history information} \end{bmatrix} \text{ will provide } \begin{bmatrix} \text{more immediate} \\ \text{access to this} \\ \text{information.} \end{bmatrix}$$

In this example, the goal might be too limiting. The goal might need to be revised to read:

Develop a sales history database to provide immediate access to sales analysis information and estimates of future product sales.

This goal now calls for a change in the way in which a company stores and processes its sales history information. Rewritten, the goal now reads:

$$\begin{bmatrix} \text{Changing the way in} \\ \text{which a company} \\ \text{stores sales history} \\ \text{information} \end{bmatrix} \text{ will provide } \begin{bmatrix} \text{more immediate access} \\ \text{to sales analysis} \\ \text{information and} \\ \text{estimates of future} \\ \text{product sales.} \end{bmatrix}$$

Determination of Appropriate Design Strategy

Several terms need clarification before we can proceed. A *goal*, as indicated in the previous section, is a broadly stated purpose. An *objective*, as we will see, is a much more concrete and specific statement of purpose. What falls between a goal and an objective is a *strategy*: the devising of a plan to allow for the eventual attainment of a goal. Accordingly, an *organizational design strategy* is devising a plan (or model) and explaining how the plan (or model) is to be implemented to allow for the eventual attainment of a goal.

Unlike goals, statements of organizational design strategy should include references to computer technology as appropriate. One statement of strategy might be to install online computer terminals and small computers at remote billing centers. This strategy helps explain how a decentralized billing system model would be implemented in an organization.

Another statement of strategy might be to use computers to integrate budget accounts and accounts payable transactions for one division of a multidivision company. This strategy helps explain how a new system design will improve the planning and control of company expenditures.

How does a company decide on an appropriate organizational design strategy? While there is no single correct answer to this question, the following areas should be considered in the development of a strategy:

1. computer-based versus manual systems,

2. centralization versus decentralization,

3. build versus buy, and

4. prototype versus fully functional system.

COMPUTER-BASED VERSUS MANUAL SYSTEM

There are several good reasons for building a computer-based system as well as several reasons that oppose such a decision. Advantages include improved processing speed, greater consistency in processing, and so forth. Remember the types of problems stated in chapter 4? In defining a problem it was suggested that analysis should begin by determining whether a system suffered from the problems of reliability, validity, accuracy, economy, timeliness, capacity, and throughput. If there were shortcomings in any of these areas, it would lend support to the design of a computer-based system.

The reasons why an organization might find it not to its advantage to go ahead with a computer-based system include

1. Rapidly changing user environment. In rapidly changing organizations, computer-based systems may be outdated before they can be implemented.

2. Major increase in equipment and technical support cost. The cost of acquiring sufficient computing capacity and hiring technical staff such as system designers may be too great for some organizations.
3. Displacement of lower-paid staff. The cost associated with the retraining of personnel displaced by a computer-based system may be too great for some organizations.
4. Work place restrictions. If, for example, a new system requires changes in work patterns and if these changes are prohibited by employee contracts, a computer-based system would be inappropriate.
5. Unresolved technical requirements. With new or unique applications, computing vendors may not be able to resolve mandated technical requirements.

CENTRALIZATION VERSUS DECENTRALIZATION

Centralization means the concentration of resources (e.g., people, computers) in an organization, while *decentralization* entails the dispersing of resources to several locations. With the advent of powerful yet relatively inexpensive computers, combined with reliable data transmission facilities, the issue of centralization versus decentralization warrants considerable study. Consider the following example. The Ready-Made Shoe Company has in times past followed a centralized approach in billing its customers. As shown in **Figure 7–1,** their approach consists of receiving customer orders at regional sales offices, followed by sending all sales orders (modified customer orders) to Ready-Made's central of-

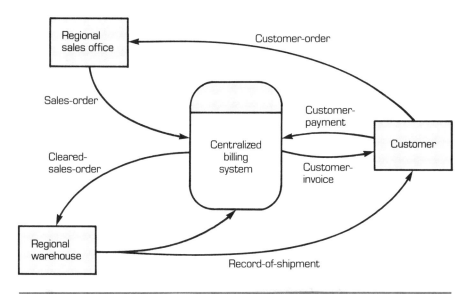

FIGURE 7–1 Centralized billing and decentralized order filling

fice. The central staff determines whether an order should be filled and identifies the regional warehouse to fill the order. This information is transmitted by a cleared sales order. The regional warehouse fills and ships merchandise to the customer, based on information received from the corporate office. Once goods are shipped, a record of shipment is sent to the customer and to the central office. On receipt of the record of shipment, the central staff prepares the customer invoice and sends it to the customer. The customer, in turn, is expected to make payment to the central office.

Suppose that Ready-Made decides to change it's existing system by developing a decentralized approach to billing. As shown by **Figure 7–2,** such a system would feature the sending of sales order information directly to a regional warehouse. On receipt of the sales order, the warehouse crew would fill the order, ship the goods, and prepare the customer invoice. As part of processing, a record of shipment and a customer invoice would be sent to the customer. In addition, a record of customer billing identifying the customer, the details of the shipment, and the details of the invoice would be sent to Ready-Made's central office. As before, the customer would be expected to make payment to the central office.

In the changeover from a centralized to a decentralized billing system, a company such as Ready-Made would be able to realize several system gains, such as improved timeliness and throughput. Ready-Made realized, for example, that it could reduce the time between the receipt

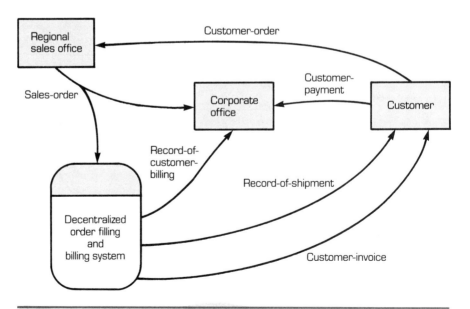

FIGURE 7–2 Decentralized order filling and billing

of a customer order and the mailing of the customer invoice by up to five days. That is, with a centralized design, the central staff took up to two days to receive, process, and send the sales order; the regional warehouse staff took up to three days to process and send a record of shipment back to the central staff. However, the new system would also be disadvantageous. What would be lost in the process would be the central staff's ability to closely monitor customer orders. With the new system, regional sales offices would be required to screen customer orders to determine which ones to fill and ship; they would be responsible for managing customer billing.

BUILD VERSUS BUY

Should an organization build a new system, using internal staff, or should it contract for a new system with an external party? With an ever-increasing number of software products available for sale, this build-versus-buy question is becoming increasingly difficult to answer. Some companies currently require their system analysts to investigate the cost of acquiring a system from an external party, for modification to meet internal needs. Organizations have discovered that when a search is successful, the time and cost of a new project can be reduced by up to 50 percent.

Suppose that the decision is made to build rather than to buy. In today's software environment, the analyst must face another decision: whether to build using a standard high-level programming language, such as COBOL, RPG II, or Pascal, or to build using a variety of software development tools, such as database management systems (DBMSs) or application development systems. Assume for the moment that a mail-order company desires a special mail list. As shown in Figure 7–3, such a list involves creating a special mail file (see **Figure 7–3a),** and printing a special mail list (see **Figure 7–3b)** from this file. If this system were built using a standard language such as COBOL, two programs would be needed: an interactive program to create the file and a batch program to print from the file.

As a development alternative, consider how software tools might be used in place of COBOL. As shown in **Figure 7–4,** several different types of software tools might be used. These include *query languages, DBMSs, application development systems, display screen support systems, report-writer systems, test data generators,* and *modeling systems.*

For example, a DBMS might be used directly to create the special mail file. Once the file is created, a report-writer system might be used to print the desired mail list.

Let's consider each of the terms shown on the figure. While our intent is not to describe any terms in detail, they should be familiar to those who work in the field of systems analysis. They include

1. query language, which is used to ask the end user a question, such as "Which type of special mail list do you wish to create?";

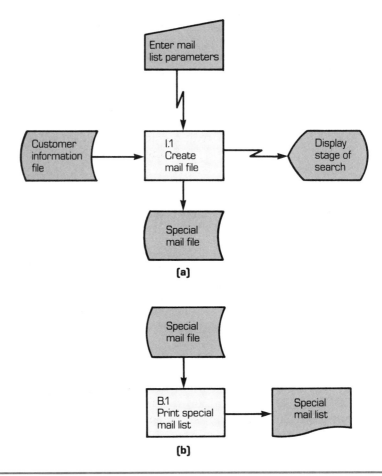

FIGURE 7–3 System flowcharts showing steps needed to create special mail file (a) and print special mail list (b)

2. DBMS or file management system, which is used for data-handling functions such as retrieving answers to user queries and creating and maintaining data files;
3. application development system, which is used to rapidly write computer software and may include standard code to simplify the programming of repetitive tasks, such as input and output editing;
4. display screen support system, which is used to write online terminal display instructions;
5. report-writer system, which is used to prepare special listings and user reports;
6. test data generator, which is used to supply sample data for testing such items as the layout of a report; and
7. modeling system, which is used to show the features of a larger-scale system.

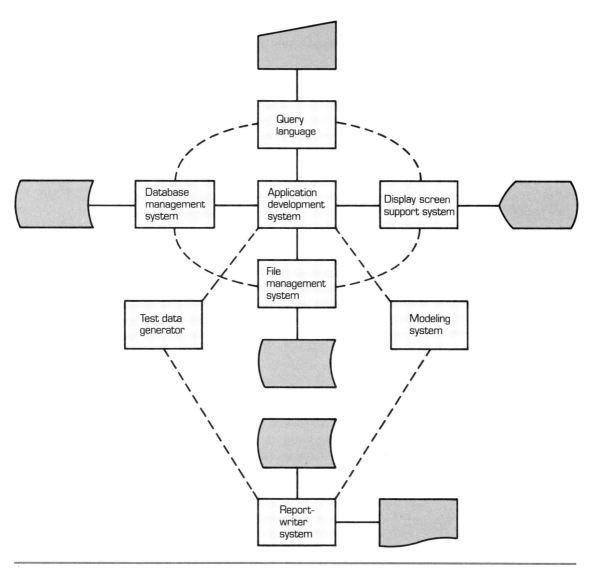

FIGURE 7-4 Examples of software development tools

PROTOTYPE VERSUS FULLY FUNCTIONAL SYSTEM

With the advent of software development tools, many organizations are tending to favor the design of operational prototypes rather than fully functional systems. A *prototype* is a version of what will be eventually required and is built with the understanding that future modifications, or so-called enhancements, will be required. In contrast, a *fully functional system* is built with a different set of expectations. Such a system is expected to satisfy most, if not all, user requirements; it is not expected to require substantial modification or enhancement.

There are several reasons why organizations tend to support operational prototypes rather than fully functional designs. These reasons include the following:

1. With any new system, user needs can be defined only partially. Users can, however, specify their processing needs more exactly once a prototype is made operational.
2. Prototype development encourages greater user involvement in the systems development process. The user participates by indicating what is liked and disliked and what will be required (in the next prototype) to achieve a satisfactory design.
3. Prototypes cost less than fully functional systems to design and implement. Since the analyst need not be concerned with meeting all user expectations, an operational prototype can often be implemented in weeks or months, as opposed to months or years.
4. Prototype design allows the systems analyst and systems designer to validate the logical and the technical system specification. With validation, these specifications serve as the blueprints for building the next version of the prototype.
5. Prototyping typically involves the use of labor-saving software development tools. Tools such as application development systems make it possible to complete software quickly; they also help the systems analysis team document the internal structure of a design.

Identification of Specific Design Objectives

Closely related to the process of goal setting and formulating an organizational design strategy are the identification and ranking of specific design objectives. As stated earlier, an objective differs from a goal in that it is more specific and immediate. A *specific objective*, for example, contains a statement of purpose and a measurable outcome. The general form of such an objective is as follows:

```
IF we accomplish our statement of purpose
   THEN we should realize a set of desired outcomes
     (ELSE) an opportunity is lost
```

Consider the following examples of specific design objectives:

- to increase the percentage of goods shipped for orders received the same day from 10 percent to 80 percent within nine months;
- to reduce the average number of overtime hours worked each week per department from 12.5 to 7.5 within six months; and
- to reduce the number of minutes needed to look up the status of a customer order from twelve minutes to thirty seconds within four months.

As these examples show, a specific objective contains a statement of purpose and a measurable outcome. Moreover, it specifies the time frame in which the objective is to be accomplished. Objectives thus clarify the concept of a user's performance definition. They specify in precise terms how a new system is expected to perform.

Some might question the lateness of setting specific design objectives. Consider this criticism: "Don't users set measurable outcomes early on, as part of finalizing project specifications?" Perhaps the best answer to this question is, yes, they may, but they seldom do because they lack sufficient knowledge. For example, users may not know that only 10 percent of orders received are shipped the same day, or that the time to look up the status of a customer order is twelve minutes. Instead, users initially provide the analyst with a general set of processing requirements, with the gut feeling that there is a performance problem. Systems analysts, in turn, investigate these requirements. Remember the purpose underlying systems analysis: to determine whether a problem exists and if it does, to recommend a logical improved solution. To this definition, we can now add the following condition: *The value of a logical, improved method of processing lies in its ability to improve performance from the user's point of view.*

Ranking of Specific Design Objectives

The ranking of specific design objectives becomes necessary if user objectives are in conflict. Assume that users agree that the following goals are essential to improving a university's library check-out procedures:

1. to reduce the time needed to verify the borrower's identification (the problem of timeliness);

2. to reduce the time needed to check out each book (the problem of timeliness);

3. to reduce the number of people required to check out books (the problem of economy);

4. to reduce the number of errors made in checking out books (the problem of accuracy); and

5. to reduce the waiting line of borrowers during peak hours (the problem of throughput)

Observe that these objectives may be in conflict. It may be possible, for example, to reduce the number of people required to check out books, provided the waiting line of borrowers is not reduced. However, it would be impossible to reduce the number of people required to check out books during peak hours and to reduce the waiting line of borrowers during the same period.

The ranking of objectives tells the analyst which objectives are mandatory and which others are highly desired. Suppose that the user and the analyst rank the five objectives shown above as follows. The mandatory objectives are reducing the number of errors in checking out books (the problem of accuracy) and reducing the waiting line of borrowers during peak hours (the problem of throughput). The highly desired objectives are reducing the number of people required to check out books (the problem of economy), reducing the time needed to verify the borrower's identification (the problem of timeliness), and reducing the time needed to check out each book (the problem of timeliness). This ranking tells the analyst that solving the problems of accuracy and throughput must be accomplished—they are mandatory—and that solving the problems of economy and timeliness, while important, are less so.

Identification of Design Constraints

The final item of the user's performance definition is the identification of the limits to be placed on a new design. As in our earlier discussion dealing with the components of a system, *constraints* limit the scope of a new design; they set conditions that must be either achieved or avoided. In the library example, for instance, design constraints might be identified as follows:

1. Development costs must be less than $75,000.

2. Maximum throughput must not exceed 1,000 books per hour.

3. The design must be compatible with plans to decentralize university computing facilities.

4. The design must be interrelated with existing online student records.

These constraints all indicate a measurable outcome that is either to be achieved or avoided. The first two design constraints indicate outcomes to be avoided: costs greater than $75,000 and volume greater than 1,000 books per hour. The second two constraints are different. These constraints specify outcomes to be achieved: a design that is compatible with plans to decentralize and that can be interrelated with existing online student records.

DESIGNING THE NEW SYSTEM

The user's performance definition identifies how the user expects a new system to perform and is the first step in answering the questions: "How should the new system be designed?" and "What user requirements will be satisfied by the new design?"

Closely coupled with the preparation of the user's performance definition is the determination of whether or not it is possible to logically model and define a new system. Such a system must be able to satisfy user goals and objectives. Moreover, such a system must be able to provide users with desired system outputs.

Designing the new system places special requirements on systems analysis team members. They are generally asked to prepare several design alternatives rather than one, before deciding on the design considered most feasible. There are five identifiable steps in the process of designing a new system:

1. determining new system outputs,
2. developing design alternatives to achieve those outputs,
3. evaluating each alternative,
4. ranking each design alternative, and
5. selecting the best design.

Determining New System Outputs

The design feature considered most important to users is the outputs to be produced by the new system. Some users will remark, "I don't care how the computer works, but I do care about the kinds of information I will receive and whether that information is timely and accurate." An inventory manager might comment, "Remember that we need several types of reports. We need a daily stock-status report, a monthly price book, a daily purchase advice report, a month-end inventory value analysis report, a month-end stock performance report, and a daily product change report."

At this point it should be clear how the analysis team determines new system outputs and inputs. For example, the data collection process is directed toward such a determination. Likewise, DFDs serve to document both inputs to and outputs from a system; they determine whether all system output requirements have been accounted for. As another example of the role played by DFDs, consider the student registration system shown in **Figure 7–5.** This context diagram documents the five main types of external information to be produced by the system, which are labeled as

1. student schedule, showing the classes a student has registered for;
2. class list, showing the students registered in a class;
3. class load, showing the number of students registered in each class for each faculty member;
4. class-faculty load, showing the number of students registered in each class for each member of each type of faculty; and
5. revised student program, showing the record of classes completed and the student's current schedule.

Developing Design Alternatives to Achieve Those Outputs

Developing design alternatives to achieve the outputs shown by a high-level DFD is easier to accomplish than it may initially appear. Consider the student registration context diagram once again (see Figure 7–5).

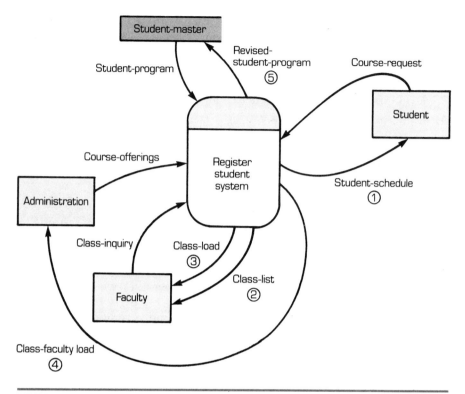

FIGURE 7-5 Student registration context diagram

Possible design alternatives in processing a student's course request include

1. allowing students to enter their course requests and to pick up their student schedules at any one of several remote locations;
2. requiring students to submit their course requests at a single, central location and to wait for their student schedules to be received in the mail;
3. allowing students to enter their course requests at any one of several remote locations, but requiring students to pick up their approved student schedules from their faculty advisers; and
4. requiring students to submit their course requests to their faculty advisers, who are required to approve and to send all requests to student registration, and to pick up their student schedules from their faculty advisers.

Of these alternatives, the context diagram adequately describes the first two design possibilities. (Exercises 7–4 and 7–5 at the end of the chapter ask you to modify this diagram to show the information flow for the third and fourth alternatives.)

Besides considering design alternatives at the uppermost levels of a system, such as those shown by the context diagram or the level-0 DFD, various design possibilities can be shown at lower levels. Consider the way in which the accounts receivable file is altered by the Prepare customer invoice transform of Mansfield's billing system (illustrated in chapter 3). The two primary outputs from this transform were the customer invoice and the invoice summary (see **Figure 7–6a**). The level-1 diagram went on to show how both of these outputs were to be created (see Figure 5–15). For writing information to the accounts receivable file, the steps were as follows (see also **Figure 7–6b**):

1. Invoice records were created by the transform for creating the invoice record. Each record was written to the invoice transaction file.

2. Invoice summary records were created by the transform for creating the receivable summary. Each summary record—that is, a condensed version of an invoice record—was written to the accounts receivable file.

There is an alternative to this design, however. As shown in **Figure 7–7**, a separate transform is not required in creating invoice summary records. Instead, the transform 1.1, create invoice record, can be altered to produce the invoice record and the invoice summary record as outputs.

Finally, when a design prototype is proposed, design alternatives are based on the number of functions to be added to a design. Suppose that you are responsible for designing a graphics software package. As the level-1 diagram shows (see **Figure 7–8),** one package might include four parts: draw picture, edit picture, save picture, and load picture. Let's consider how this prototype might work. In the drawing or editing of a picture, input consists of signals transmitted from a graphic input device to a computer. The computer stores the in-process picture in internal memory and transmits the draw commands to a color monitor. To save a picture, the completed picture is written to picture storage, where it is stored under a unique file name. This file name is also used to load a previously stored picture.

Decomposing transform 1.1, draw picture, into its main parts demonstrates how design alternatives come about. As shown below, the first design alternative might be limited to three draw-picture functions, while a second design alternative—a more functional alternative—is limited to five draw-picture functions.

Graphics Software Package—Design Alternative 1
    1. Draw straight lines
    2. Draw orthogonal lines
    3. Draw circles

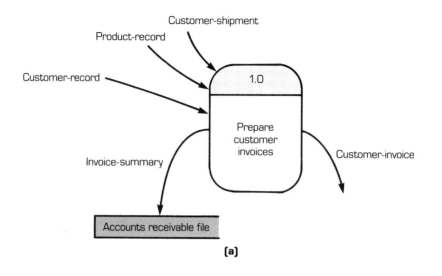

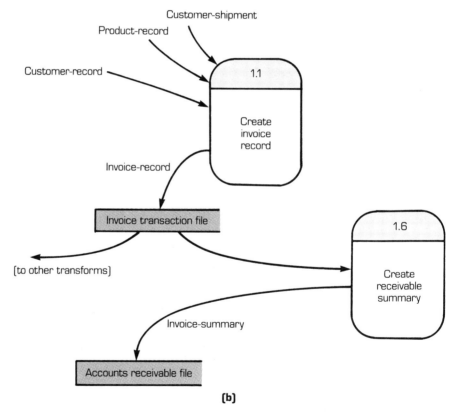

FIGURE 7–6 Abbreviated level-0 and level-1 DFDs showing the main outputs from preparing customer invoices (a) and the steps leading to the processing of invoice summary information (b)

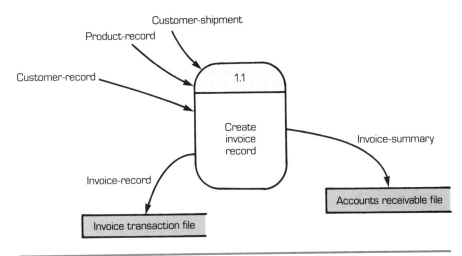

FIGURE 7-7 Alternative processing design in writing invoice summary information to the accounts receivable file

Graphics Software Package—Design Alternative 2
1. Draw straight lines
2. Draw orthogonal lines
3. Draw circles
4. Draw freehand
5. Set pen color.

Evaluating Each Alternative

Once decision alternatives have been documented, the work of evaluation begins. Before starting this task, the analyst might decide on major evaluation categories and select criteria appropriate to each. For example, an analyst might select system requirements, system performance, and system costs as three major evaluation categories. Then, as shown on **Figure 7-9,** the analyst might devise several evaluation criteria for each of these categories to form a table. The completed table represents a *design evaluation matrix.* This kind of matrix is used to rate and provide a weighted score for design alternatives.

Consider next the evaluation criteria shown on Figure 7-9. *System requirements criteria* call for an analysis of the impact of each design alternative. Typical system requirements criteria include the impact of the design on organizational policy, personnel, equipment, and physical space. The requirements for management control and employee training are also to be evaluated. *System performance criteria* concern how well the system is expected to perform once it becomes operational. As shown, six of the seven criteria described in chapter 4 are once again pertinent. Questions of timeliness and of capacity, for instance, may be

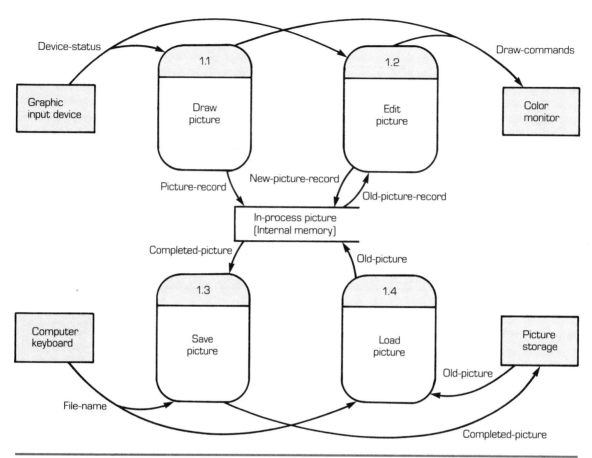

FIGURE 7–8 Graphics software level-1 diagram

especially important in evaluating design alternatives. Timeliness examines the *response rate* of each alternative, while capacity examines the *growth potential* of each alternative. *System costs criteria* include the costs of developing and operating the new system. In addition, return on investment and payback are generally shown. As stated in chapter 1, return on investment requires that the costs of the new system be compared with the projected savings.

Following the identification of appropriate evaluation criteria—there exists a wide variety of choices—the analysis team must decide on an evaluation scheme appropriate for each. The general rule is: *Whenever possible, use objective measures rather than subjective measures.* For example, **Figure 7–10** shows a partially completed design evaluation matrix for two design alternatives. As illustrated, design alternative 1 is projected to have the capacity to handle 20,000 accounts, with a throughput rate of one hundred accounts per hour; design alternative 2

Evaluation Criteria \ Design Alternatives		Alternative 1	Alternative 2	
Requirements				
Policy				
Personnel				
Equipment				
Space				
Management control				
Training				
Performance				
Reliability				
Validity				
Accuracy				
Timeliness				
Capacity				
Throughput				
Costs				
Development				
Operation				
Return on investment				
Payback				
Total Score				

FIGURE 7–9 Design evaluation matrix

Evaluation Criteria / Design Alternatives		Alternative 1	Alternative 2	
Requirements				
Policy				
Personnel				
Equipment				
Space		80 square feet	80 square feet	
Management control				
Training				
Performance				
Reliability				
Validity				
Accuracy		less than .05%	less than .05%	
Timeliness				
Capacity		20,000 accounts	40,000 accounts	
Throughput		100 accounts/hour	250 accounts/hour	
Costs				
Development		$50,000	$75,000	
Operation				
Return on investment		16.0%	16.8%	
Payback				
Total Score				

FIGURE 7–10 Partially complete design evaluation matrix

is projected to have a capacity to handle 40,000 accounts, with a throughput rate of 250 accounts per hour. Both design alternatives are shown to be identical in terms of accuracy and required space. System development costs for alternative 2 are 50 percent higher than for alternative 1; however, the return on investment is only 5 percent greater for alternative 2.

When evaluation criteria must be evaluated subjectively, a summated scale is recommended. Consider the following: Estimate the chances of finding the personnel required to complete each design alternative.

Alternative 1

Very Good	Poor	Fair	Good	Very Good
(1)	(2)	(3)	(4)	(5)

Alternative 2

Very Good	Poor	Fair	Good	Very Good
(1)	(2)	(3)	(4)	(5)

As discussed in the next section, a summated scale simplifies the rating of evaluation criteria.

Ranking Each Design Alternative

There are times when the evaluation of a design leads to an obvious best alternative, making the ranking of alternatives unnecessary. For example, suppose that a school is committed to decentralizing student registration. Returning to the alternatives in processing a student's course request, a commitment to decentralize rules out the second alternative, requiring students to submit their course requirements at a single, central location. Suppose next that a faculty vote is taken to determine whether students should be required to pick up their schedules from their advisers. A faculty representative summarizes: "The vote indicates that students should not be required to pick up their schedules from their advisers." Thus, this policy decision drops alternatives three and four from further consideration, leaving only the first design alternative. The question then becomes: Should this alternative be approved or should work begin to consider still other alternatives?

When an evaluation of several alternatives does not lead to an obvious best choice, a weighted ranking is recommended. Such a ranking involves

1. assigning weights to each evaluation category and to each criterion within each category;
2. rating each criterion by assigning a number from a possible range, such as the numeric range shown by the summated scale;
3. scoring each criterion by multiplying the rating number times the criterion weight; and
4. scoring each design alternative and each category within each alternative by adding the weighted scores.

Figure 7–11 shows such a weighted design evaluation matrix. As indicated, the three main categories were weighted fairly evenly: 30 percent for system requirements, 36 percent for system performance, and 34 percent for system costs. The rating scale ranges from 1 to 5, and is identical to the summated scale. A rating of 1 indicates very poor or very low; a rating of 5 indicates very good or very high. The column score was calculated by multiplying the criterion weight times its rating. Thus, a weight of 5 and a rating of 4 leads to a column score of 20.

Selecting the Best Design

This final step is the easiest, provided the scoring of each design alternative was done fairly and clearly shows one design to be superior to all others. As Figure 7–11 illustrates, alternative 2 is ranked higher than alternative 1, though the difference in total score is only thirty points. The evaluation category scores are as important as the total score. As shown, alternative 2 ranks best as measured by system performance and system costs; alternative 1 ranks best in terms of system requirements. The interpretation of this ranking leads to the conclusion that alternative 2 is the best design; however, it will require greater system resources and be more difficult to implement than alternative 1.

SOCIAL, TECHNICAL, AND ORGANIZATIONAL IMPACT ANALYSIS

The process of designing the new system leads to selecting the best design from several design alternatives. However, feasibility analysis does not end here. It is not sufficient to move directly from the logical to the physical design of a system. What remains are the overriding questions of system performance and system acceptability. More specifically, the analyst must ask whether the model can really deliver, and if it can, what its social, technical, and organizational impact will be.

The first question is perhaps the easier to answer, provided a structured approach is followed in analysis. A structured approach produces three payoffs: a new kind of logical reasoning, methods for changing a system, and new insights as to how parts of a revised system are to be knitted together. By fully using a structured methodology, the systems

Evaluation Criteria / Design Alternatives		Alternative 1		Alternative 2		
	weight					
Requirements	30	Rating	Score	Rating	Score	
Policy	[5]	4	20	4	20	
Personnel	[5]	4	20	3	15	
Equipment	[5]	4	20	2	10	
Space	[5]	5	25	3	15	
Management control	[5]	3	15	3	15	
Training	[5]	3	15	4	20	
			115		95	
Performance	36					
Reliability	[6]	4	24	3	18	
Validity	[6]	4	24	3	18	
Accuracy	[6]	5	30	5	30	
Timeliness	[6]	2	12	4	24	
Capacity	[6]	2	12	4	24	
Throughput	[6]	3	18	4	24	
			120		138	
Costs	34					
Development	[4]	5	20	3	12	
Operation	[10]	2	20	4	40	
Return on investment	[20]	3	60	4	80	
Payback						
			100		132	
Total Score			335		365	

FIGURE 7–11 Weighted design evaluation matrix

analysis team can determine in large part whether or not the proposed model can really deliver.

The impact of the new design is more difficult to evaluate. For example, **Figure 7–12** shows that task, structure, technology, and people are interrelated in an organizational system. If we introduce something new in one part of the system, the changes are often felt in other parts—often in places in which no one had expected them to be felt. This does not mean that organizational change is impossible, however. What it does mean is that the analyst must know how a proposed system will affect the tasks, structure, technology, and people within an organization.

Task Analysis

Positions or jobs within an organization are made up of a collection of tasks. A *task* consists of building or designing things or of providing a service. Thus, whenever you decide to accept a position with a company, what you are really doing is agreeing to perform a prescribed set of tasks.

Tasks can be described as ranging from relatively precise, repetitive, and programmed to poorly defined, unique, and loosely structured. With programmed tasks, an individual is generally given little freedom in per-

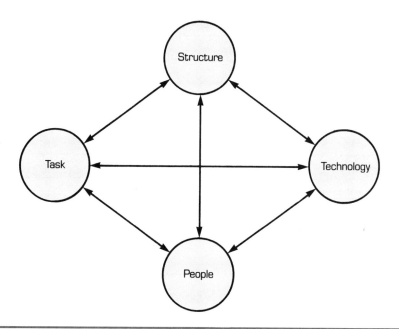

FIGURE 7–12 Basic parts of a large organization

forming a job; with loosely structured tasks, the opposite holds: individuals enjoy considerable freedom and personal autonomy in performing a job.

Task analysis requires the systems analysis team to determine how the tasks of an organization will be changed by a new system. The team must consider three types of change:

1. the elimination of existing tasks,

2. the creation of new tasks, and

3. the modification of tasks (e.g., making them more or less programmed).

The analysis team must also devise some method of documenting these types of change.

One method for conducting a task analysis involves studying a person's job to show how a proposed system change will alter that person's prescribed set of tasks. Consider the revision of the accounting clerks job, shown earlier as standard procedure A/R 394 (see figure 5–11). Consider also that a proposed system specifies the online processing of customer payments, compared with manual methods, and that the clerk will be responsible for entering payment data into the computer.

Figure 7–13 illustrates the task analysis worksheet, showing the tasks currently performed and the proposed set of tasks. A comparison reveals that three new tasks are now specified and that two current tasks require some modification.

By documenting tasks before and after a design change, the analysis team can answer questions such as "How will the new design change the jobs in the accounting department? Are such changes feasible? Will the proposed changes be acceptable to the users?" In our example, the team might conclude that it is reasonable for accounting clerks to handle the new set of tasks, provided that they receive adequate training. What kind of training? The task analysis worksheet also identifies training requirements. It indicates that an accounting clerk must be trained to use a computer terminal. The clerk will be expected to enter data, verify data, correct entered data, and print bank deposit slips.

Technical Analysis

The technology of a system consists of the tools that enable people or other machines to perform tasks. The function of these tools is to simplify the performance of a task or a set of tasks. Given this brief overview of the concept known as technology, we might change our concept of a job with a company. Rephrased, we might say that when we accept a position with a company what we are really doing is agreeing to perform a prescribed set of tasks using the available technology.

Technical analysis requires the analysis team to determine how new

Task Analysis

Department: Accounts receivable	Reference: See A/R 394	Abstract: Explains the steps to be taken in processing customer payments
Subject: Processing of customer payments	Issued: January 1, 19xx	
Position: Accounting clerk	Prepared by: Pamela Bennett	Page: 1 of 1

	Tasks Before Change		Tasks Following Change
	1. Opens mail		1. Opens mail
	2. Reviews payment for accuracy		2. Reviews payment for accuracy
	3. Completes customer payment stub as necessary		3. Completes customer payment stub as necessary
E	4. Prepares cash-control slip	N	4. Key enters clerk details into the computer
E	5. Sums total cash payment received	N	5. Visually verifies all key entered data
E	6. Forwards copy of cash-control slip to data entry department	N	6. Prints bank deposit slip
		M	7. Prepares bank deposit
M	7. Prepares bank deposit		8. Makes bank deposit
	8. Makes bank deposit	M	9. Forwards copy of bank deposit slip to cash-management supervisor
M	9. Forwards copy of cash-control slip and bank-deposit slip to cash-management supervisor		

Key
E eliminate task
N new task
M modify task

FIGURE 7–13 Task analysis worksheet

technology will change people's jobs within an organization. Much like task analysis, technical analysis considers three types of change:

1. the elimination of existing technology,
2. the creation of new technology, and
3. the modification of existing technology.

In addition to the study of these three types of change, the team must ask questions regarding availability and acceptability. These questions include: Is the technology indicated by the proposed system available? If so, will it be acceptable to the users of a system?

To illustrate the concepts underlying technical analysis, let's return to the processing of customer payments (see Figure 7–13). The accounting clerk, as indicated, is expected to use a computer terminal to enter and process payments. Technical analysis begins with the study of the workplace to determine how new technology will change existing tasks and work patterns. Such an analysis might use a technical worksheet (see **Figure 7–14**). This worksheet contains a before-and-after section, to document the changes in tools required in performing a job. As illus-

Technical Analysis

Department: Accounts receivable	Reference: See A/R 394	Abstract: Explains the technical impact of the new customer payment posting system
Subject: Processing of customer payments	Issued: January 1, 19xx	
Position: Accounting clerk	Prepared by: Pamela Bennett	Page: 1 of 1

Technology Before Change	Technology Following Change
Desk calculator Paper, pencil Typewriter Bank forms	Computer terminal Paper, pencil Typewriter Bank forms

Technological Assessment

Mental: More required initially to learn new system; however, less later on.

Skill: Less. Computer will sum amounts previously summed by hand.

Responsibility: Same. Bank deposit must be verified before it can be sent. Computer will help in error checking.

Supervision: Same. Bank deposit slip will continue to be forwarded to supervisor. Accounting clerk position is a nonsupervisory position.

Physical: Additional eye strain with new system.

Overall Evaluation:

New system lowers the mental and skill requirements of the position. Responsibility, supervision, physical requirements are not substantially changed.

FIGURE 7–14 Technical analysis worksheet

trated, the new tool important to the proposed system is a computer terminal; it replaces an old tool, a desk calculator. The second section of the worksheet provides a *technological assessment* of the proposed change. As indicated, the impact of the change in technology is more striking. The new technology is shown to lower the mental and skill requirements of the position. This, in turn, leads one to question whether the accounting clerk will appreciate the new system once it becomes operational. The clerk might attempt to forestall the new system if the job requirements were perceived as initially exciting but more tedious later on.

Organizational Analysis

Weaving new technologies with modified tasks into an acceptable balance characterizes *organizational analysis*. In this instance, the analysis team must determine how the task and structure of an organization must change as a result of a proposed system change. Organizational analysis is not a simple undertaking. It is often impossible to make everyone happy about a proposed change; new technologies cannot always be fitted neatly into an established organization.

Several strategies are available to the analysis team in performing an organizational analysis. Four such strategies are

1. decentralizing system tasks,
2. job enlargement,
3. task sharing, and
4. tool sharing.

DECENTRALIZING SYSTEM TASKS

In recent years, decentralization of system tasks has made system change much more acceptable to the users of a system. With decentralization, people responsible for the processing of data are given the authority to perform data processing activities. An online system might be proposed that requires the credit officers of a company to modify customer credit limits directly, for example. This might represent a substantial change from the past, when all credit change notices were passed along to the data entry department for computer processing. Would the credit officers prefer having authority over what is entered into processing and what is not? After all, if a mistake is made, the credit department takes the blame, not some data processing unit like data entry. Numerous examples from industry suggest that people do prefer to have direct authority over matters for which they are held responsible. Moreover, the practice of equating authority with responsibility is a basic tenet of management theory.

JOB ENLARGEMENT

Instead of using system change as a means of making jobs increasingly automatic or programmed, system change offers the opportunity to make jobs larger and richer for the end user. This process of improving job quality is known as *job enlargement,* or in some instances *job enrichment.* You might ask, "How can this be done, when the introduction of new technology acts to simplify an existing set of tasks?" Consider the accounting clerk example once again.

Left unchanged, the new set of tasks proposed for the accounting clerk would make the job less challenging or exciting. This fact was documented further by the technical analysis. However, instead of leaving the job more automatic or programmed, suppose the analysis team decided to enlarge the job by adding new tasks. For instance, the following two new tasks might be added to the task analysis worksheet in Figure 7–13:

10. analyzes daily receipts totals and week-to-date totals and determines percentage relationships; and
11. determines week-by-week and day-of-the-week customer payment trends.

As this example suggests, job enlargement provides the opportunity to bring tasks before and after system change into balance. The analysis team need not try to convince the accounting clerk by saying "Sure, your job will be easier, but you will learn to adjust." Rather, the team can say, "With the new system some parts of your job will become easier and other parts will become more difficult. On balance, the job is much like it was before we recommended any type of change."

TASK SHARING

The concept of task sharing is similar to job sharing and entails the spreading of favorable and not-so-favorable tasks among employees. Whenever possible, the systems analysis team will attempt to share the unpleasant features of a proposed system. Why? Because if each person is asked to do a little, a large, unpleasant task often appears to be smaller than it is.

Consider the review of a computer-printed file listing. Suppose the listing consists of 300 pages and people are asked to read each page in an attempt to discover processing errors. Without task sharing, one or two people might be required to read the entire listing, month after month, assuming of course that they decided to remain with the organization. With task sharing, thirty or more people might be asked to review a section of the listing—such as ten pages or so. While these thirty people may dislike their assigned task, they may realize that it's the fairest way of assigning such an unpleasant task.

TOOL SHARING

Tool sharing is similar to task sharing and involves the dispersing of a tool among employees. With data processing, the tool might be any one of a number of software tools designed for the computer. For example, *electronic mail* is currently available on several computers. With electronic mail, clerks, blue-collar workers, supervisors, and top-level managers can send messages from one computer terminal to another terminal, provided the terminals are linked by software to form a network. You might ask, "Would electronic mail be an essential part of a proposed system?" Actually, it might or might not be. Instead of being an essential part of a proposed system, it might represent an add-on feature. With tool sharing, it becomes possible for the analysis team to state: "If we implement the proposed system, we also plan to add the following computer tools to be shared by the user staff."

The Mansfield, Inc., Case Study

Once satisfied with the job of documenting the logical current inventory system, Roger Bates pressed on with the design of the proposed new system. Specifically, he decided to move ahead with the development of a more integrated billing and receivables system.

"With our new system," Roger claimed, "we should be able to resolve the problems of the current system. And while the new design might appear to be similar to the current system, it will actually contain a number of new features. Let me explain."

At this point Carolyn and Neil looked at each other, wondering what Roger had in mind. They both hoped that Roger would also be receptive to some of their ideas. After all, weren't alternative designs worthwhile?

Roger began by skipping over the context and level-0 DFDs, beginning instead with a modified level-1 DFD (see **Figure 7–15**). He used this diagram to emphasize the first two points of his presentation.

"First, we are going to change the way in which we update our files. Instead of initially writing records to the invoice transaction file and later writing summary records to the accounts receivable file, we are going to do both at the same time. In this way, we will make sure that the customer account number is correct for billing and for receivables.

"Second, we will modify our customer account number sequence by providing each customer with a unique account number, discontinuing the use of postal code. This step is critical: it will help us clear up most of our update problems. With the new system, when customers move, their account numbers will stay the same.

"Third, we will move to cycle billing. One scheme is to process one-third of our customer accounts on each of three days during the month; for instance, on the fifteenth, the twentieth, and the twenty-fifth. Why these dates? Be-

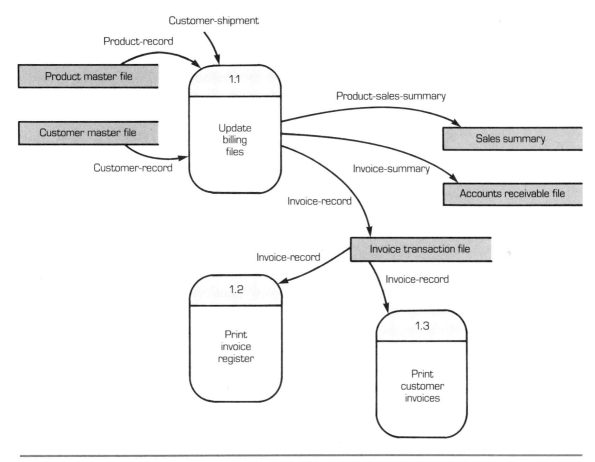

FIGURE 7–15 Level-1 DFD showing Mansfield's proposed new system

cause we need to get bills in our customers' hands before the first of the month to ensure prompt payment.

"Fourth, I am recommending that a new person be transferred to billing on a part-time basis to help them with cycle billing—at least in the short run. I realize that the current staff problem is critical; however, once cycle billing begins, the part-time person most probably will not be needed.

"Fifth, we will speed up the processing of customer payments. Let me use a system flowchart to show how this posting will work (see **Figure 7–16**). As indicated, each time we post a customer payment, we update the accounts receivable file directly. To begin processing, we verify the customer account number. This step helps us avoid billing the wrong customer and identifies customer account errors. With new manual procedures in place to help us discover who correct customers are, we should be able to avoid billing customers when they have paid. Notice that after we update the accounts receivable file we also write the details of the payment to a customer payment transaction file.

236 Chapter 7: Feasibility Analysis

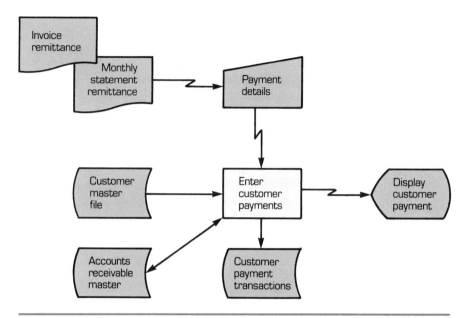

FIGURE 7–16 System flowchart showing process of entering customer payments

We print the contents of this file, as the current system does, to produce the customer payment cash register.

"Well, what do you think?" Roger asked. "Do you see any flaws with the new design?"

After some discussion, Carolyn and Neil were able to agree with the concepts outlined by Roger. "I see only one apparent weakness," remarked Neil.

"What's that?" queried Roger.

"Reducing the time to change customer and product record changes by marketing and manufacturing is still a problem," commented Neil.

"I'm working on that one," said Roger. "By placing terminals in marketing and manufacturing and by training their staffs to enter changes online, we should be able to process both customer and product changes once they become known."

Carolyn also had a question: "What's the purpose of the sales summary file?"

"Didn't think you would miss that one," Roger remarked. "The sales summary file (see Figure 7–15) will allow us to do an analysis of sales by salesperson and by product line. I will fill you in on the details of this important new feature once all logical requirements are fully specified."

Finally, Carolyn asked Roger how the new system would be organized. Roger explained.

"We will need to write eighteen to twenty computer programs. A tentative program processing menu (see **Figure 7–17**) indicates that we will need one interactive program to update the billing files, one to enter customer payments, and one to enter accounts receivable adjustments. We will need two master file update programs: one to update the customer master file and one to update

A. INTERACTIVE PROGRAMS
1. Update billing files
2. Enter customer payments
3. Enter accounts receivable adjustments
4. Update customer master file
5. Update product master file
6. Display customer account
7. Display product account
8. Display receivables account

B. BATCH PROGRAMS
1. Print customer invoices
2. Print customer monthly statements
3. Print invoice register
4. Print cash payments register
5. Print monthly statements register
6. Print customer master file register
7. Print product master file register
8. Print customer master file adjustments
9. Print product master file adjustments
10. Print accounts receivable master file adjustments

FIGURE 7–17 Program processing menu for Mansfield's proposed billing and receivables system

the product master file. At a minimum, we will need three display programs to display customer accounts, product accounts, and receivable accounts.

"Besides these eight interactive programs," he continued, "at least ten print programs are necessary. The first two will print customer invoices and customer monthly statements. We also need to print five file registers: an invoice register, a cash payment register, a monthly statement register, a customer master register, and a product master register. Last, we must design three programs to show any adjustments made to the customer, product, or accounts receivable master files."

SUMMARY

This chapter has dealt with the question of whether it is possible to evolve an acceptable logical design. The first concern dealt with the issue of whether the design would meet or exceed user performance requirements. Clear goals, objectives, and design constraints were viewed as essential to the definition of these requirements.

The analysis team must be able to communicate to user groups the performance expectations of a new system once it becomes operational. This step is clarified with specific objectives.

The analysis team must also clarify organizational design strategy. Four decision-prompting issues considered in the development of such a strategy are computer-based versus manual systems, centralization versus decentralization, build versus buy, and prototype versus fully functional systems.

The second concern of this chapter dealt with developing design alternatives and performing a technical assessment of each alternative. The process of developing design alternatives begins with determining new system outputs followed by developing alternative ways of achieving those outputs. Once this is done, an evaluation of each alternative is required to decide which alternative is most feasible.

The development of design alternatives requires considerable work. Typically, the analysis team must design a set of DFDs and document each alternative system. Fortunately, the tools of structured analysis simplify this time-consuming undertaking. The practicing analyst learns to rely on DFDs in particular to develop workable design alternatives and prototypes.

The evaluation of design alternatives requires the selection of evaluation categories and criteria important to each. The purpose of evaluation is to measure objectively, whenever possible, key system requirements, features, and costs. A design evaluation matrix is used to illustrate which design alternative is rated as best. The matrix helps the analysis team make a final system recommendation.

The third concern of this chapter was with the social assessment of the proposed new system. Social, technical, and organizational impacts of the new system are the final test of overall system feasibility. Techniques important to analyzing the impact of the new system include task analysis, technical analysis, and organizational analysis. Four strategies designed to lessen the impact of a new system are decentralization of system tasks, job sharing, task sharing, and tool sharing.

REVIEW QUESTIONS

7–1. What are the main purposes of the formal system feasibility study?

7–2. Name the various parts of the user's performance definition.

7–3. How does a goal differ from an objective?

7–4. What is an organizational design strategy?

7–5. What does the build-versus-buy decision involve?

7–6. What is a prototype and how does it differ from a fully functional system?

7–7. Explain why organizations tend to support the development of operational prototypes compared with fully functional systems.

7-8. What are the two parts of a specific objective?

7-9. What is the difference between a mandatory and a highly desired objective?

7-10. How do constraints influence a design?

7-11. List the steps important to designing the new system.

7-12. Name the three main types of system criteria used in design evaluation.

7-13. What is task analysis? Technical analysis? Organizational analysis?

7-14. What strategies are available to the analyst in performing an organizational analysis?

7-15. How does task sharing differ from tool sharing?

EXERCISES

7-1. Rewrite the following goal statement to explain how a system change leads to a business improvement.
   Goal Statement: Tie together budgeting with accounts payable to improve the planning and control of company expenditures.

7-2. Rewrite the following goal statement and indicate whether it is a well-written or poorly written statement. What conclusion can you draw from your analysis?
   Goal Statement: Acquire personal computers to improve managerial response to company problems.

7-3. Consider this statement: "Suppose a company decides to begin with six months of sales history and build a sales database slowly." Is this an example of a goal, a design strategy, an objective, a constraint, or a description of new system outputs? Explain the reason for your choice.

7-4. Modify the student registration context diagram shown in Figure 7-5 to depict the situation described in alternative 3, which allows students to enter their course requests at any one of several remote locations, but requires them to pick up their approved student schedules from their faculty advisers. Show only the revised course-request, approved-student-schedule loop.

7-5. Modify the student registration context diagram shown on Figure 7-5 to depict the situation described in alternative 4, which requires students to submit their course requests to their faculty advisers, who must approve and send all requests to student registration, and to pick up their student schedules from their faculty advisers. Show only the revised course-request, approved-student-schedule loop.

7-6. Complete the design evaluation matrix shown on page 240. Based on the highest total score, which is the best alternative? Which alternative do you consider best? Explain your answer.

Evaluation Criteria \ Design Alternatives	weight	Alternative 1		Alternative 2	
		Rating	Score	Rating	Score
Requirements	32				
Policy	[4]	5		4	
Personnel	[5]	4		4	
Equipment	[4]	3		5	
Space	[6]	3		4	
Management control	[7]	5		3	
Training	[6]	3		4	
Performance	34				
Reliability	[6]	5		4	
Validity	[5]	3		4	
Accuracy	[6]	5		3	
Timeliness	[7]	5		4	
Capacity	[5]	3		4	
Throughput	[5]	3		4	
Costs	34				
Development	[5]	5		3	
Operation	[15]	2		4	
Return on investment	[14]	5		4	
Payback	[—]				
Total Score					

WORLD INTERIORS, INC.—CASE STUDY 7
DESIGNING THE NEW SYSTEM

Introduction

Following careful study of WI's current method of processing, John Welby reached the same conclusion as Rena—namely, that solving the company's administrative problems would require a much improved computer system. He based this view on several findings:

1. The current system was expensive to operate and maintain and, moreover, had never been fully used.
2. The current system did not address major data processing problems, such as reconciling customer orders with vendor payable or keeping track of problem orders.
3. The current system kept all customer order and mailing label information online, though offline storage was acceptable in many instances.
4. The current system did not calculate in-state sales tax, determine the amount of sales tax payments, or calculate sales tax credits.
5. The current system was not designed to store retail store information.
6. The current system could not analyze either direct or retail store sales or estimate the effects of advertising and promotional programs.

System Design Modifications

Changing the current design of processing required close examination of the types of data to be stored online. There was little reason for keeping customer order information online, for example, provided the order had been filled by the manufacturer and the customer appeared to be satisfied with the shipped merchandise. In addition, there was little need to store mailing label information online, especially if that information was rarely displayed, modified, or used. Conversely, there were reasons for creating an online open-orders file and an online supplier file. John defined the open-orders file as follows:

"An *open-order file* would be similar to the old customer order history file. It would store 1) all orders taken, but not shipped by the manufacturer; and 2) all orders taken, filled, and shipped within the past six months. Finally, it would replace the pending orders file."

John went on to define the contents of the new supplier file.

"The *supplier file* would be similar to the old supplier product file. It is needed to check the correctness of keyed data and to add supplier information to customer orders. It would also be used by marketing to determine the relative success of the various advertising and promotion programs. We plan to store product and sales history information in this file. Each time an item is ordered we plan to update the sales history portion of this file."

In putting together new system specifications, John concluded that the new system would be required to:

1. check the correctness of each customer order,
2. calculate in-state sales tax,
3. keep customer order information online for six months past the time that WI received the order from the customer,
4. match customer invoice numbers against corresponding numbers shown on the factory invoice and entered into processing,
5. determine in-state sales tax payment amounts and due dates,
6. store historical mailing label information online, and
7. store sales summary totals online by product code for each supplier.

Program Processing Menu

The program processing menu for the proposed order processing system helped identify the functions to be performed by the new system. The first four programs were to be used on a daily basis and were entitled:

1. ENTER CUSTOMER ORDERS,
2. PRINT DAILY ORDER REGISTER,
3. PRINT CUSTOMER INVOICES, and
4. DISPLAY OPEN ORDERS.

John explained the proposed system: "The first program is the most critical. This program will be designed to check all line-item numbers, finish numbers, and product prices—for each item ordered by the customer. It will add line-item names to each item ordered by extracting information from the supplier file. However, it will not update the supplier's sales history. This will be done later when we enter factory invoices."

Besides a change in files, there were other significant design changes. For example, envelopes with windows were to be used instead of mailing labels. John commented, "This eliminates the need to maintain the labels file. The address printed on the invoice will show through the window on the envelope."

In addition, since the open-order file replaced the pending invoice file, the third copy of the invoice (see the current system) could be either dropped or printed and placed in an offline file called the factory orders file. It was decided to place an invoice copy in such a file.

CASE ASSIGNMENT 1

Revise the level 1.2 transform, prepare customer invoice, to explain the new method to be followed in producing invoices. As before (see chapter 5), the input to 1.2 is a cleared customer order; the output consists of three copies of the customer invoice. In your analysis, limit the DFD to four transforms; the titles of these should be the same as the four computer programs to be used on a daily basis listed previously.

CASE ASSIGNMENT 2

Prepare system flowcharts for these first four programs to show the physical features of computer processing. Use standard symbols in drawing each flowchart. If sort routines are needed, show these as separate programs.

Problem Orders

The handling of problem orders was never built into WI's initial system design. "We missed on that one," stated Rena. "We never dreamed that the number of problem orders would increase in proportion to company growth."

There are several reasons for problem orders: incorrect sizing by the factory, incorrect finish, goods damaged in transit, and manufacturer's defects. In addition, some customers never received their goods, while others wound up with the wrong orders. Regardless of the cause, WI attempted to manually track all problem orders. Admittedly, the company was not doing a very good job.

In an attempt to improve the management of problem orders, WI established a customer service center. The purpose of this center was to receive and process all customer complaints. New steps in processing included receiving a complaint from the customer and setting a problem flag for orders stored on the open-order file. A problem order remained on the open-order file until it could be resolved. Accounting was responsible for the actual resolution of a problem order. Once the problem was resolved, accounting instructed the service center to clear—that is, to remove the flag from—the open order. Once cleared, the open order could be paid and, if older than six months, removed from the file.

Besides setting and clearing problem flags, the customer service center was required to complete a customer service sheet. This sheet specified the nature of the customer complaint. The original copy of the sheet was generally sent to the manufacturer. Three other copies were prepared: one stayed with the service center, one was sent to the customer, and one was sent to accounting. Accounting waited for a service sheet response from the manufacturer before taking further action.

Extended Program Processing Menu

The additions to the revised processing menu for WI's order processing system included several new programs:

5. ENTER FACTORY INVOICES,
6. PRINT FILLED-ORDERS REGISTER,
7. ENTER/CLEAR PROBLEM ORDERS,
8. PRINT PROBLEM-ORDERS REGISTER,
9. PRINT CUSTOMER SERVICE SHEETS,
10. RESET OPEN-ORDERS FILE, and
11. PRINT CUSTOMER ORDER HISTORY.

The first of these seven programs would be designed to perform two important file updates: to change the status of an order stored on the open-order file by showing that an order had been filled, and to change sales and profitability totals stored on the supplier file. Besides these updates, all filled orders were to be written to a filled-orders file.

For processing to begin, the customer invoice number stored on the open-order file was to be matched against the customer invoice number shown on the factory invoice (see **Figure 7–18**) and entered into processing. In addition, factory item numbers, finish numbers, and costs were to

Factory name and address				**FACTORY INVOICE**	
SOLD TO: World Interiors, Inc. 1919 Fairways Drive Wilsonville, OR 97000		SHIPPED TO: Esther Peters 1986 Wilson St. Brooklyn, NY 11200			
Customer Invoice Number: 22345　Date: 5/04/xx	Shipped via: UPS　From: HOBO—37			Refer to → Invoice No.　When remitting 686313	Invoice Date: 5/27/xx
Stock Number	**Description**	**Quantity**	**Finish**	**Price**	**Amount**
H37　F352	Vertical Blind	1	100	62.25	62.25
H37　V352	Valance	1	100	6.75/ea	6.75
Account Number: 74286		2 Total Quantity			69.00 Invoice Total
Terms: 2% 10, net 30.* Past-due balances will be assessed a service charge of 1.5% per month (18% per annum)					Page 1—Last page
* A 2% discount will be allowed if invoices dated 1st–15th are paid by the 25th and invoices dated 16–31st are paid by the 10th of the following month.					

FIGURE 7–18 Factory invoice form

be entered and compared with stored numbers and totals. As a special part of processing, the program accumulated the total sales tax for filled in-state orders.

The next program was to be designed to print the filled-orders register. The register showed all orders in which invoice numbers matched and factory information was in agreement with customer order information. After printing, the register was approved by the data processing staff. Once approved, it was forwarded to the accounts payable section. The payables staff merged the filled-orders information with other outstanding payables. Payables processing, another design area facing John, would consist of selecting factory invoices to pay and writing vendor-voucher checks. Checks were to be mailed to the manufacturer.

The next two programs were to be designed to enter and clear problem orders and to print the results of this processing step. The next program was needed to print the specifics of problem orders on customer service sheets.

The last two programs were to be used to remove customer orders from the open-order file and to print the results of this processing step. As stated earlier, orders were to remain on the open-orders file for six months, provided the order had been filled and was not classified as a problem order.

CASE ASSIGNMENT 3

Modify the level-0 DFD prepared earlier (see chapter 3) to show a new transform entitled, "3.0 Process problem orders." Include in this new drawing the open-orders file. Remember that the new open-orders file replaces the pending customer invoices file.

CASE ASSIGNMENT 4

Decompose the level 2.0 transform prepared for case assignment 3 to show the steps important to entering factory invoices and to printing the filled-orders register. Inputs to processing should include the factory invoice and information stored on the open-orders file. Include in this DFD the transforms of entering the factory invoices, printing the filled-order register, and processing accounts payable.

CASE ASSIGNMENT 5

Decompose the level 3.0 transform prepared for case assignment 3 to show the steps important to entering and clearing problem orders, printing the problem-orders register, and printing customer service sheets. Inputs to processing should include a customer complaint or accounting's decision to clear a flag. The clearing of a problem order typically follows a satisfactory response from the manufacturer.

CASE ASSIGNMENT 6

Prepare system flowcharts for programs 5 through 11 to show the physical features of computer processing. Use standard symbols in drawing each flowchart. Include sort routines as separate programs.

8

Logical Design Specification

INTRODUCTION

A well-conceived and -conducted feasibility analysis study should lead to the selection of the best new system—the system to propose to management for its approval. This system must meet or exceed user performance requirements, produce required outputs, and have a favorable technical and social impact. What remains for the analysis team is the task of finalizing the logical design specification. More specifically, the team must prepare the preliminary technical design specifications, combine those technical specifications with the logical design, and develop the schedule for system design and implementation. Once all of this is completed, the document that results—the logical design specification—is presented to top management and user groups for their approval.

Observe that a completed logical design specification combines the logical aspects of a design with preliminary technical requirements. While you might think that this is odd, such mixing is done for a variety of reasons. For example:

1. Technical design requirements, if known, must be specified in terms that can be understood by the systems designer and users.
2. Ideas for a successful technical design must be passed from the systems analysis team to the systems design team, top management, and users.
3. Operational aspects of a design must be specified far enough in advance to allow management and user groups to judge a proposed design by its logical and technical features.
4. Resources needed in support of the technical aspects of a design must be clarified in order for management and user group support to be obtained.

This chapter first considers these preliminary design requirements. Following this, the contents of the logical structured specification are considered. The chapter ends with a look at how a planning schedule is developed for systems design and implementation. When you complete this chapter, you should be able to

- describe system design requirements and indicate why a preliminary version of these requirements is included in the logical design specification;
- understand the difference between transaction-centered and transform-centered designs;
- prepare a structured specification, using many of the tools presented in earlier chapters;
- design a detailed time schedule, showing the critical activities of design and implementation; and
- prepare a detailed project budget.

PRELIMINARY SYSTEM DESIGN REQUIREMENTS

The mixing of logical and preliminary system design features in the logical design specification suggests that the organization of the document should contain two main parts: the proposed logical system and system design requirements. The *proposed logical system* describes the functional view of the proposed system. As we have seen in previous chapters, this functional view is prepared through the use of the tools of structured analysis. It includes DFDs, a data dictionary, data store descriptions, and transform descriptions.

System design requirements provide a different set of design specifications. Included in this part of the document are five sets of preliminary design requirements:

1. system organization requirements,
2. input/output requirements,
3. data file and database requirements,
4. computer program requirements, and
5. processing control requirements.

A more complete discussion of each of these topics is found in chapters 9 through 13. As an introduction to those chapters, let's consider what the analysis team must do to frame a set of preliminary design specifications.

System Organization Requirements

One of the most difficult aspects of systems analysis and design is determining how a system is to be packaged. *Packaging* is the process of indicating how the various components of a system are to be organized as a set of technical units, in which the units represent computer programs or steps in a computer program.

The steps involved in packaging, as discussed briefly in chapter 3, consist of

1. dividing a system into jobs,
2. dividing each job into job steps, and
3. illustrating how each job step is to be designed.

When these steps are completed, the final form of organization for a software product can be determined.

An obvious question is: "How does an analyst begin this process of packaging?" Actually, there is little agreement on this subject. Most analysts acknowledge that a system will be organized one way if its design is transform centered and another way if its design is transaction centered. Consequently, let's consider these two design characteristics as a starting point.

Figure 8–1 illustrates a transform-centered design. In such a design, data flows to and from a single central transform, or from an area of central transform. This central transform acts as a dividing point between system inputs and outputs. A transaction-centered design, in contrast, is quite different. As **Figure 8–2** illustrates, one transform, called a transaction center, receives information from a source and uses this information to activate different data paths.

Two more realistic examples will clarify the differences between a transform- and a transaction-centered design. **Figure 8–3** illustrates the context diagram prepared for updating a customer master file. The four inputs to processing consist of new customer information, name and address changes, customer deletions, and an old customer record (which applies when either changes or deletions are made to customers currently on file). Outputs from processing include a customer change register, which reports all changes made to the customer master file, and either a revised customer record or a new customer record.

Because this design does little more than impart a new form to the customer master file, it is classified as a transform-centered design. The entire design can be organized as a single job (e.g., a single computer program), with several distinct job steps. An analyst might specify:

Program Name
Update Customer Master File

Required Job Steps
1. Add new customer information
2. Make name and address changes
3. Delete customer account
4. Display customer record

Figure 8–4 illustrates an accounts payable system to describe a transaction-centered design. This system is triggered by the receipt of an invoice (or bill) from a vendor. Once received, the bill is either paid or

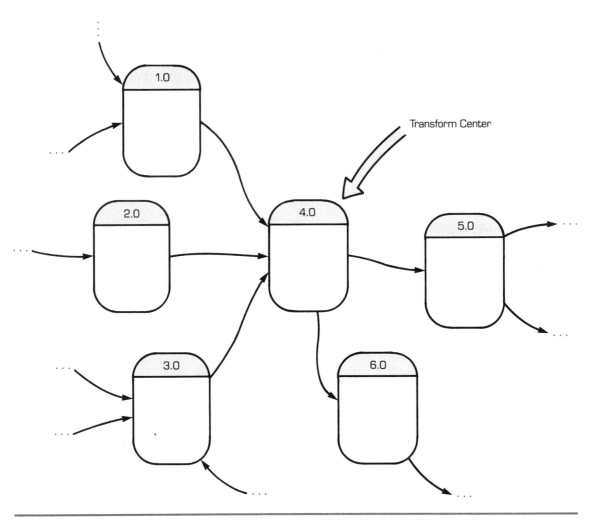

FIGURE 8–1 Transform-centered design

stored in a pending (e.g., a to-be-paid) bills file. When the bill is paid, a personal check is written, the paid bill is placed in a file of bills paid, and the new account balance is written in the personal checkbook register. When a bill is stored, it remains in the file of pending bills until the decision is made to pay the bill. This decision removes the pending bill from the file of pending bills and transfers it to the file of bills paid.

A transaction-centered design usually depends on user decisions or on some external stimuli. As the description of the accounts payable design suggests, the user must decide which bill to pay, when to pay it, and in what amount. Most analysts generally propose several jobs (e.g., computer programs) rather than one in a transaction-centered design be-

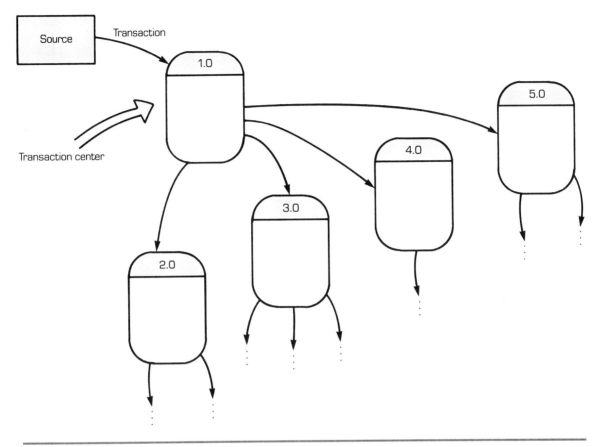

FIGURE 8–2 Transaction-centered design

cause of this dependence on user decisions. In the accounts payable design, the analyst might show a combination of interactive and batch programs, much like the following:

```
Program Menu
I.1  Enter vendor invoices (bills)
I.2  Enter dollars for payment of bills
I.3  Display bills remaining to be paid
I.4  Make vendor payment
       .              .
       .              .
       .              .
B.1  Print personal checks
B.2  Print personal check register
B.3  Print list of all bills to be paid
       .              .
       .              .
       .              .
```

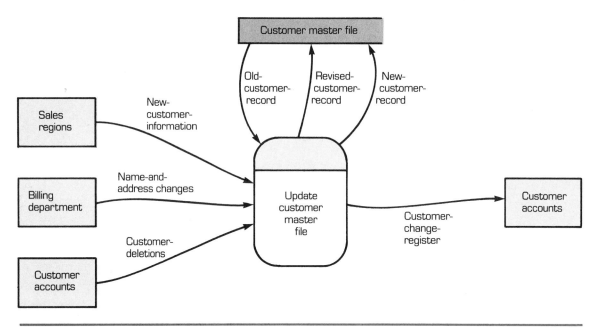

FIGURE 8–3 Updating a customer master file: an example of a transform-centered design

As this list suggests, at least four interactive programs will be needed to enter vendor invoices into the system, to enter dollars for payment of bills into the system, to display bills remaining to be paid, and to make a vendor payment. Three batch programs are also required: one to print personal checks, one to produce a register of all printed checks, and one to print a listing of all bills to be paid.

Given this rather lengthy introduction to the concept of packaging, the question remains: What system organization requirements might the systems analyst be expected to include in the logical design specification? Realistically, the analyst can be expected to suggest how the system might be divided into jobs and how some jobs might be divided into job steps. Accordingly, the analyst generally includes in the logical specification a program processing menu and system flowcharts to highlight the features of each program shown on the menu. As will become apparent later, however, it is the systems designer, not the systems analyst, who decides how each job step is to be structured. It is the designer's responsibility to transform the logical DFDs into an organized set of technical units.

Input/Output Requirements

Defining preliminary input/output (I/O) requirements for a proposed system is easier than determining packaging requirements; however, documenting this second set of requirements can take considerable time. Typ-

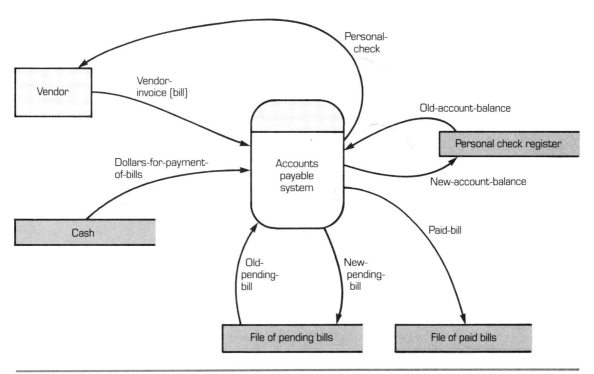

FIGURE 8-4 Accounts payable system: an example of a transaction-centered design

ically, users insist on reviewing the I/O forms proposed for a new system before they will give their approval to begin the design of such a system. Why? First, users need to know how outputs will look in advance to determine whether the types of information they have specified will actually be produced by the new system. Second, users need to know the work requirements of their staffs who will enter data into the system and handle processed output.

The analysis team attempts to address these user concerns by preparing sample drafts of system outputs and inputs. In preparing I/O requirements, the team will usually prepare samples of output terminal screens and computer-printed documents, and input terminal screens (including interactive dialogues) and data entry documents.

Figures 8–5 through 8–8 illustrate samples of I/O layouts. In this instance, the system is a computer terminal scheduler for a highly used, multistation computer. As shown in **Figure 8–5,** an output of this proposed system is a daily schedule display. It shows the number of computer terminals that have been reserved for the day, beginning at 0830 and ending at 2330. **Figure 8–6** illustrates a second system output, the 0830 hourly schedule. This display contains space to show the names of the persons who have reserved a computer terminal on the date specified.

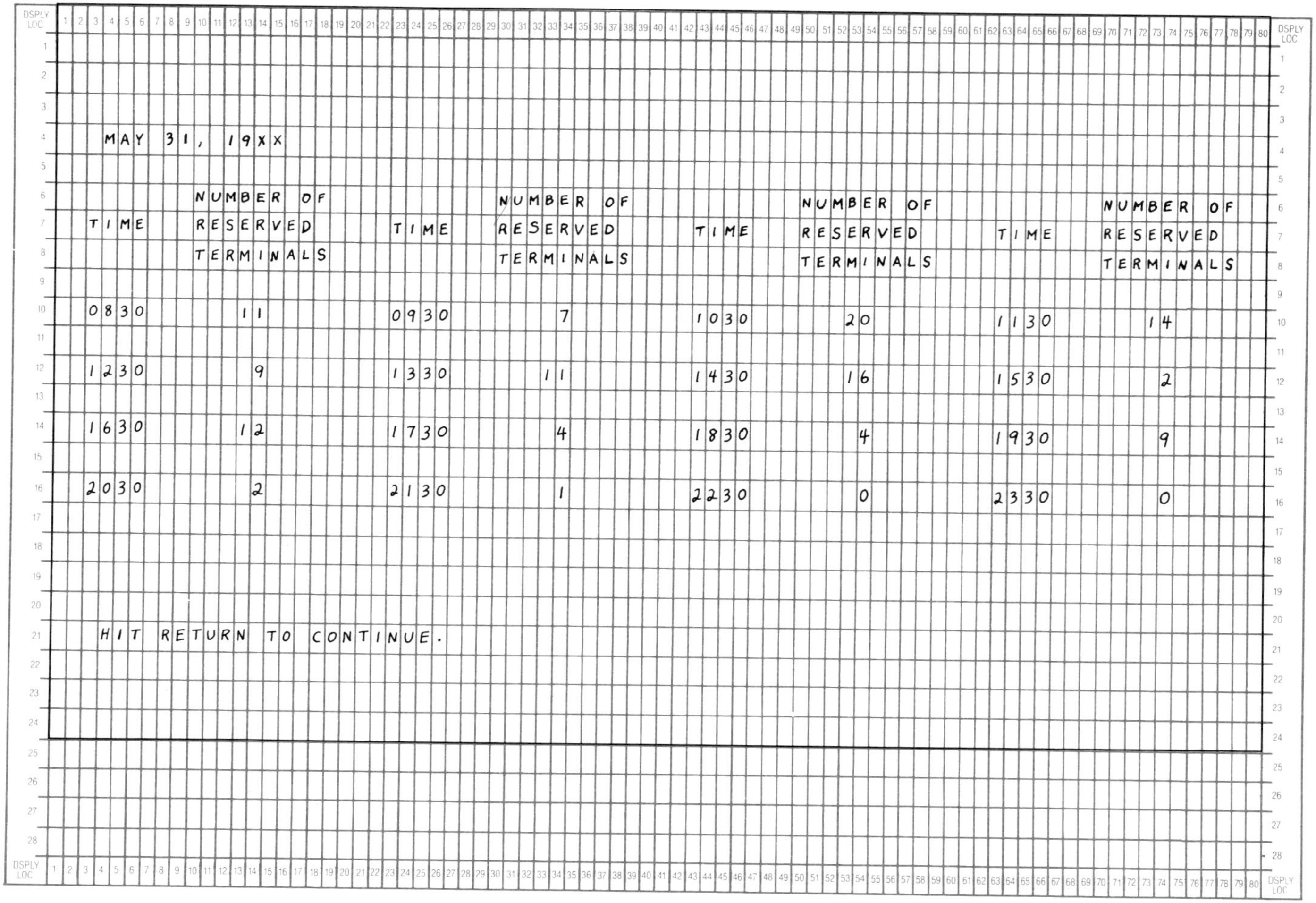

MAY 31, 19XX							
TIME	NUMBER OF RESERVED TERMINALS	TIME	NUMBER OF RESERVED TERMINALS	TIME	NUMBER OF RESERVED TERMINALS	TIME	NUMBER OF RESERVED TERMINALS
0830	11	0930	7	1030	20	1130	14
1230	9	1330	11	1430	16	1530	2
1630	12	1730	4	1830	4	1930	9
2030	2	2130	1	2230	0	2330	0

HIT RETURN TO CONTINUE.

FIGURE 8–5 Proposed system output: display of daily schedule

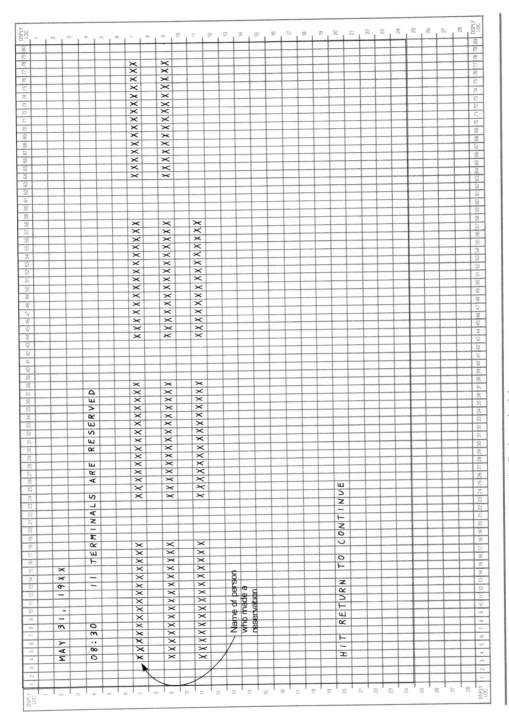

FIGURE 8–6 Proposed system output: hourly display schedule

Figure 8–7 illustrates an input display menu. This menu tells the user that the computer terminal scheduler will limit input to five choices. Users of the NAX scheduling system will be able to display the entire schedule, display their personal schedule, make a reservation, cancel a reservation, or exit from the system. **Figure 8–8,** a second input display, suggests an input dialogue to be followed in producing an hourly schedule. As indicated, the dialogue may contain acceptable and unacceptable computer-printed messages. This system, for example, will not accept the input of a month in which no times are scheduled (to avoid attempting to display an empty schedule), nor will it accept an alphanumeric response when a numeric response is appropriate.

The documentation of I/O requirements in the logical design specification is often limited to a listing of each requirement, followed by a sample layout of all major input documents and screens and all major output documents and screens. When a project moves to the design stage, the systems designer must decide whether these initial I/O samples require modification. In addition to I/O samples, the design specification may also include volume and frequency estimates. For example, the analyst may be expected to estimate

1. the number of lines to be printed in producing each document and the number of documents to be printed;
2. the number of characters to be key-entered for each source document and the number of documents to be processed;
3. the time frame in which reports are to be printed; and
4. the time frame in which source documents are to be received and processed.

Finally, the I/O section may include an analysis of staffing and equipment requirements. If the new design calls for the training of ten computer operators and the acquisition of ten computer terminals, this information must be added to the logical specification.

Data File and Database Requirements

As with I/O requirements, the analyst is often required to list the main physical files required by a system and to document their logical contents. Fortunately, much of this work should have already been completed, through the preparation of data store descriptions. Suppose that a proposed design features a customer master file. The analyst would begin to define this file as follows:

```
Customer-master-file = {customer-record}
```

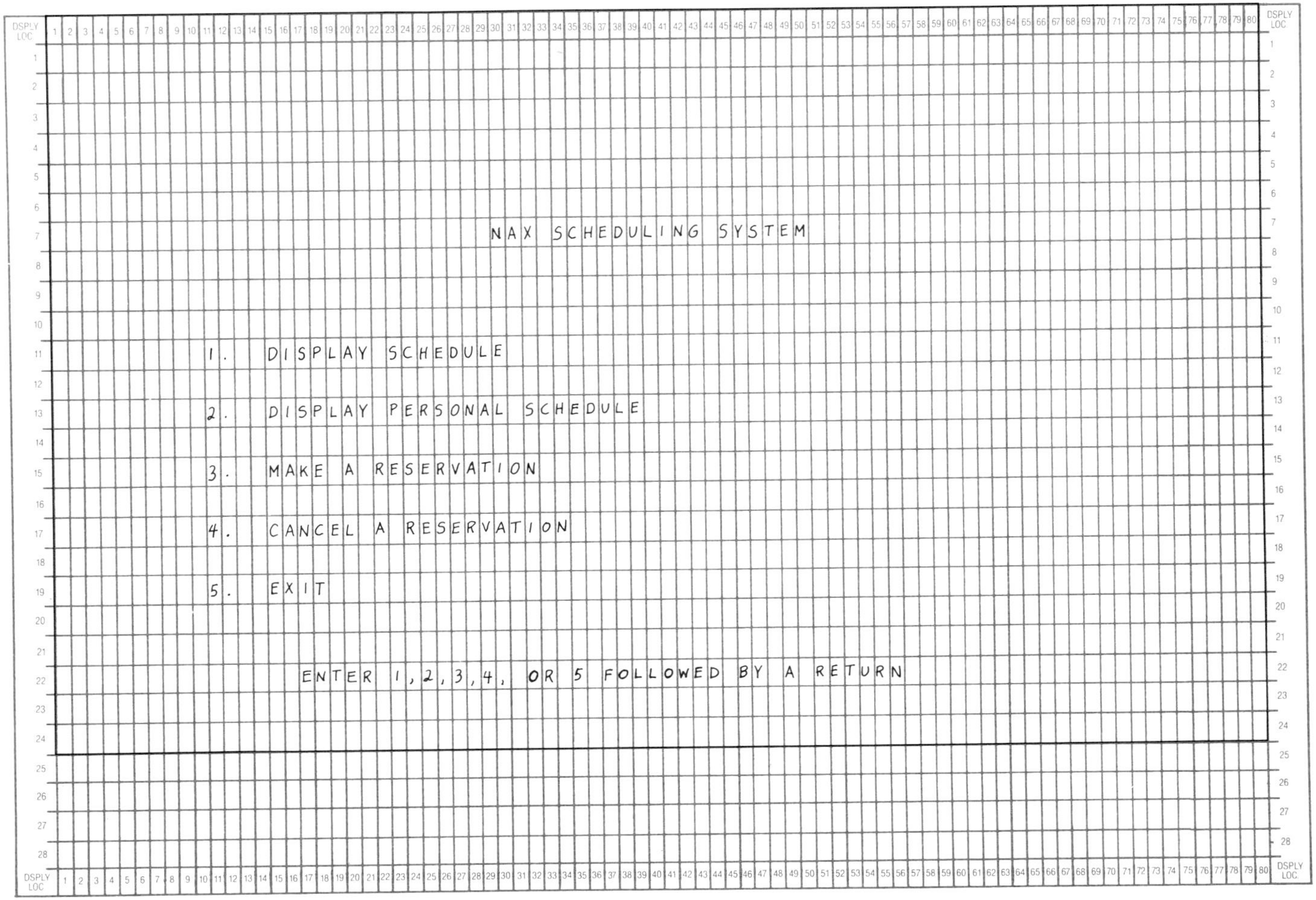

```
                    NAX SCHEDULING SYSTEM

          1. DISPLAY SCHEDULE
          2. DISPLAY PERSONAL SCHEDULE
          3. MAKE A RESERVATION
          4. CANCEL A RESERVATION
          5. EXIT

                    ENTER 1, 2, 3, 4, OR 5 FOLLOWED BY A RETURN
```

FIGURE 8–7 Proposed system input: input display menu

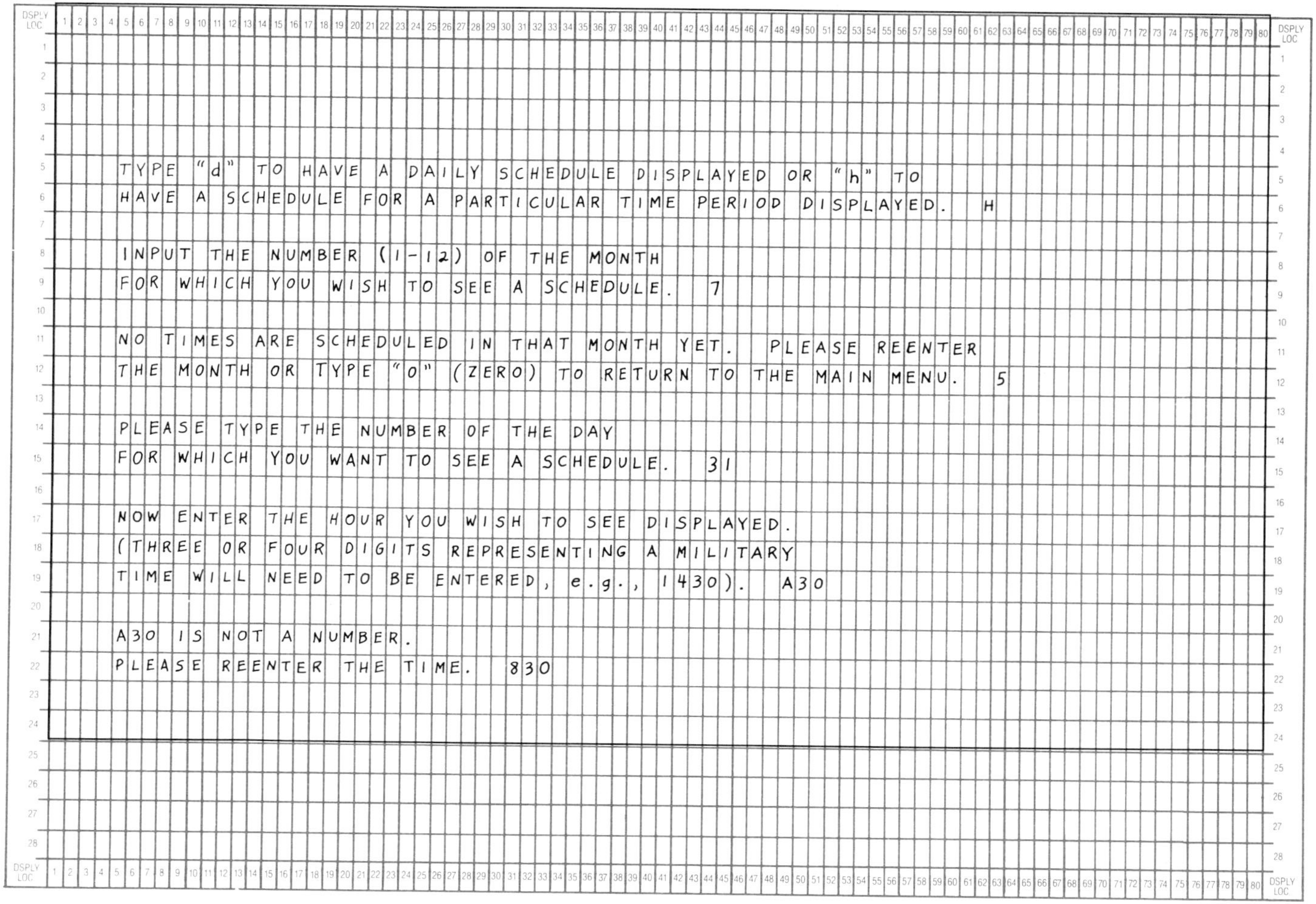

FIGURE 8–8 Proposed system input: input dialogue

```
 1  TYPE "d" TO HAVE A DAILY SCHEDULE DISPLAYED OR "h" TO
 2  HAVE A SCHEDULE FOR A PARTICULAR TIME PERIOD DISPLAYED.  H
 3
 4  INPUT THE NUMBER (1-12) OF THE MONTH
 5  FOR WHICH YOU WISH TO SEE A SCHEDULE.  7
 6
 7  NO TIMES ARE SCHEDULED IN THAT MONTH YET.  PLEASE REENTER
 8  THE MONTH OR TYPE "0" (ZERO) TO RETURN TO THE MAIN MENU.  5
 9
10  PLEASE TYPE THE NUMBER OF THE DAY
11  FOR WHICH YOU WANT TO SEE A SCHEDULE.  31
12
13  NOW ENTER THE HOUR YOU WISH TO SEE DISPLAYED.
14  (THREE OR FOUR DIGITS REPRESENTING A MILITARY
15  TIME WILL NEED TO BE ENTERED, e.g., 1430).  A30
16
17  A30 IS NOT A NUMBER.                                     830
18  PLEASE REENTER THE TIME.
```

In decomposing this equation, the analyst would also define the contents of each customer record, much like the following:

```
Customer-record = Customer-number + customer-name-
                  and-address + customer-credit +
                  salesperson-number
```

Next, the analyst might document the entire contents of this master file by using a nested listing. As shown by **Figure 8-9,** such a listing includes the name of the data store and all stored attributes, including all key attributes and pointer attributes.

In addition to showing the contents of each data store, the analyst must document volume and frequency considerations, much like the documentation for I/O requirements. For instance, the analyst might need to provide estimates for

1. the number of records to be placed in each data store;
2. the number of records to be added to, modified, or removed from each data store over a predefined time period (such as a one-week period);
3. the frequency with which updating of a file is expected to take place (e.g., daily, weekly, monthly);

```
Customer-master-file
    Customer-record
        Customer-number
        Customer-name-and-address
            Sold-to-name
            Ship-to-name
            Sold-to-address
                Street-address
                City-address
                State
                Postal-code
            Ship-to-address
                Street-address
                City-address
                State
                Postal-code
        Customer-credit
            Credit-limit
            Credit-rating
        Salesperson-number
```

FIGURE 8-9 Nested listing showing contents of the customer master file

4. the response time considered acceptable between each request for stored data and its return; and
5. the file backup considered appropriate for records stored on computer files.

Computer Program Requirements

Similar to data file and database requirements, the work of documenting computer program requirements should have been completed, in large part, through the preparation of transform descriptions. What remains is the identification of *processing constraints, unit controls,* and *unit test plans.* As an example, suppose that a functional primitive is identified as "1.1.2 Display User Menu." The structured English written for this transform might be simple enough and read as follows:

```
CLEAR computer terminal display screen.
DISPLAY processing menu choices.
DISPLAY select-from-menu prompt.
```

To complete this logical description, the analyst adds three short statements to describe processing constraints, unit controls, and unit test plans. Consider these three statements:

Processing constraints: Menu choices and select-from-menu prompt instructions should require fewer than twenty-two lines. The computer terminal display screen has a screen width of eighty characters and accepts twenty-four lines of input.

Unit controls: The use of a counter is advised to determine whether the proper number of lines is displayed.

Unit test plan: A visual test is needed to determine whether the proper menu and the correct number of lines are displayed.

While these three descriptions require additional time to complete—especially if they are prepared for each functional primitive—they are invaluable in helping to evolve the systems design. Processing constraints, for instance, advise the designer of equipment and staff limitations. Moreover, these constraints can be artificially set, thus acting as *built-in constraints.* Suppose that we need to provide a design that allows a user to escape from a computer program. One solution is to set an upper limit of five tries to enter a correct response. The built-in constraint would read:

```
IF Count = 5
    Clear the video display screen.
```

Likewise, *built-in unit controls* can be added to a design. A counter, combined with the appropriate control message, illustrates a built-in control. The structured English might read:

```
IF Count = 5
    Clear the video display screen.
    Display day-of-month termination message.
    Display options menu.
```

This description tells the designer to set a counter, clear the screen if the count is reached, display a message to tell the user why the screen was cleared, and display a menu allowing the user to escape.

Finally, the unit test plan section describes possible tests to determine whether a programmed module is working correctly. Some unit test plans are visual tests, others are built in, and still others are experimental. An example of an experimental test plan might be

1. Enter invalid menu selections until the count limit is exceeded.
2. Following this, the program should
   a. clear the screen,
   b. display a termination message, and
   c. display an options menu.

Processing Control Requirements

By describing unit controls and unit test plans for each functional primitive, the analysis team can provide the designer with a clear picture of how internal programmed procedures might be written and tested. What remains is the design of processing controls of the system as a whole. Transaction-centered designs, for example, require an *audit trail* to verify the correctness of processing totals at several stages in processing. Consider the accounts payable system once again, from a control point of view. **Figure 8–10** shows that if $500 is entered into processing and $200 for personal checks is returned to vendors, then $300 must be added to the file of pending bills. Likewise, the $200 representing the dollar amount of bills paid must be added to the file of bills paid; the new account balance must be adjusted as follows:

```
new-account-balance = old-account-balance - $200
```

With transform-centered designs, processing controls for the system as a whole are somewhat easier to develop; however, they also tend to provide less useful diagnostic information. Consider the update of the customer master file once again (see Figure 8–3). A systemwide processing control involves the recording of all updates on the customer-change register. This register helps ensure that the file changes processed were

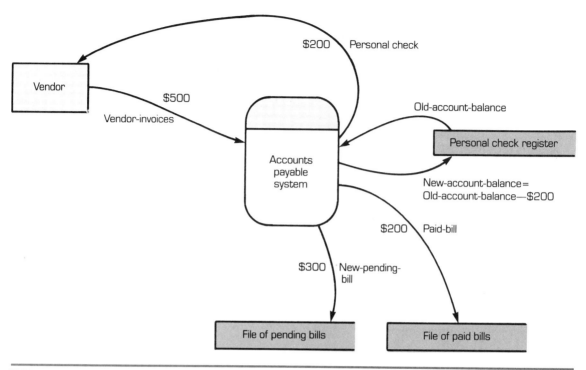

FIGURE 8-10 Designing an audit trail for the accounts payable system

correct; unfortunately, it does not guarantee that the physical file actually stores what it is supposed to store.

Design Assumptions

Asking the systems analysis team to describe preliminary system design requirements assumes several things. It assumes that the team knows a great deal about the design process and that systems designers will want to have design suggestions passed along. It also assumes that team members understand the nature of the design environment. For example, they must know such things as the availability of computing equipment, screen sizes, and so forth. In addition, it assumes that the probability of a proposed logical design being accepted is high; otherwise, the work required to add preliminary system design requirements would not be worthwhile. Finally, we need to add a note of caution. The addition of system design requirements does not mean that these requirements must be strictly followed by the systems design team. Rather, what the analysis team is asked to do is to frame a preliminary set of design specifications. In the final analysis, the design team must work with users to determine an acceptable technical design.

PREPARING THE STRUCTURED SPECIFICATION

Once preliminary systems design specifications have been determined, the task of finalizing the logical design specification begins. This specification serves to define the work of the systems analysis team. It should describe

1. the current logical system, together with the problems that the analyst set out to solve;
2. the results of the feasibility study, beginning with user's performance definitions, continuing with design alternatives and the selection of the best design, and ending with the impact of the new design on the organization;
3. the logical requirements of the proposed new system; and
4. the preliminary system design requirements, following the guidelines presented in the previous section.

Figure 8–11 illustrates a suggested table of contents for the logical system specification. With the possible exception of the last three headings, the outline should contain no surprises.

    I. Executive summary
   II. Current logical system
  III. Feasibility study
       A. User's performance definition
       B. Processing alternatives
       C. Proposed design
       D. Impact of the proposed design
  IV. Proposed logical system
       A. Data flow diagrams
       B. Data store description
       C. Transform descriptions
       D. Data dictionary
          1. Data flow definitions
          2. Data store definitions
          3. Transform definitions
   V. System design requirements
       A. System organization
       B. Input/output
       C. Database and data file
       D. Computer program
          1. Processing constraints
          2. Unit controls
          3. Unit test plans
       E. Processing control
  VI. Schedule and budget
 VII. Hardware and software
VIII. User's guide

FIGURE 8–11 Suggested table of contents for logical design specification

Executive Summary

The executive summary describes the problem investigated and the purpose of the study. It lists the processing alternatives and briefly considers the proposed design and its impact on the organization. Remember the reasons for summarizing the purpose and the results of the study *before* considering any details? Much like the executive summary prepared for the project specification (see chapter 4), the executive summary here introduces managers to the significant recommendations reached by the study; it motivates managers to read the balance of the study report, which by this time is quite lengthy.

Current Logical System

This description is an extension to the system described by the project specification. Besides showing the main system components and how these components are interrelated, it should specify the problems discovered by the analyst—including those discovered during data collection and analysis and data organization and documentation. To aid in this presentation, the analyst should use a context diagram and a level-0 diagram, at the very minimum. These diagrams help document the problems associated with the current system.

Feasibility Analysis

As discussed in the last chapter, the feasibility analysis describes the various logical designs that might be used in altering the current system and specifies which logical system is considered the best or most feasible. This section is often difficult to prepare, especially if several alternatives have been considered. One reporting format is to summarize the possible design alternatives, describing in detail the structure of the most feasible design in the next section of the report, and placing the details of the alternative designs in an appendix to the report. Another useful format is to clearly state the planning assumptions made in conducting the feasibility study. In this way the analysis team is somewhat protected from criticisms such as, "Why didn't you consider this alternative?"

Following the guidelines of the project specification, the feasibility section should clarify the action to be taken (the proposed new system), the probable risks (the costs and the probability of success of the new design), and the implications to be realized (the benefits and side effects) once the new system becomes operational. Decision makers require such clarification in their determination of whether to proceed with systems design; designers need this information to appreciate the risks and benefits of the proposed system.

Proposed Logical System

The main part of the logical design specification typically involves a complete description of the proposed logical system. As indicated by Figure 8–11, the proposed logical system is fully documented, including a complete set of DFDs, data store descriptions, and transform descriptions. All words used in the DFDs and data store descriptions are included as entries in the data dictionary; all keywords used in writing the transform descriptions are placed in the data dictionary. When completed, the proposed logical system description tells what the system will do, is graphic, is top-down (is partitioned from top to bottom), and is nonredundant. With such a reporting format, users and designers alike can quickly form a complete picture of what the proposed design can be expected to do.

System Design Requirements

System design requirements provide the analyst's view of how the proposed system might be organized (e.g., packaged), together with a preliminary set of technical specifications—I/O, database and data file, computer program, and processing control specifications. To aid in this presentation, program processing menus (showing possible job steps), system flowcharts, sample I/O layouts, database and data file layouts, and DFDs modified to show audit trails are recommended. Some analysts prefer to combine transform descriptions with computer program design requirements. This would modify the outline shown on Figure 8–11 as follows:

  C. Transform descriptions
     1. Structured English
     2. Processing constraints
     3. Unit controls
     4. Unit test plans

More important, this revised outline begins to show how the logical and technical aspects of a system design begin to blend.

Schedule and Budget

This section describes the project management portion of the specification. It compares estimated times and costs, prepared for the project specification, with the actual amounts required by the systems analysis team and provides detailed time and cost estimates for the design and implementation stages of systems development. Since resource requirements tend to be more substantial during these next two stages of development, a visually aided presentation is encouraged. For example, both time and cost schedules are placed in this part of the specification. Time schedules are similar to the project planning and control charts considered earlier (see chapter 1). Cost schedules include a revised estimate of direct and indirect costs and a recalculation of return on investment (also see chapter 1).

Since much more is known about the project following systems analysis, the revised time and cost schedules prepared at this point should show less variation than those contained in the earlier project specification. As discussed later in this chapter, estimates that are close to actual figures are mandated by many data processing organizations. If user groups and managers learn to trust the time and cost figures presented by the data processing staff, they are more likely to approve the continuation of systems development efforts.

Hardware and Software

So far little has been said about the hardware and software required for a project. This was not an oversight, because it is often the systems designer who finalizes hardware and software requirements.

What the systems analysis team is expected to do is make their findings known regarding hardware and software, much like the findings suggested for systems design. At the very minimum, the team is expected to address such questions as:

1. What types of hardware and system software will be needed to transmit, transform, and store the data specified by the logical design?
2. Will current hardware and system software be able to handle the specified processing requirements, or will hardware or systems software need to be purchased?
3. What are the sources, the availability, and the cost of hardware or systems software? Can hardware or systems software be leased, or must it be purchased?
4. Is application software available, and if so, what are the sources and the costs of buying rather than building parts of the proposed design?
5. What are the potential problems of buying rather than building the design?
6. What are the likely hardware, systems software, and application software interfacing problems? Will existing computer programs need to be rewritten to run on new equipment?

These and similar questions suggest that the hardware and software section of the structured specification should contain two sections: one for hardware and another for systems and application software. For the hardware specification, the analysis team is advised to begin with the minimum computer configuration that will produce all system outputs. It is then possible to modify this configuration to show what steps will be needed to handle any predicted growth of the system over a two- to-four-year period. Generally, the configuration presented should be broken down into required units, such as units of main memory, disk drives, tape drives, line printers, and computer terminals; it should specify the required speed of disk and tape drives, line printers, and computer terminals.

For the software specification, the analysis team should indicate the operating system and the programming languages to be supported, any required utilities, and file or database software requirements. The application software section should describe the features of any product that is to be purchased and its cost. In addition, the add-on costs of modifying purchased application software should be estimated and included in this section of the specification.

User's Guide

A final section to be included in the structured specification is the user's guide. Some of you might ask: "Why deal with the user's guide at this stage? Why not wait until the design is better defined, or even wait further until after the design has been tested and partially implemented?" Actually, there are several good reasons for beginning to suggest ways in which the user's guide might be organized at this early stage. Since by this time the analysis team members have already spent considerable time working with user groups, it is often possible for them to specify concisely the user's training needs along with instructions for the eventual use of the proposed system.

Consider the scheduling system discussed earlier (see Figures 8–5 and 8–8). The analysis team designing this system might determine that two types of user materials will be needed in support of the technical design: a user's guide, for the person using the terminal scheduler; and an administrator's guide, for the person responsible for the setup and maintenance of the scheduler. **Figure 8–12** outlines the steps important to each guide. The user's guide represents what is commonly known as *user documentation*. It deals with how the individual will use the system once it becomes operational. The administrator's guide is one type of *operations documentation*. It explains how the software is to be operated and maintained—that is, it includes provisions for maintaining system files and changing system parameters.

DESIGN SCHEDULE AND BUDGET

In preparing the schedule and budget section of the structured specification, the analysis team is usually advised to 1) include a summary of the projected-to-actual times and costs for systems analysis and 2) work with the systems designer to prepare the systems design and implementation schedule and budget. The first of these steps is easy to complete, provided the analyst begins with a project planning control chart and

Volume 1: User's Guide

1. Introduction to the computer terminal scheduler
2. How to use the manual
3. How to schedule time
    a. Reviewing the current schedule
    b. Reserving terminal time
    c. Canceling a reserved time
4. How to confirm your schedule
5. Error messages and what they mean

(a)

Volume 2: Administrator's Guide

1. Loading the computer terminal scheduler
2. Printing user schedules
3. Updating master files
4. Changing system parameters
5. Error messages and what they mean

(b)

FIGURE 8–12 User guide outlines for the user (a) and the system administrator (b)

has collected actual time and cost data. **Figure 8–13,** for example, compares projected to actual times and costs. The project, as indicated, took somewhat longer to complete than expected; however, costs were initially kept in check. Toward the end of the analysis, both time and cost exceeded the projected figures.

If the cost of systems analysis is greater than expected, there is a likelihood that the cost of systems design and implementation will also be greater than initially projected. Why? Generally the process of analysis reveals that the project is more complicated than it initially appeared. Even if costs are not higher than expected, two detailed schedules, one showing projected design times and another showing design costs, are often prepared for the structured specification. These schedules express the most current estimates of the time and resources needed to complete the proposed project.

Preparing the Detailed Time Schedule

The analyst prepares the detailed time schedule by first constructing a network chart and then transforming it into a chart that resembles a Gantt chart. A *Gantt chart* is a chart form named after Henry L. Gantt. It shows each project activity as a line or color-coded bar. Consequently, it is much easier to read and use than a network chart.

Let's consider a scheduling problem similar to the problem discussed in chapter 1 to demonstrate how a Gantt chart is prepared. Suppose we

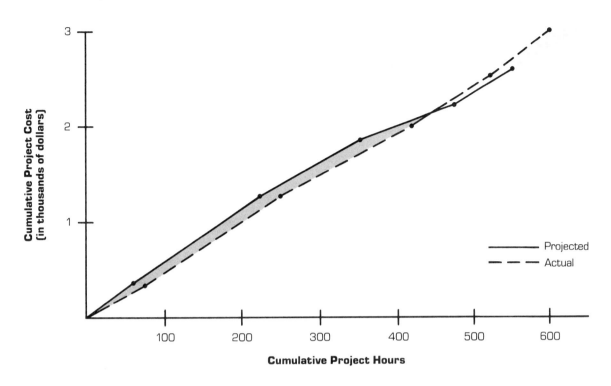

FIGURE 8–13 Over-budget and -time project planning and control chart

know that eight systems design activities are to be completed. As shown in **Figure 8–14**, these activities begin with the design of system output and conclude with the writing of the technical specification. Suppose we also know the node relationships between activities. These facts make it possible to draw the resulting network charts.

Before we continue, we should clarify the meaning of activity D on the figure, entitled *dummy activity*. This activity takes zero time to complete and represents a constraint on the system. In the example, activity F, the design of I/O computer programs, cannot begin before activity A

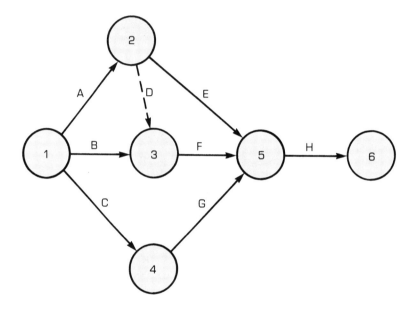

Activity	Node Relationship	Description	Time (in months)
A	1–2	Design of system output	2.08
B	1–3	Design of system input	2.50
C	1–4	Design of system files	2.50
D	2–3	(Dummy activity)	—
E	2–5	Design of system controls	1.88
F	3–5	Design of I/O computer programs	2.50
G	4–5	Design of file processing computer programs	2.50
H	5–6	Writing of technical specification	.63

FIGURE 8–14 Network chart with a dummy activity

(the design of system output) is completed. This constraint is thus quite reasonable. We need to know how system output will look before we can design the programs to produce it.

Next, let's determine the *critical path* through the network, or the minimum time needed to complete the project. We will begin by representing the first three activities of the network as horizontal bars on a Gantt chart (see **Figure 8–15a**). As indicated, activities B and C take the same time to complete, while activity A takes less time to complete. If we were to end our analysis at this point, we could conclude that there are two parallel critical paths, B and C, and that the minimum time to complete the project is 2.5 months. Activity A is not critical. Rather, activity A contains .42 (2.5 − 2.08) months of slack—that is, we could begin activity A .42 months later than activities B and C and still complete the project in 2.5 months.

Let's add the remaining activities to complete the Gantt chart. As shown by **Figure 8–15b,** the longest path, the critical path, is represented by the activity chain B, F, and H. The dotted line shown in sequence A–E represents slack, which is now increased from .42 months to 1.25 months (5.0 − 3.75). Sequence C–G is also shown to contain slack. This path's total slack time is .25 months (5.0 − 4.75).

At this point we might question: "What happens if activities are started as late as possible. Will the Gantt chart look the same?" The answer is no, as shown by **Figure 8–15c.** The connecting line from activity A to B (shown as D) indicates that activity A cannot be delayed by more than .42 months, because of the constraint set by Activity D.

Finally, let's redraw the network to show the expected time (te) for each activity, and the earliest possible time (T_E) and the latest possible time (T_L) for each event. This revised chart is called a *critical path method* (CPM) *diagram,* because it shows the critical path of the network (see **Figure 8–16**). However, because a CPM diagram is difficult to read, we also might prepare a *critical path predecessor table* (also see Figure 8–16). This table shows all events, predecessor events, earliest and latest start times, and slack, which is defined as $T_L - T_E$.

The way in which the earliest and latest start times are calculated can be traced once the CPM diagram and the predecessor table have been completed. T_E is calculated by adding the expected activity time (te) to the predecessor event time. When there are several activities, T_E is defined as the highest of these computer values. For example, event 5 has the following possible T_Es:

$$T_E(5) = te + T_E(2) = 1.67 + 2.08 = 3.75$$
$$T_E(5) = te + T_E(3) = 2.50 + 2.50 = 5.00$$
$$T_E(5) = te + T_E(4) = 2.25 + 2.50 = 4.75$$

Of these three, the highest time of 5.0 months is selected.

T_L is calculated by subtracting the expected (activity) time (te) from a subsequent event time. When there are several activities, T_L is defined

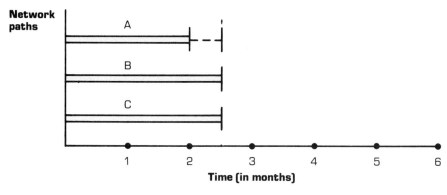

(a) First three activities

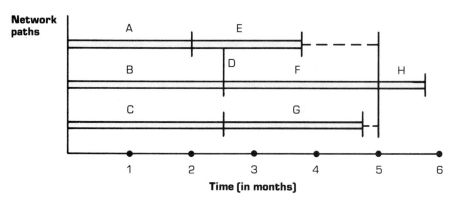

(b) Earliest possible start time (EPST)

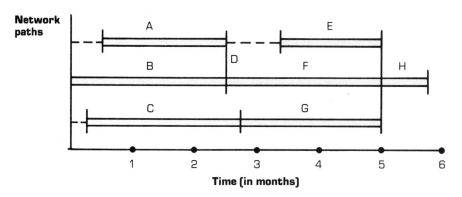

(c) Latest possible start time (LPST)

FIGURE 8–15 Gantt charts showing first three network activities (a), the completed network and earliest start times (b), and the completed network and latest possible start times (c)

Design Schedule and Budget **271**

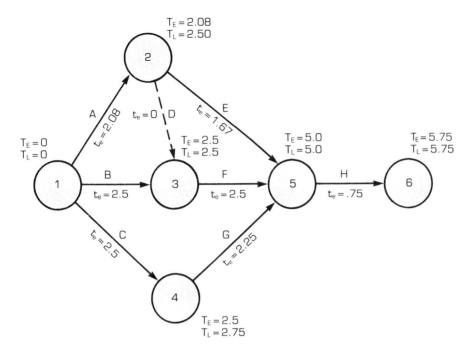

Event	Predecessor Events	T_E	T_L	Slack
1	—	0	0	0
2	1	2.08	2.50	.42
3	1,2	2.50	2.50	0
4	1	2.50	2.75	.25
5	2,3,4	5.00	5.00	0
6	5	5.75	5.75	0

FIGURE 8–16 CPM diagram and predecessor table

as the lowest of these computer values. For example, event 2 has the following possible T_Ls:

$$T_L(2) = T_L(5) - te = 5.0 - 1.67 = 3.33$$
$$T_L(2) = T_L(3) - te = 2.5 - 0 = 2.50$$

Of these two, the lowest time of 2.5 months is selected.

Preparing the Detailed Project Budget

Once the detailed time schedule has been completed, the work of preparing the detailed project budget becomes greatly simplified. As shown in **Table 8–1,** direct labor costs are based on the time required to complete the various design activities. Let's consider how one of these projected costs is calculated.

Activity A requires 2.08 months. These hours, multiplied by a staff size of 1.5, yield 525 projected hours, in which each person works an average of 168 hours per month.

Next, suppose that a systems designer is expected to work full-time on activity A and a senior programmer is expected to work half-time. Suppose also that the systems designer is paid $16 per hour and that the senior programmer is paid $13 per hour. The projected cost for the design of system output then becomes

$16 (168 hours/month × 2.08 months) + $13 (84 hours/month × 2.08 months) = $7,875

Besides helping to determine the direct labor costs for the design of a project (that is, the costs that can be directly associated with design ac-

TABLE 8–1 Design cost spreadsheet for direct labor

Activity	Description	Time (in months)	Staff Size	Projected Hours	Projected Costs
A	Design of system output	2.08	1.5	525	$ 7,875
B	Design of system input	2.50	1.5	630	9,450
C	Design of system files	2.50	2.0	840	14,700
D	Dummy activity	0	0	0	0
E	Design of system controls	1.67	1.0	280	5,600
F	Design of I/O computer programs	2.50	2.5	1,050	18,375
G	Design of file processing computer programs	2.25	1.5	565	8,475
H	Writing of technical specifications	.75	2.0	250	4,375
Totals		14.25		4,140	$68,850

tivities), detailed project budgets generally include an estimate of indirect labor costs, computer costs, and project overhead.

Indirect labor costs are costs that cannot be measured directly, but nonetheless should be accounted for. The cost of the clerical staff in support of a project represents an indirect labor cost.

Computer costs are the costs associated with the use of computing equipment. Generally a flat machine rate is charged for computer time, such as $100 per hour. The analyst must determine the number of hours of machine time needed for each design activity and multiply this number by the machine rate.

Project overhead involves the cost of administering the project. The systems manager's salary, for instance, would be included in this category. So might expenses for office space, supplies, and so forth.

Collectively, the detailed total budget might look like the following:

Project XY Design Costs:
Direct labor costs (1)	$68,850
Indirect labor costs (2)	10,327
Total labor cost	$79,177
Computer costs (3)	1,500
Total labor and equipment cost	$80,677
Projected overhead (4)	7,918
Total design costs	$88,595

Supporting notes would accompany each figure shown in the budget, such as the following:

Notes:
1. See the design cost spreadsheet for a breakdown of direct labor costs.
2. Indirect labor costs are based on 15 percent of direct labor costs.
3. Computer costs are estimated at fifteen hours multiplied by $100 per hour.
4. Project overhead is estimated at 10 percent of total labor cost.

The schedule and budget section for the logical design specification cannot be treated casually by the systems analyst. Because many design projects require extensive company resources, this is one section that managers responsible for approving projects should read carefully. What happens when managers believe that costs are too high? When this occurs, tradeoffs in design are generally required. A *tradeoff* involves giving up part of the design in return for completing the project faster and at a lower cost. For example, the logical specification might include the design of several management reports, some of which are critical to the success of the project, and others of which are classified as nice to have.

If the latter reports are excluded from the design, total design hours could be reduced, say by 200, with a corresponding reduction in cost of $3,743. This reduction is calculated as follows:

Direct labor (200 hours at an average cost of $14.50/hour)	$2,900
Indirect labor (.15 of direct labor cost)	435
Total labor cost	$3,335
Computer cost (.05 of 15 hours × $100)	75
Total labor and equipment cost	$3,410
Project overhead (.10 of total labor cost)	333
Total project cost reduction	$3,743

This process of dropping items from the design would continue until either design costs were acceptable or the proposed design were rejected as too expensive.

Besides permitting tradeoffs, a detailed estimate of time and costs enables managers to set aside resources for the balance of systems design, before any work is done. Many systems analysts consider this budgeting of design funds to be the only way in which they know that their logical system specification has been accepted. In effect, managers have communicated to the analysis team, "We have set aside the funds needed to design the proposed system. Let's move ahead."

The Mansfield, Inc., Case Study

Roger, Carolyn, and Neil worked together in writing the structured specification describing the new billing and receivables system. They also called in John Seevers, the lead systems designer, to critique and generally help them with their work.

Roger reminded the group of the task to be completed. "We cannot lose sight of who will be reading our specification," he remarked. "Besides our boss (Ken Westerling), we might expect two or more vice-presidents to read our report. I know for a fact that David Orring (the vice-president of finance and accounting) plans to read the report, and since the proposed system contains features of interest to marketing, we might expect Tony Chung to be interested in our recommendations."

Structured Specification Outline

In preparing the outline for the structured specification, Roger insisted that the group prepare an executive summary section first, followed by a description of the current logical system, the results of the feasibility study, and the description of the proposed system. They were also to work closely with John Seevers in preparing their preliminary design specifications, the detailed sched-

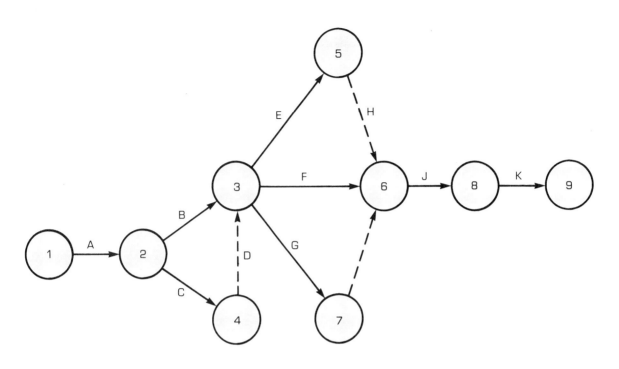

Activity	Node Relationship	Description	Time (in weeks)
A	1-2	System organization	2
B	2-3	Design of system output	4
C	2-4	Design of system input	2
D	4-3	(Dummy activity)	0
E	3-5	Design of system files	8
F	3-6	Design of interactive programs	15
G	3-7	Design of batch programs	10
H	5-6	(Dummy activity)	0
I	7-6	(Dummy activity)	0
J	6-8	Design of system controls	3
K	8-9	Writing of technical specifications	4

FIGURE 8–17 Network chart for the design of Mansfield's proposed system

ule and budget, hardware and software requirements, and the preliminary version of the user's guide.

Carolyn and Neil questioned Roger about the feasibility study section. "Did we conduct such a study?" Carolyn asked. "Or did we pick only the design alternative that looked best?"

Roger responded: "Actually, I did look at various alternatives, including using an outside service facility to handle all of our billing and receivables. My analysis led me to conclude that revising our current system was the most economical thing to do."

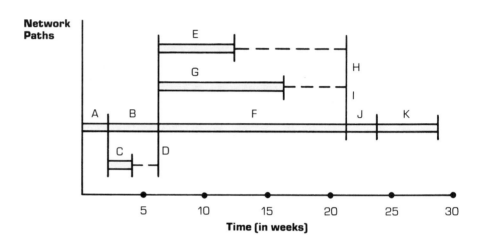

Event	Predecessor Event	T_E	T_L	Slack
1	—	0	0	0
2	1	2	2	0
3	2,4	6	6	0
4	2	4	6	2
5	3	12	21	9
6	3,5,7	21	21	0
7	3	16	21	5
8	6	24	24	0
9	8	28	28	0

FIGURE 8–18 Gantt chart and predecessor table for Mansfield's proposed system

Detailed Schedule

John Seevers was asked to work with Carolyn to produce a detailed schedule for the proposed billing and receivables system. **Figure 8–17** illustrates the network chart that followed from this joint effort. Carolyn explains: "We decided to spend the first two weeks organizing the design team and finalizing the structure of the new system before moving on to work on more detailed design steps. Once we're past the planning stage we will start work on output and input design, completing all of these designs before concentrating on the design of system files and computer programs."

She continued: "We divided the program design work into two categories: interactive programs and batch programs. Since we must design approximately twenty programs in total, the total time of twenty-five weeks to complete all program design seems reasonable."

John added a final note: "After all file and program design work is complete, we plan to design system controls and write the technical specification."

Figure 8–18 illustrates the Gantt chart and the predecessor table that follows from the network chart prepared by John and Carolyn. "With activities which do not lie on the critical path," he explained, "we show a fair amount of slack. This will give the design team some flexibility. We should be able to give the younger staff a chance to work on batch program and file design, while keeping them away from the critical path."

SUMMARY

The logical design specification is the major design document produced by the systems analysis team. It explains the logical features of the proposed system; it describes how user requirements can be met following its scheduled implementation. In addition to describing the proposed system, the specification provides five sets of preliminary design requirements, indicates how these technical requirements are related to the logical design, and illustrates the detailed time and cost schedule for systems design and implementation.

Systems design requirements, though preliminary, consist of system organization, I/O, data file and database, computer program, and processing control suggestions. Of these, system organization helps clarify how a system is to be packaged, while I/O documents the forms and screens appropriate to entering data into processing or in displaying and printing the results of processing. Data file and database requirements depict volume and frequency-of-use considerations, while computer program requirements identify processing constraints, unit controls, and unit test plans. Processing control requirements specify the necessary steps for auditing the correctness of processing.

Writing and presenting the logical design specification represent the end product of systems analysis. The specification provides an in-depth description of the favored system, as well as a historical record of the project—explaining which actions were taken and why, and which steps led to project conclusions.

Preparing the detailed design schedule and budget for inclusion in the logical design specification inspires forward thinking about the technical design stage of systems development. This section represents the project management portion of the specification. The detailed time schedule is often used to determine the minimum time needed to complete the design and implementation of the project (the critical path), while the detailed cost schedule examines funds expended to date and projects the budget for the remainder of the project.

This chapter concludes the systems analysis portion of the text. Four principles emphasized by this chapter and reinforced by the previous five help summarize what we mean when we speak of systems analysis as a structured and managed process. These principles are:

1. Systems analysis consists of a planned set of activities, beginning with the identification of user problems and ending with the preparation of the logical design specification.

2. Systems analysis employs a structured methodology in its practice to provide structured descriptions of current and proposed systems.

3. Systems analysis calls for the identification of preliminary systems design, hardware and software, and user guide requirements, even though the final specifications for each will be determined in later stages of development.

4. Systems analysis is a carefully managed endeavor; it includes such practices as the step-by-step comparison of planned and actual times and costs for determining systems requirements, collecting current and proposed system data, organizing and documenting those collected data, conducting the feasibility study, and preparing the logical design specification.

REVIEW QUESTIONS

8–1. What types of specifications are contained in the preliminary systems design requirements section of the logical design specification?

8–2. Explain what is meant by the packaging of a system.

8–3. What is the function of the central transform in a transform-centered design?

8–4. What does a transaction-centered design depend on?

8–5. How are I/O requirements documented in the logical design specification?

8–6. How are computer program requirements described in the logical system specification?

8–7. What is the purpose of an audit trail?

8–8. What are the main sections of the logical design specification?

8–9. What two types of user's guide materials are specified by the systems analyst?

8–10. How does a Gantt chart differ from a network control chart?

8–11. What is a dummy activity?

8–12. What types of cost are shown on a detailed project budget?

8–13. What is meant by a tradeoff in design?

EXERCISES

8–1. Review the employee pay system described in chapter 3 and answer the following:

- What type of transaction triggers this system (see Figure 3–4)?
- What job steps would be required for the program, Update Employee Master File (see Figure 3–20)?
- What job steps would be required to print the payroll register (see Figure 3–19)?

8–2. Suppose that you have placed five overtime pay records in a file. Suppose further that the number of overtime hours in this sample is equal to fifteen and that the overtime pay total, once calculated, should equal $250. Design a test plan to check the following procedure, given the information stated above.

```
Procedure
DO-WHILE there are overtime-pay records
    Read a record.
    IF end-of-file is reached
        THEN Display a termination message, over-
            time-hours-total, and overtime-pay total.
        ELSE Add overtime-hours-worked to over-
            time-hours-total.
    Add overtime-pay to overtime-pay total.
    END-IF
END-DO-WHILE
```

8–3. Suppose that a procedure is to be designed to enter quantity-shipped totals into processing. In structured English this procedure appears as

```
Procedure
Enter quantity-shipped.
Multiply quantity-shipped by item-cost to arrive
 at item-extended-cost.
```

First, add built-in controls to this structured English procedure to make it impossible to enter a quantity-shipped total that is zero or less or greater than 999. Second, list the key terms in this revised procedure to be placed in the data dictionary.

8–4. Determine T_E and T_L for the two network charts shown below.

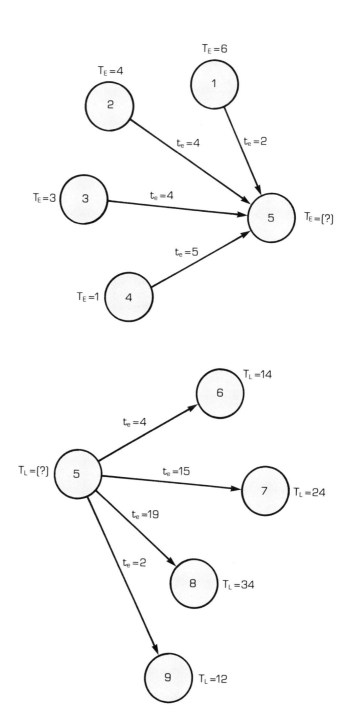

8–5. Construct a predecessor table for the network diagram shown below. What is the critical path?

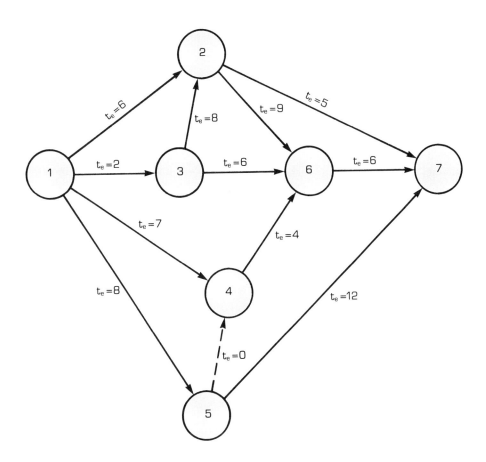

8–6. Determine the total estimated cost for a project based on the cost table shown on page 282. In addition to knowing the time (in months) and the staff size for each activity, you know that
- each staff member works 168 hours per month;
- the designer is paid $20 per hour and the programmer is paid $15 per hour;
- indirect labor cost is 10 percent of direct labor cost;
- computer time is $5,000; and
- project overhead is 20 percent of total labor cost.

Activity	Description	Time (in months)	Staff Size Designer	Programmer
A	Design of system input and output	1.5	1.0	.5
B	Design of system files	1.0	1.0	—
C	Design of system controls	.75	1.0	—
D	Design of computer programs	4.0	1.0	3.0
E	Writing of technical specifications	.5	1.0	1.0

8–7. Suppose the costs for the project outlined in exercise 8–6 are too high. The analyst states that the project cost can be reduced by more than $14,500 if the time to program the new system is reduced by one month (from four months to three) and the cost of computer time is reduced by 10 percent. Is the analyst correct? Prove your answer.

WORLD INTERIORS, INC.—CASE STUDY 8
DESIGNING PROJECT CONTROL CHARTS

Introduction
WI's decision to design an improved method of processing customer orders and analyzing retail store sales was combined with the decision to acquire a larger computer. Before putting these decisions into action, however, Rena and Ray insisted that John Welby develop a method of project control. Such a method would divide the total work effort into stages. They insisted that each stage would have to be unique. This would clarify each objective (mission) and the point at which it could be evaluated (its milestone).

Project Control Chart
In response to the Logens' request, John began to break down the activities important to systems design. He began with a listing of activities, followed by their relationship to one another, their description, and their time and cost to complete (see table, page 283).

CASE ASSIGNMENT 1
Prepare a critical path method (CPM) diagram showing the paths that make up the design network. For each node show T_E, the sum of the expected activity times, and T_L, the latest allowable time. For each activity, show te, the expected activity time. Finally, show the critical path on your completed network.

CASE ASSIGNMENT 2
Prepare a table of CPM calculations showing, for each node, the predecessor node, T_E, T_L, and the slack. Prove that the critical path shown on the network diagram is the same as the critical path indicated by the table.

CASE ASSIGNMENT 3
Prepare a Gantt chart from the network diagram prepared for assignments 1 and 2. Once again, prove that the critical path shown by the chart is the same as the critical path on your network diagram.

World Interiors, Inc.—Case Study 8

Activity	Node Relationship	Description	Time (in months)	Cost
A	1–2	Program specifications	2	$6,000
B	2–3	Program coding	3	7,000
C	3–12	Program testing	2.5	6,000
D	12–13	System testing	2	5,000
E	2–4	Database specifications	1	$3,000
F	4–5	File conversion	2	5,000
G	5–10	Database testing	2	5,000
H	1–6	Equipment specifications	1.5	$4,000
I	4–6	(Dummy variable)	0	0
J	6–7	Vendor selection	1.0	2,500
K	7–10	Installation and testing	1.5	4,000
L	2–8	Procedure specifications	1	$3,000
M	8–9	Training specifications	1	3,000
N	9–10	Preinstallation training	2	5,000
O	9–11	System documentation	1.5	4,500
P	11–12	Final user documentation	1	3,000

9

System Organization

INTRODUCTION

THE acceptance of the logical design specification by users and managers leads to the second major stage of systems development—namely, the stage of systems design. The main work of systems design lies in translating logical design requirements into a software plan. This plan, called the *technical design specification,* documents the physical features of processing. It shows how computer programs are to be designed and how all input, output, files and databases, and processing controls are to be made operational.

The technical design specification is much more exacting than the logical design specification. For example, it features structure charts instead of DFDs. A *structure chart,* as defined in chapter 3, shows the internal organization, or structure, of a computer program. The greatest virtue of such charts is that they greatly simplify the writing, testing, and understanding of computer programs. In addition, the technical specification contains pseudocode rather than structured English in describing processing procedures. *Pseudocode,* or false code, is a method of specifying the logic to be followed in the design of a computer program. Although resembling actual source code, it does not conform to the syntactical rules of any particular programming language. Finally, the technical specification provides mockups of input and output screen and report layouts, the specific contents of physical computer files, and the types of processing controls. All parts of the specification thus show the exact features of processing. In this way, the technical design specification differs from the logical specification, which is limited to suggested features of processing.

This chapter begins with the subject of the design process and the development of a preliminary technical design. The preliminary design involves the question of packaging, which as stated in the last chapter indicates how the various components of a logical system are to be organized as a set of technical units. A preliminary design also considers the hierarchical structure of a system. By examining the hierarchy of a system, the designer can break down a complex system into manageable pieces. The main objective of this chapter is to provide an overview of the design process and to describe how hierarchical input-process-output (HIPO) charts and structured walkthroughs assist designers in formulating preliminary design specifications. When you complete this chapter, you should be able to

- describe what is meant by a preliminary functional design;
- transform either a transaction-centered or a transform-centered DFD into a hierarchical design;
- state the main objectives of the HIPO technique and construct a set of HIPO diagrams;
- describe how structured walkthroughs help designers explain the details of a design; and
- conduct a structured walkthrough.

THE DESIGN PROCESS

Pressman[1] defines the design process as consisting of four distinct steps: preliminary design, detailed design, coding, and testing. The *preliminary design* consists of devising a design that exhibits a hierarchical organization, consists of modular parts, and permits stepwise refinement (e.g., decomposition) of different levels and parts. The purpose of such a preliminary design is twofold:

1. to produce a design document, showing how a software product is to be structured, and
2. to allow for a review of the design before work on the detailed design is begun.

In contrast to the preliminary design, the *detailed design*, provides a blueprint for the coding and testing of computer programs and system components. It consists of a detailed specification with sufficient information to allow a programmer to produce usable software.

Coding, the third step, is the process of translating the detailed design into machine-readable instructions. A programmer might use a higher-level programming language, such as Pascal or COBOL, to prepare these instructions, which when complete become a computer program.

The fourth step is *testing*, which involves the execution of coded instructions on a computer in an attempt to discover whether all instructions are accurate and complete, and correctly placed in sequence. Testing is performed to uncover errors or bugs in a computer program.

Figure 9–1 shows the interrelationships between the four steps in the design process, coupled with a final package called the *maintenance package*. Logical design requirements serve as the main input to the preliminary design. This initial package is prepared by a design group to provide a high-level functional design. Following the review of the preliminary design, a detailed functional design is prepared. This second design leads to a complete, accurate, and easy-to-understand design specification. The coding and testing package represents the first attempt to implement the design. Design reviews at this point include program

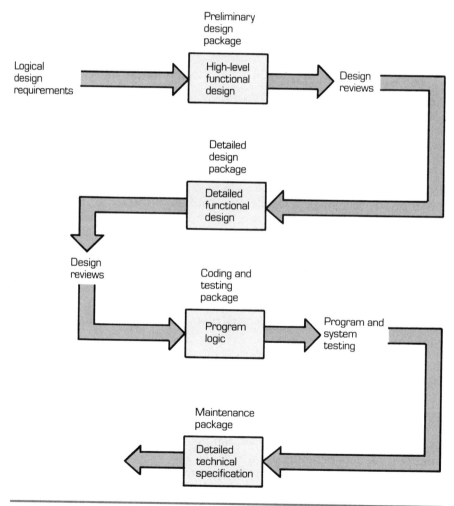

FIGURE 9–1 Four types of design packages

testing to determine whether programs work as expected and *system testing* to determine whether the system performs as expected. Finally, the maintenance package consists of materials describing how changes are to be made to the system once it becomes operational. Since the people responsible for maintaining a system are often not those who were responsible for its design, documentation must exist so that maintenance personnel can understand the upkeep of the system.

Preliminary Functional Design

Preparing the preliminary functional design involves making two assumptions, namely that a system can be organized into a hierarchical structure of functions and that the scope of the system can be depicted by the uppermost levels of such a structure. Once again, consider the uppermost levels of Mansfield's simplified payroll system (see **Figure 9–2**). The scope of this system is defined by three top-level functions: update the hours worked, print the payroll register, and print employee paychecks.

By now you may be asking, "How are top-level functions defined? Is there a particular methodology that helps the designer transform a DFD into a hierarchical structure of functions?" The next two sections deal with these questions. The first section examines evolving a structure from a transform-centered design; the second examines evolving a structure from a transaction-centered design.

EVOLVING A TRANSFORM-CENTERED DESIGN

Suppose our task is to transform the DFD shown on **Figure 9–3** into a hierarchical structure. Since this DFD represents a transform-centered design, our first step is to identify the area of central transform. Following this, our second step is to construct a level-0 structure chart by lim-

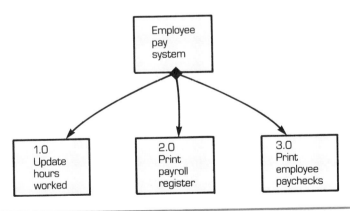

FIGURE 9–2 Payroll level-0 structure chart defining the scope of the system

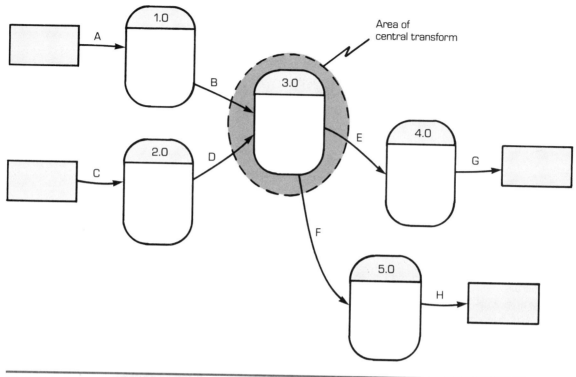

FIGURE 9–3 DFD showing the area of central transform

iting the functions shown on the chart to those corresponding to the area of central transform. **Figure 9–4** illustrates a level-0 structure chart to demonstrate how the area of central transform is organized as a structure of functions. For the preliminary design, these functions can be limited to input, process, and output descriptions, which are shown as:

1. *Get* (in<u>p</u>ut) a data flow.
2. *Make* (pro<u>c</u>ess) something using one or more data flows.
3. *Put* (out<u>p</u>ut) a data flow.

With this understanding, Figure 9–4 shows that the system will first get B and D, the two data flows that enter the area of central transform. Next, the system will make E and F, the two data flows that leave the area of central transform. Last, the system must put E and F once they are made.

Figure 9–5 shows the expanded structure chart, beginning with level-0 and showing all lower levels. Once again, the functions are limited to gets, makes, and puts. This expanded chart clearly illustrates the underlying principle of a transform-centered design: the uppermost level describes the structure of the area of central transform.

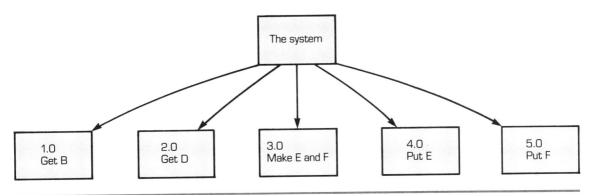

FIGURE 9–4 Level-0 structure chart showing the main gets, makes, and puts of a system

Often, the area of central transform will consist of several transforms shown by the DFD. **Figure 9–6** provides such an example. In this example—let's call it an order filling system—the first transform suggests a get function while the last two imply put operations. Transforms 2.0, 3.0, and 4.0, however, all suggest make functions. Verbs like *determine, compute, add, evaluate, process,* and so forth indicate that some form of make operation is required.

Figure 9–7a illustrates the level-0 structure chart derived from the logically defined order filling system. By including several transforms within the area of central transform, the top-level structure chart continues to look surprisingly simple. However, as shown in **Figure 9–7b,** the structure of the central transform becomes more complex once the top-

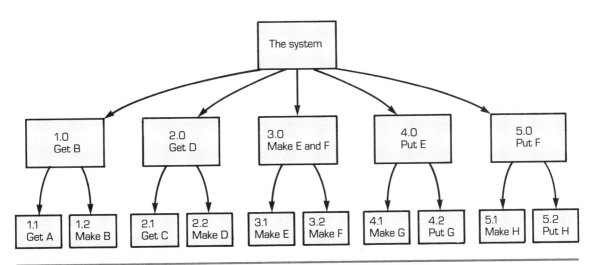

FIGURE 9–5 Expanded structure chart showing the complete functional hierarchy

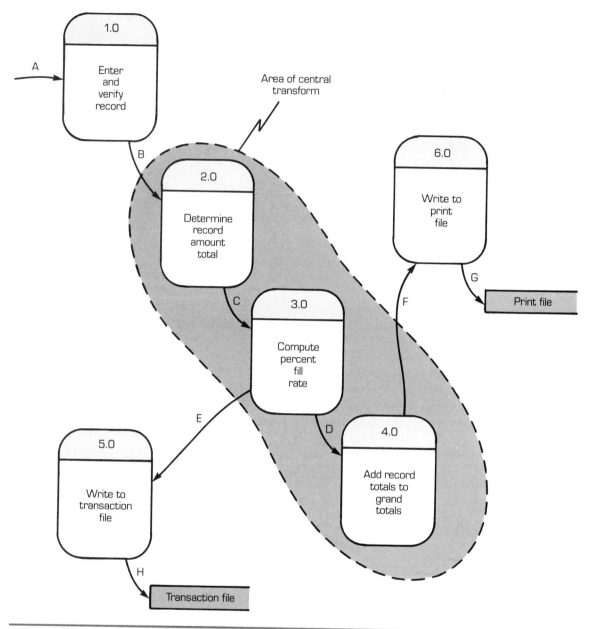

FIGURE 9–6 DFD showing a more complex area of central transform

level system is decomposed. In this system, C must be made, followed by D and E. Only then can the system make F.

Figure 9–8 attaches data couples to the modules shown on Figure 9–7b to test the logic of this level-2 structure chart. A *data couple*, as

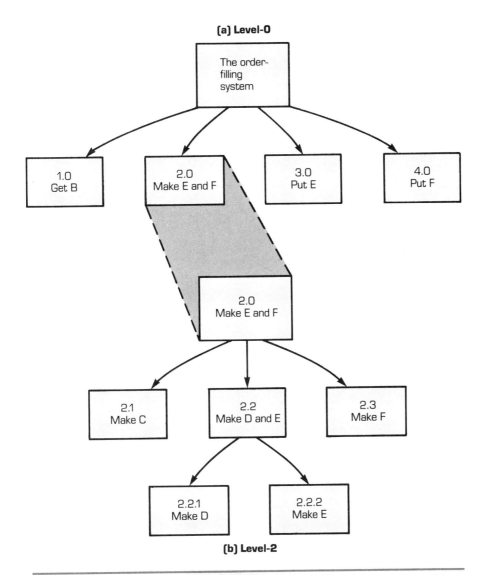

FIGURE 9–7 Structure chart showing the level-0 module (a) and the modules within the area of central transform (b)

stated in chapter 3, is represented by a small arrow with a white tail. It defines the information to be passed between a high-level module, known as the boss, and a subordinate module, known as a worker. The boss module, 2.0 in this instance, receives data couple B, which it then sends to worker 2.1. Worker 2.1 uses B to make C, which is returned to the boss module 2.0. The boss accepts C and sends it to a second worker, 2.2. This worker receives C and returns D and E, which are sent by two

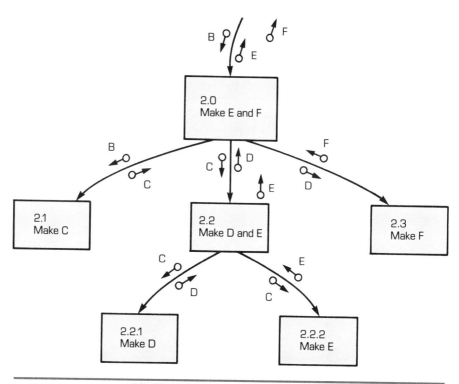

FIGURE 9–8 Area of central transform showing both modules and data couples

subordinates of 2.2, modules 2.2.1 and 2.2.2. Finally, the boss (module 2.0) needs a third worker, 2.3, to receive D, make F, and return F.

EVOLVING A TRANSACTION-CENTERED DESIGN

The process of transforming a transaction-centered design into a hierarchical structure of functions is much like the process associated with creating a transform-centered design. Consider the DFD shown in **Figure 9–9**. The transaction center, as indicated, receives input A from a source; it transforms A into B and places B in a data store. Once B is stored, various actions can take place. For example, four transforms can be used to make G, H, I, and J.

Figure 9–10a shows the level-0 structure chart for this transaction-centered design. The main system might be called a *driver program*, or *superordinate module*. As shown, the main system activates one of several subsystems. The diamond-shaped symbol designates that the main system must select from one of several possible actions—that is, it is possible to invoke one of several lower-level modules.

Figure 9–10b shows the structure of subsystem 1, which is limited to get A, make B, and put B. This figure should lead you to conclude that

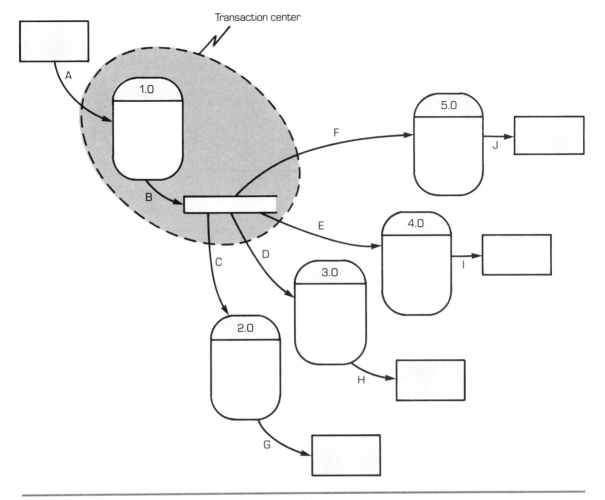

FIGURE 9-9 DFD showing components of the transaction center

subsystem 1 is nothing more than a simple transform-centered design and that transform-centered designs are contained within a transaction-centered design.

Finally, the horizontal arrow leading from 1.3 to 1.1 in Figure 9–10b reminds us that subsystem 1 is to be represented as an iterative process. Modules 1.1, 1.2, and 1.3 are processed repeatedly, such as once per new transaction. This process might be described as follows:

```
REPEAT
     1.1  Get A
     1.2  Make B
     1.3  Put B
UNTIL a signal is received to stop.
```

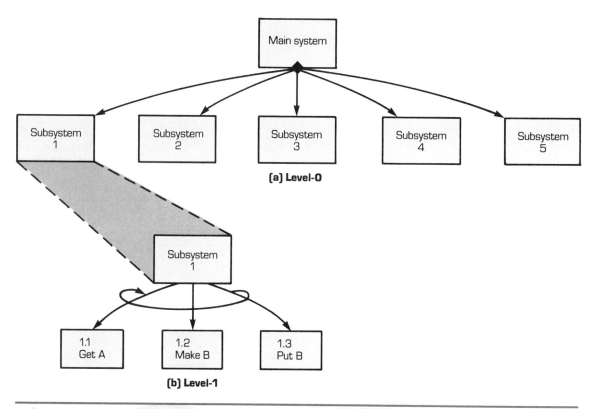

FIGURE 9–10 Structure chart showing the level-0 subsystem (a) and the modules within the transform center (b)

Systems Design: The Process

Creating top-level structure charts from transaction- and transform-centered logical designs is a main step in preparing the preliminary design. They begin to show how program structure is related to the flow of data in a system. They also help separate data input and output functions from data processing functions.

Nonetheless, the overall process of systems design consists of more than preparing structure charts. As shown in **Figure 9–11,** the design process begins with system organization—at least in its preliminary form. Once the structure of the software design is known, the more detailed work of I/O design, file and database design, program design, and processing control design begins.

Activities that run parallel to detailed design steps also enter into the process. As the figure shows, three separate specifications, the equipment specification, the test specification, and the user-interface specification, must also be completed. The initial forms of all three specifications were found in the logical design specification. What the systems

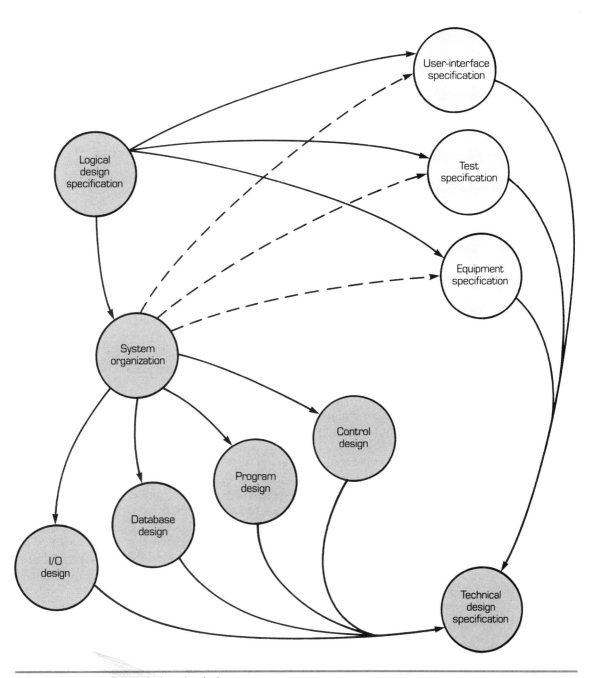

FIGURE 9-11 Systems design: the design process

designer must do is refine and expand these specifications, as necessary, to make each one as concrete as possible.

Let's consider briefly the purpose and contents of each of these separate specifications:

1. *Equipment specification*—an organized plan for obtaining the hardware and software required in support of a new design. While the plan will vary from one project to another, it typically contains three sections: equipment needs analysis, vendor selection procedures, and installation and testing procedures. An *equipment needs analysis* involves the determination of whether existing hardware and software can support the new system or whether additional hardware and software are needed. *Vendor selection procedures* deal with steps to be taken when hardware and software must be acquired. These procedures specify how requests for information from vendors are to be written; they also indicate how bids from vendors are to be obtained and evaluated. *Installation and testing procedures* deal with testing new equipment once it arrives from the vendor. Some specifications call for full field testing—that is, the equipment must handle typical day-to-day processing without error. The equipment must pass this test before it can be considered operational.

2. *Test specification*—an organized plan for testing a new system. A complete test specification should include test plans and policies, a test strategy, test points and test runs, criteria for determining successful test runs, names of personnel to test the system, and time and cost projections for testing. Most designers are aware that testing requires considerable time to plan and complete. Accordingly, a test strategy generally involves a combination of unit testing (e.g., the testing of individual program modules), subsystem testing, and main system testing. In addition, the new system will be tested by users as one of the final steps in deciding whether it is acceptable.

3. *User-interface specification*—an organized plan for helping end users understand the new system. This plan should cover user education policies and guidelines, user education strategy, user reviews of the design, training needs and programs, and user documentation. Of these materials, user reviews of the design are generally the most crucial. They require the user to determine whether program functions are designed correctly and whether they are consistent with user needs outlined in the logical design specification.

CONSTRUCTING HIPO CHARTS

Design teams usually begin their preliminary analysis by trying to visualize the hierarchy of functions to be built into a design and to specify the inputs to and outputs from each function. One set of materials developed to encourage this type of design work is called HIPO (hierarchical/input-process-output). The main objectives of this technique are to provide[2]

1. a method of illustrating the hierarchical structure of system functions;
2. a visual description of inputs to each function and outputs produced by each function; and
3. an extended description of processing, as required by each major function.

The HIPO technique uses three types of diagrams to document the structure of a system: a visual table of contents, overview HIPO diagrams, and detail HIPO diagrams.

Visual Table of Contents

The first step in the design process is to create a *visual table of contents* (VTOC) to document the main functions to be performed by a new system, the relationships between functions, and the hierarchical structure of the new system. Sound familiar? A VTOC, as we will soon see, closely resembles a top-level structure chart.

Figure 9–12 illustrates a VTOC prepared by John Seevers in the design of Mansfield's payroll system. "The top-level module, 1.0, shows the overall function of the system (to prepare a record of hours worked)," John stated. "The next level shows main subfunctions, which are to get the employee timecard, make the hours-worked record, write the hours-worked record, and display the hours-worked record. Besides

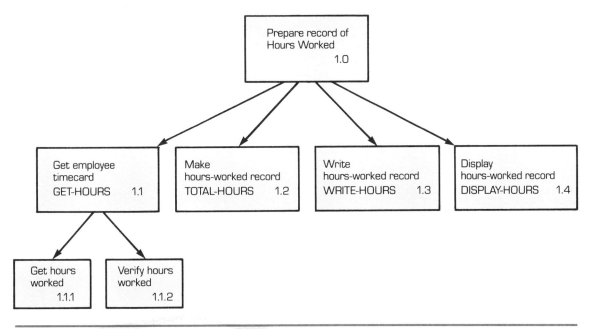

FIGURE 9–12 Visual table of contents (VTOC)

labeling all modules (using a top-down numbering notation), we identify each program or job step name. For example, 'Get employee timecard' is identified as job step GET-HOURS."

Overview HIPO Diagrams

Overview HIPO diagrams are similar in design to system flowcharts and thus help designers describe how a hierarchical structure is to be physically implemented. **Figure 9–13** illustrates an overview HIPO diagram. As shown, the worksheet contains an input, process, and output section. The input and output sections feature flowcharting symbols to show pro-

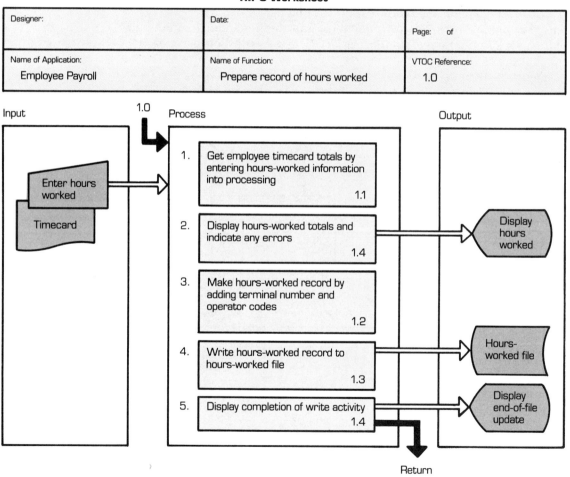

FIGURE 9–13 Overview HIPO diagram

cessing requirements, such as magnetic disk storage or information display by computer terminal. The process section describes each main program function or job step. For example, the module named "Get employee timecard," which was shown on the VTOC, is further described as "Get employee timecard totals by entering hours-worked information into processing." We will let John explain further.

"The makeup of HIPO worksheets requires that we use a variety of special-purpose flowcharting symbols. White arrows, such as the white arrow from 'enter hours worked' to process 1, or the white arrow from process 2 to 'display hours worked,' are *data management arrows*. They show that groups of data will be moved to and from a particular point in processing. Black arrows, such as the arrow entering at the top of the process box and the arrow exiting the process box, are *control arrows*.

"Notice how processing steps are clearly indicated by overview HIPO diagrams," John added. "Figure 9–13 shows that five job steps are invoked by module 1.0, beginning with 'Get employee timecard totals by entering hours-worked information into processing.' The figure also shows that control is returned to module 1.0 following the execution of all five job steps."

Detail HIPO Diagrams

Although detail HIPO diagrams are much like overview HIPO diagrams, their purpose is quite specific: to explain how functions work, when each function is identified by the next higher-level HIPO diagram. **Figure 9–14** shows the first page of a two-page description of the function to get the employee timecard (as points of reference, see module 1.1 on Figure 9–12 and process 1 on Figure 9–13). As illustrated, the detail HIPO diagram breaks down a single function into specific processing steps, which in this example consist of specific data entry steps (enter employee identification number, enter regular hours worked, enter overtime hours worked, and so forth). Following each appropriate or inappropriate entry by the user, a message is displayed.

Besides describing the sequential steps of processing, a detail HIPO diagram shows internal control steps, documents important processing parameters, and references error messages. Notice how Mansfield has designed its validation checks. John comments: "After an employee identification number (EMP-ID) is entered, it is checked for alphabetic characters (see Figure 9–14). Why do we do this? Because EMP-ID should only contain numeric characters."

John provides a second example: "After we check the EMP-ID for alphabetic characters we conduct a range test to determine whether the number is less than 10,000. In order to do this, we must set a parameter limit, which we store as ID-LIMIT."

Detail HIPO diagrams are often difficult to interpret in that they contain a wide variety of control options, parameters with little-understood

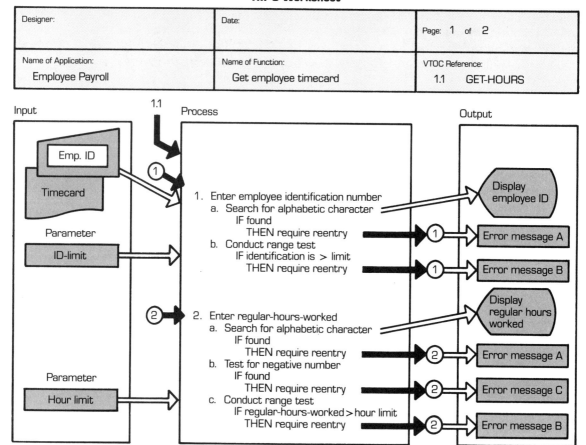

FIGURE 9-14 Detail HIPO diagram

names, error messages, and special-purpose displays. To aid in the understanding of these various design properties, an extended description is generally appended to each detail diagram. **Figure 9-15** illustrates such a description. As shown, an extended description consists of a set of notes explaining how control options work and the meaning of error messages. Error message A, for instance, clarifies that the employee identification number must be numeric, not alphabetic or alphanumeric.

Advantages and Disadvantages

HIPO documentation serves several purposes. Its advantages include the following:

1. It is an excellent top-level planning tool. Equipped with a complete set of HIPO materials, the designer or a review team can evaluate

Notes	Routine	Label	Reference

Control options:
 	
1. This routine is entered when
    (a) an alphabetic character is found or
    (b) EMP-ID is greater than ID-LIMIT.
2. This routine is entered when
    (a) an alphabetic character is found,
    (b) a negative number is discovered, or
    (c) regular-hours is greater than HR-LIMIT.
 	
Error Messages:
 	
A. Alphabetic character was entered, please reenter [variable name]
B. Range test limit exceeded, please reenter [variable name]
C. Negative number, please enter positive number.			

FIGURE 9–15 Extended description

and refine a design before initiating work on more detailed design steps.

2. It supports the development of a functional design. The contents of the process section show processing functions rather than coded statements.

3. It partitions a system into modules, follows a hierarchical organization, is graphic, and offers a strategy for transforming a hierarchy into an input-process-output (IPO) sequence. The interplay between a VTOC and overview and detail HIPO diagrams simplifies the process of showing how a design is to be implemented. This capability is perhaps the most important advantage of the HIPO documentation.

4. It simplifies program coding by showing specific program functions. For example, besides showing that an employee identification number must be entered into processing, a detail HIPO diagram tells the programmer to test the entered number to determine whether it contains an alphabetic character and falls within an acceptable range.

5. It helps show the control structure of a computer program. One item of particular concern to a programmer is how the various functions of a program are to be executed, especially when a processing error is discovered. Returning to our previous example, suppose that an invalid employee number were discovered. The programmer might ask: What should be done next? Should the program loop back to allow another number to be entered? Should the program stop? Should a bell or buzzer be sounded? Or should the program move to the next function, looping back to correct errors after all data have been entered into processing? As Figure 9–14 indicates, the control structure needed in handling an invalid employee error is quite clear. When an invalid number is discovered, an error message is displayed and the program loops back to allow another number to be entered.

6. It simplifies program maintenance. The HIPO documentation speeds up the process of tracing through a design to determine which code to modify. Suppose that a decision is made to change the value of all employee identification numbers by adding another digit to each number. Before beginning this task, the maintenance programmer would study all detail HIPO diagrams to determine exactly where and when employee identification numbers were used in processing and how they were checked for correctness.

7. It facilitates meaningful design reviews. Because the HIPO documentation is broken down into manageable pieces, the design team can easily understand the system. This increases the likelihood of discovering design flaws and omissions.

There are also disadvantages associated with HIPO documentation:

1. It requires special notation. While this is not a particularly serious matter, the HIPO technique does require special symbols and diagrammatic notation.

2. It requires considerable time for preparing and modifying diagrams, especially when all documents are prepared by hand. This disadvantage is much more serious than the first and cannot be easily dismissed.

3. It is difficult to diagram complex processing functions, especially those with elaborate branching and control structures. A severe criticism of HIPO documentation is that it cannot be used to describe some processing environments, such as programs that contain interactive dialogue or programs that feature complex data structures, pointers, or recursive routines.

STRUCTURED WALKTHROUGHS

Maintaining a dialogue with users during the process of systems design is often difficult. Some designers have even broken off all communication with users once the logical specification has been approved. Their thinking is, "Well, after all, users did approve the logical design. All we have to do is give them a system that is similar." Later, these same designers express surprise when users maintain, "The system they dumped on us was certainly not what we expected."

Wise designers make it a point to involve users at several stages in the design process. In particular, the earlier stages of design are considered especially critical. For example, it is far better to have users identify flaws in design early on, especially before coding has begun.

Structured walkthroughs offer one way of involving users directly in the design process. A *structured walkthrough* is an organized step-by-step tracing through of a design by a group of people. It is conducted to test the overall quality of a design (e.g., is it accurate, complete, and consistent with what was expected?). It thus differs from a mere "show and tell," in which designers demonstrate to users what they have done.

Types of Structured Walkthroughs

While a structured walkthrough can be done at any time in the design process, designers have come to realize that three types of walkthroughs are often needed: preliminary design walkthrough, detailed design walkthrough, and pseudocode (or source code) walkthrough. A *preliminary design walkthrough* tests the quality of the preliminary design, while a *detailed design walkthrough* tests the quality of the design just before the start of coding. A *pseudocode walkthrough* is the most detailed of all. It involves checking the internal logic of modules that are scheduled to be translated into computer code and, in some instances, checking the logic of modules that have been translated into computer source code.

Why are there three types of walkthroughs? A review of Figure 9–1 helps explain why. As indicated, once the preliminary design package is complete, design reviews are in order. The preliminary design walkthrough is one such design review. As also indicated, once the detailed design package is complete, a second set of design reviews is scheduled. The detailed design walkthrough and the pseudocode walkthrough are examples of this second design review.

Conducting a Structured Walkthrough

Walkthroughs differ depending on the participants of the test group. For now, let's assume that we will limit our attention to conducting a walkthrough of a preliminary design. Let's assume further that both an internal and external review will be conducted. An *internal review* involves a technical group review, where the review team consists of systems analysts, programmers, and other designers. This group must determine whether design steps are clear and technically accurate. An *external review* involves a user group review, made up of a team of users and managers. This group must determine whether the design will support user needs, as initially specified by the logical design.

To begin the walkthrough, the technical group typically collects all available design documentation: logical design requirements, project plan, a complete set of structure charts, and overview and detail HIPO diagrams. It then schedules meetings with the designer to walk and talk through all of these documents, testing the preliminary design for technical and functional accuracy. Finding the design acceptable, the technical group passes the design documentation to the external review group. This group is also required to walk and talk through all the documentation. However, the objective of this second review differs from the first. The external review group is asked to test the design for its ease of understanding and functional accuracy.

An example will help demonstrate how a structured walkthrough might work in practice. Suppose we wish to determine whether the structure chart representing the printing of the payroll register is accurate (see **Figure 9–16**). We might begin by reviewing the designer's view of the logical DFD describing the process. As shown in **Figure 9–17,** the designer has altered the DFD by adding areas of central transform. Next, we might compare this drawing with the top-level structure chart, determining whether all steps in processing have been accounted for and whether the design is acceptable—that is, whether it is accurate and complete.

Consider Figure 9–16 once again. The internal review group might ask, "How does the designer know that there are two main steps in processing (calculating employee pay and writing the employee register)?" The group members might note that the DFD provides the clue. They would then conclude: "The current payroll file placed between transforms 2.3 and 2.4 on the DFD indicates that records will be batched before the payroll register is printed." The group might then ask, "Is such a file necessary, or would it be possible to write the payroll register

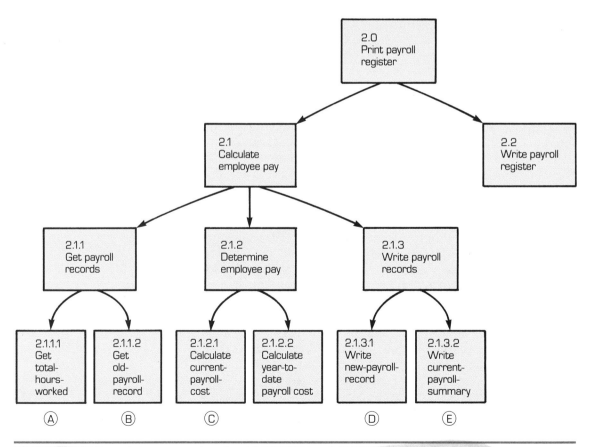

FIGURE 9–16 Expanded structure chart for the payroll system's level-2.0, print payroll register

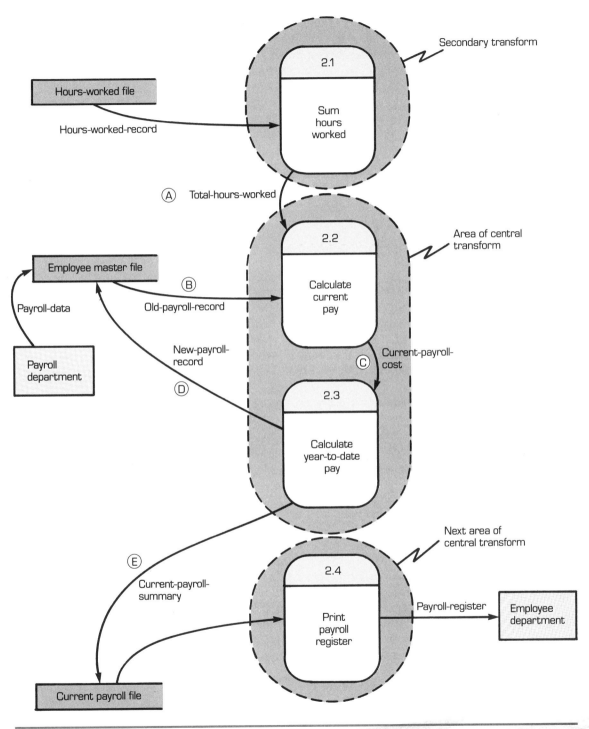

FIGURE 9–17 DFD of the payroll system showing areas of central transform and coding data flows with capital letters

directly?" The designer comments, "It is possible to print the register directly; however, this would be inefficient. Processing of module 2.1, calculating employee pay, would be delayed while the computer waited for the printer to finish printing the current payroll summary for an employee."

Finding satisfactory answers to its technical questions, the internal review group must consider issues related to functionality. Members might ask, "Are top-level functions of the technical design consistent with the top-level functions of the logical design?" In this instance the group might ask the following questions:

1. Does the technical design sum hours worked, calculate current pay, calculate year-to-date pay, and print a payroll register?

2. Does the design accept hours worked and old payroll information as input and produce as output new payroll and payroll register information?

As the preliminary structure chart indicates, all processing functions can be accounted for; however, *there is not a one-to-one correspondence*. The structure chart contains modules 2.1.1, 2.1.2, and 2.1.3 (getting payroll records, determining employee pay, and writing payroll records). None of these modules is shown as a function specific to the logical design. Rather, the logical design (see Figure 9–17) expresses these functions as sum hours worked (2.1), calculate current pay (2.2), and calculate year-to-date pay (2.3).

Once the top level of the design is understood, the review group can determine whether all logical data flows are accounted for. This test, while time-consuming and involved, is perhaps the most important test of a preliminary design package. It requires the review team to match each logical flow with a corresponding get, make, or put activity. Consider Figures 9–16 and 9–17 once again; the letters A through E have been added to show correspondence between the two diagrams. On the DFD, A represents the data flow for total hours worked, while on the structure chart A appears next to a get function, "Get total-hours-worked." Likewise, the letter C represents the data flow for current payroll cost on the DFD; on the structure chart, the letter C is represented by a make function: calculate (make) current payroll cost.

Conducting tests of data flow correspondence is helpful to the review team for several reasons. Besides determining whether all data flows can be accounted for, the review team can determine whether module names have been selected correctly. Continuing the previous example, the review team might conclude that module 2.1.2 (on Figure 9–16) is labeled improperly. The title of this module, they maintain, should be changed to "Calculate payroll costs." Tests of data flow correspondence also determine whether data flows have been labeled correctly. In this instance, review team members might conclude that module 2.1.3.2's phrase "current payroll summary" is confusing. They say that this data flow should be changed to "new payroll summary." In addition, the test helps the team determine whether the modules shown on the structure chart are *cohesive*—that is, whether they are designed to accomplish a single function, as opposed to a variety of functions. In chapter 12 the concept of cohesion will be discussed in detail. For now, our concern is that lowest-level module names appearing on a structure chart suggest that a single get, make, or put function is evident. For example, module 2.1.2.1, calculating current payroll cost, appears to be limited to a single make function. If the module were changed to storing and deriving current payroll cost, a single function would not be self-evident. The name then suggests that this module will be designed to make and put current payroll costs.

Finding structure charts to be accurate and complete, the review group might move on to examine overview and detail HIPO diagrams. **Figure 9–18** illustrates an overview HIPO diagram describing the process of printing the payroll register. In reviewing this design, the team would determine whether design steps were clear, complete, and consistent

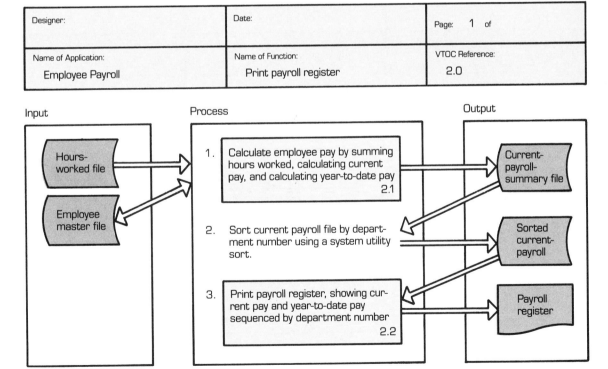

FIGURE 9–18 Overview HIPO diagram describing the print payroll register function

with the steps shown on the structure chart. The team would also determine whether inputs and outputs to processing were accurate. As the figure illustrates, printing the employee register involves getting records from two files, the hours-worked file and the employee master file, and putting records in two files, the current payroll summary file and the employee master file. A temporary file called the sorted current payroll file is also indicated. This file results from sorting the current payroll file by department number. Moreover, a sort program does not need to be designed, since a system utility program (a program designed to sort records into a desired sequence) already exists.

A review of a detail HIPO diagram (see **Figure 9–19**) allows the review team to consider other issues. For example, the team might question the addition of two outputs, "indicate every ten writes," in calculating employee pay. The designer comments: "Suppose we have 1,000 hours-worked and employee master file records to process. Without these special displays, processing would begin and end without any communication between the computer and the computer operator. With

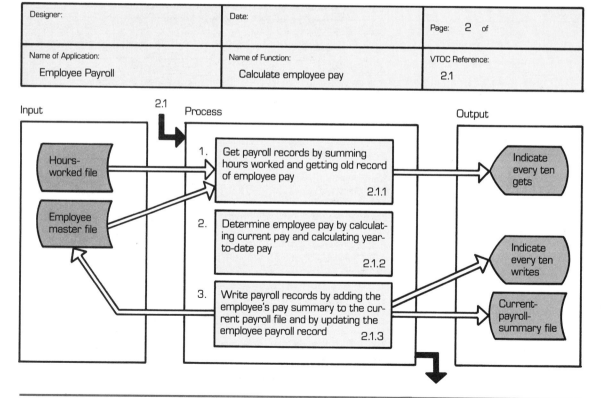

FIGURE 9–19 Detail HIPO diagram describing the calculate employee pay function

special displays, communication improves. The operator is informed of each successful processed set of records, where a set is defined as ten records. Should processing fail at some point between the beginning and end, these special displays tell the operator approximately where and when the failure occurred. This information is needed to pinpoint the source of the problem."

Even with this explanation, review team members might still be critical of the design. They state, "We agree with the need for special output displays, but why two when one would be adequate?" The designer might side with the review team and add, "Yes, I suppose only one output display (indicating the successful completion of every ten writes) is necessary."

After all the documentation has been reviewed, the review team is usually required to submit its suggested changes to the designer. A *walkthrough summary,* such as the one shown on **Figure 9–20,** is a formal way of showing each design item to be corrected. In completing this form, the review team must properly reference each suggested change, by structure chart, DFD, or IPO diagram. The designer, in turn, must respond to each item listed. If the designer accepts the review team's suggestion, the explanation of action taken is often limited to changes made. If, however, the designer disagrees with a suggestion, the reasons for not making a change must be clearly stated.

Walkthrough Summary

Review team: _____ Date of walkthrough: _____
         Walkthrough Coordinator
                                     Date of walkthrough response: _____
Design Team: _____
                                              Employee payroll
         Project design manager   Project: _____

Item. No.	Ref. No.	Item to Be Corrected	Date of Action	Action Taken
1	SC 2.0	Change name of module 2.1.3.2 to "Write new-payroll-summary"	1/23	Name changed
2	DFD 2.0	Change data flow, letter E, to "new-payroll-summary"	1/24	Name changed
3	IPO 2.1.1	Eliminate output display shown as: "Indicate every ten gets."	1/25	Change mode

FIGURE 9–20 Walkthrough summary sheet

Why Conduct a Walkthrough?

As we have seen, a structured walkthrough involves the step-by-step tracing through of a design by a group of people. Its purpose is to test the overall quality of a design. Why conduct a walkthrough? By now several advantages can be suggested. They include the following:

1. It is economical; design errors are less expensive when caught early on.
2. It improves communication; people outside the design team are asked to consider and take responsibility for the quality of a design.
3. It helps fine-tune a design; review teams can often make small but meaningful changes to a design.
4. It keeps the design team honest; designers realize that their work will be evaluated.
5. It helps the designer catch major errors; such a review increases the probability that no major functions will be omitted or structured incorrectly.

The structured walkthrough also has its disadvantages, including the following:

1. It can be expensive; the staff costs of conducting an internal and external review of a design are often quite high.
2. It may foster excessive criticism; the review team may not limit itself to making constructive criticism.
3. It can result in group conflict; the review team might be split, with one subgroup wanting to make changes that conflict with the desires of another subgroup.
4. It can take too much time; the review team may take forever to determine whether a design is acceptable.

Because of these and other disadvantages, management supervision of the walkthrough process is essential. A top-level manager must establish the rules to be followed by each review team. These rules should include what actions to take if group conflict arises, if criticism of the design becomes excessive, or if the time and expense needed to complete the reviews become excessive.

The Mansfield, Inc., Case Study

John Seevers felt that it would be best if he first framed a preliminary design of the new system, before beginning any work on a detailed design. "Best approach the new system one step at a time," John thought. "This way, I can review my findings with Roger to make sure that we haven't made a major mistake."

In putting together his design team, John selected Pamela Abbott and Walter Ruckman, two experienced designers. In addition, he asked that Carolyn Liddy be permitted to stay on with the project. "Pamela and Walter are creative designers and are probably the best technical people we have," stated John. "Carolyn will add continuity to the project. She is familiar with user requirements and should be of great help in writing I/O specifications and in working with user groups during the implementation phase."

System Overview

After close examination of the design requirements specified by the analysis group, the members of the design team agreed that the new system should be menu based, with users selecting the program to be processed by the computer. John explained how the system would work.

"Consider **Figure 9–21**," he said. "As shown, we favor a two- to three-deep menu-based system. After logging on, the user will be introduced to the billing and receivables system and asked to choose among several interactive programs or several batch programs (or to end the session if desired). If a branch to the interactive portion of processing is selected, a second menu asks the user to select from one of five types of interactive processing, to branch to the batch portion of processing, or to end the session. Notice that we grouped update master files (4.0) and display account (5.0) programs. If a user selects

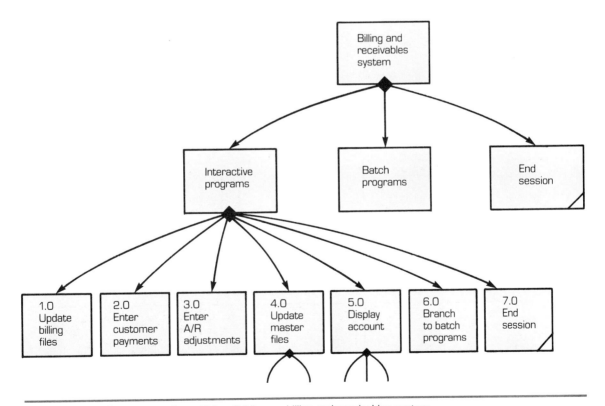

FIGURE 9–21 System overview of Mansfield's new billing and receivables system

either of them, a third menu asks the user which type of update or display he or she desires. Notice also that this design follows the program processing menu developed by the analysis team (see Figure 7–17). Thus, there should be no surprises."

Visual Table of Contents

As part of the preliminary design process, John asked his staff to prepare VTOCs of all the programs shown on the systems overview chart. **Figure 9–22** shows the VTOC prepared for entering customer payments. We will let Pamela explain this design.

"We considered the main functions of this program to be limited to key remittance data, get customer information from the customer master file, update the customer's accounts receivable record, and write the customer payment to the customer payment transaction file," she said. "Organized in this way, none of the processing steps should be difficult to design."

In review, Carolyn questioned: "What do you mean by verifying the customer account (see 2.2.2 and 2.3.2), and why is there a diagonal line shown in the lower right-hand corner for these modules?"

Pamela replied, "Verifying the customer account consists of asking the user or data entry operator to visually review the customer's name to determine whether it matches the name shown on the remittance. Remember, we only

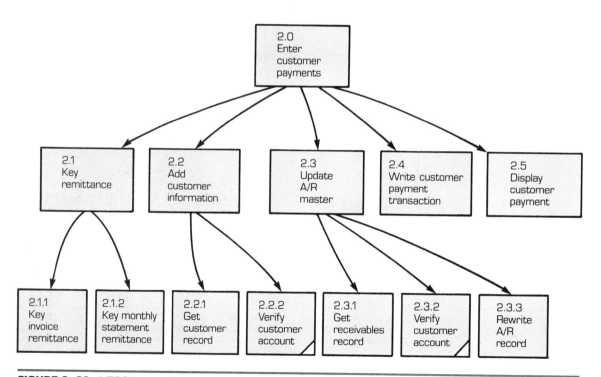

FIGURE 9–22 VTOC interactive program 2.0, enter customer payments

key in the customer number, retrieving the customer name from the customer master and the accounts receivable master file."

She continued: "The lower right-hand corner slash indicates that modules 2.2.2 and 2.3.2 are identical in function. In processing we invoke the same module twice."

HIPO Overview Diagram

Pamela illustrated the overview HIPO diagram drawn for program 2.0 to show the relationships between inputs, outputs, and processing (see **Figure 9–23**). She explained: "The HIPO diagram helps show how verifying customer name is made operational. Steps 2 and 3 indicate that after we retrieve the customer name from the file, it is displayed. Thus, this display differs from the display required for keyed data."

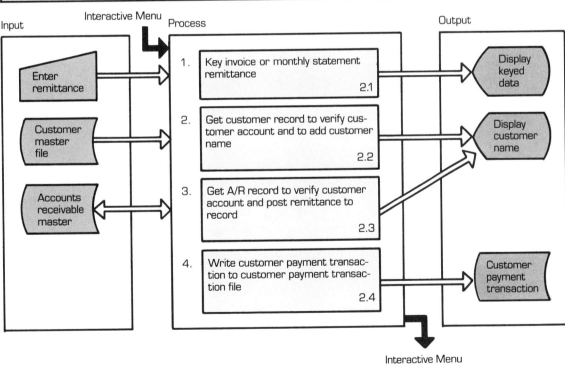

FIGURE 9–23 Overview HIPO diagram of program 2.0

She continued: "Steps 2, 3, and 4 show file processing requirements. Data are read from the customer master file in step 2, read from and written to the accounts receivable file in step 3, and written to the customer payment transactions file in step 4. Other than this sequencing of file processing steps, the instructions needed to handle customer payments are quite straightforward."

SUMMARY

Once the logical design specification is approved, work can begin on the preliminary design specification. This specification is the first part of the systems design process. The other three parts are preparing the detailed technical design, coding the design, and testing the design. Once these parts are completed, a maintenance package must be prepared. This package consists of materials describing how changes are to be made to the new system once it becomes operational.

Several parallel activities accompany the systems design process. The systems design team must refine and expand the specification for obtaining hardware and software, for testing the software design, and for helping users understand the new system.

The preliminary design requires the design team to determine which parts of the logical design represent transaction or transaction centers. Once the general makeup of the design is understood, the team can establish the scope of the design project and describe the systems software as a hierarchy of functions.

The HIPO technique is useful in developing the preliminary design of a new system. The first part of this technique—creating the VTOC—illustrates the hierarchical structure of system functions. The second part—preparing overview and detail HIPO diagrams—provides a visual description of system inputs, a written description of each process, and a visual description of system outputs. Because of their input-process-output sequencing, HIPO diagrams are also known as IPO diagrams.

The HIPO technique features a functional and a top-down approach. This includes providing a top-level overview of a system and partitioning to show lower levels of detail.

While there are several advantages to using HIPO techniques, there are also several disadvantages, which include: the charts and diagrams take considerable time to complete, are difficult to modify, and are difficult to create when complex processing functions are required. Accordingly, some designers combine HIPO techniques with other techniques, such as structured English and pseudocode. Consider the structured English example shown below as another way of explaining the first part of the process of getting employee hours:

```
Enter employee-identification-number.
    IF employee-identification-number contains
    alphabetic characters
```

```
            THEN display ''Alphabetic character was entered,
                please reenter.''
            OTHERWISE continue.
   END-IF
   IF employee-identification-number exceeds ID limit
            THEN display "Employee ID was too large, please
                reenter."
            OTHERWISE continue.
   END-IF
```

Based on this comparison, which method do you think is best? This is the question faced by many designers.

Structured walkthroughs, the step-by-step tracing through of a design by a group, are recommended to test the overall quality of a design. Just as there are two types of design—preliminary and detailed—so too are there corresponding types of walkthroughs. A third type of walkthrough—a pseudocode walkthrough—is performed just before the design is coded.

Walkthroughs are conducted by internal and external review groups. Internal reviews are more technical and serve to determine whether design steps are clear and accurate. External reviews are user oriented. They test the design for technical and functional accuracy.

Walkthroughs are conducted because it is economical to catch design errors in advance, improve communication, fine-tune a design, keep the design team honest, and catch major errors. To avoid the disadvantages associated with walkthroughs, close supervision of each walkthrough is strongly advised.

REVIEW QUESTIONS

9–1. What does the technical design specification document? What items are featured in the make-up of this design specification?

9–2. How does the detailed design of a system differ from the preliminary design of a system?

9–3. What does a maintenance package consist of?

9–4. Name the two assumptions that are made before the preliminary functional design is prepared.

9–5. What does the uppermost level of a structure chart show when a designer is working on a transform-centered design? What does it show when a designer is working on a transaction-centered design?

9–6. What types of specifications are prepared in parallel with the technical design specification?

9–7. What does the acronym HIPO stand for?

9–8. What are the types of HIPO diagrams and how are they used in documenting a system?

9–9. What do white arrows signify on a HIPO diagram? What do black arrows signify?

9–10. Which feature of the HIPO documentation is probably the most important?

9–11. Define the meaning of the term *structured walkthrough*.

9–12. What is the difference between an internal and an external structured walkthrough review group?

9–13. Name several advantages of a structured walkthrough. Name several disadvantages.

EXERCISES 9–1. Suppose that you are shown the following DFD transform and the corresponding structure chart. What errors were made in constructing this chart? Redraw the chart to remove the errors.

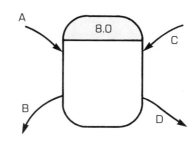

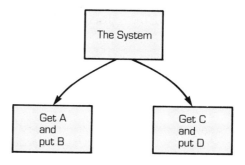

9–2. Suppose a systems designer decides to modify the DFD showing the processing of records for the order filling system as shown below. Draw a complete set of structure charts (e.g., level-0, level-1, level-2, and so forth) to describe this revised system. Also, add data couples to your level-2 structure chart (the level that describes the modules within the area of central transform).

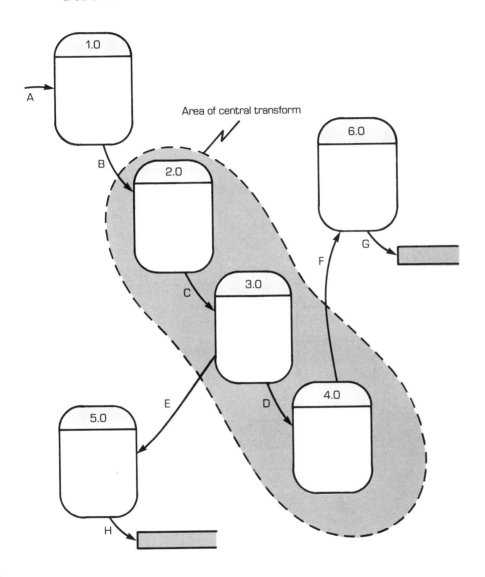

9–3. Suppose that you are required to transform the transaction-centered DFD shown below into a set of structure charts. First, draw a set of structure charts (level-0, level-1, level-2, and so forth) to describe all subsystems of this system. Second, redraw the level-1 structure chart to show all data couples that are important to the transaction center.

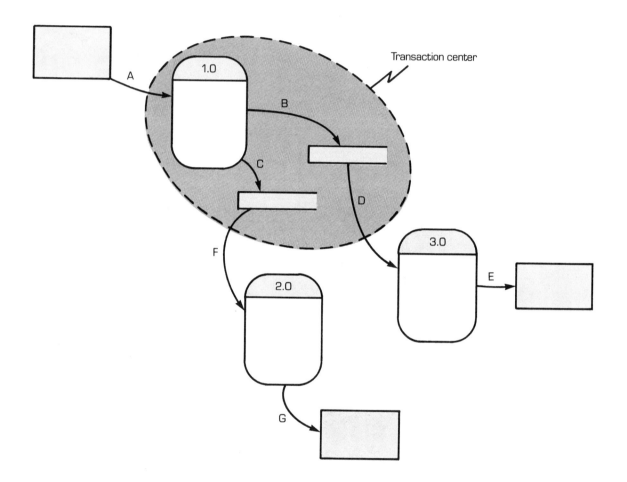

9-4. Suppose that you are handed the following DFD. Draw a level-0 structure chart or VTOC showing the uppermost modules of this system. Also draw an expanded structure chart (e.g., an expanded VTOC) showing the complete functional hierarchy.

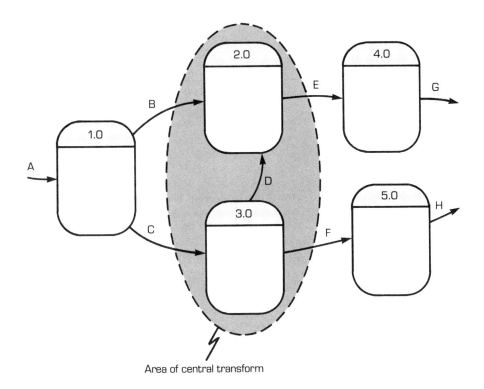

Area of central transform

WORLD INTERIORS, INC.—CASE STUDY 9
DESIGNING HIPO CHARTS

Introduction

John Welby decided to design a VTOC and hierarchical input-process-output (HIPO) charts for each computer program needed by the revised order processing system. "The interactive programs in particular will be the most difficult," he reflected. "Checking the correctness of the customer order, followed by checking the correctness of the factory invoice, may be particularly troublesome. Two special problems are processing sales tax information and checking to make sure that the profit for each customer order is acceptable."

The Correctness Problem

Checking the correctness of the customer orders and of factory invoices was never considered by the designers of WI's current system. They had assumed that the order details and the customer payment would be checked by hand and that only correct information would be entered into processing. They also assumed that all factory invoices were correctly processed by the manufacturer. John thought that both of these assumptions were dangerous. Consequently, with the new design, he planned to use the computer to determine whether the item number, finish number, and sales total recorded by the customer on the customer order were correct and, later on, whether the item number, finish number, and unit price entered on the factory invoice were correct. He also wanted to determine whether the customer had correctly added all line-item charges and, for in-state orders, had correctly calculated the sales tax. Later on, he wanted to know if the factory costs for individual items had been calculated correctly. As policy matters, John decided that 1) a customer order would not be written to the open-order file until the amount due and the customer payment were the same and 2) an open order would be treated as a problem order if factory charges appeared to be incorrect.

The Sales Tax Problem

The processing of state sales tax payments was an area that was never considered by the designers of WI's initial computer system. By law, WI was required to pay sales tax within ten days following the shipment of goods by the manufacturer to the customer. The company was also required to apply for a sales tax credit within ten days for any funds returned to the customer.

In practice, WI's administrative procedures were hampered by the existing computer system, since the company could not store the sales tax collected from a customer. Because of this shortcoming, WI had been forced to pay in advance its in-state sales-tax payments. Rena wanted this practice changed as quickly as possible.

John remarked, "With the new system, things will be different. When we receive the factory invoice, we will have ten days from the date of the invoice to pay sales tax."

The Profit Margin Problem

The checking of profit margin information was a third area never tackled by WI's first design team. The profit margin set by WI ranged from a low of 6 percent on ready-made items to 15 percent on custom-made items. Before payment was made on a factory invoice, a check of the actual profit margin (selling price less factory cost) was desirable. This check would test the invoice's accuracy. If the test revealed that the profit margin was low, the open order would be flagged for later review as a potential problem order.

Preparing the First HIPO Chart

John began his HIPO charting by drawing the level-1.0 VTOC for Program 1.0, enter customer orders (see **Figure 9–24**; see also the first flowchart prepared for case assignment 2 in chapter 7). John remarked, "First, I need to enter specifics of the order into processing. Most of this in-

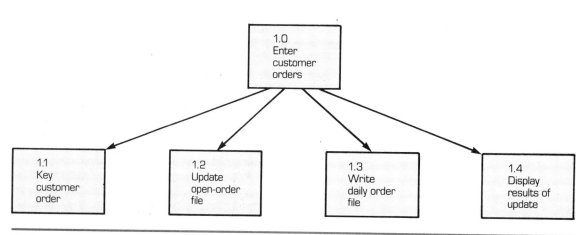

FIGURE 9–24 Level-1.0 VTOC

formation will need to be entered directly from the terminal keyboard; however, some information can be written to processing from the supplier file. After this is done, I will check the customer information to determine whether it is correct. If it is correct, I will add the new order to the open-order file and write the open order with sales totals to the daily order file."

CASE ASSIGNMENT 1

Complete the VTOC shown on Figure 9—24. Your design should include the following modules:

1. get supplier information,
2. calculate sales tax,
3. match sales and payment totals,
4. write open-order record,
5. check line-item charges, and
6. create open-order record.

These modules are not listed in correct order and will require some reorganization. You may add modules to your design. However, explain your reasons for any additions.

CASE ASSIGNMENT 2

Using the VTOC prepared for program 1.0 (entering customer orders) as a guide, draw the corresponding HIPO overview diagram. Inputs and outputs shown should be the same as those shown on the first system flowchart prepared for assignment 2 of chapter 7. Assume that two displays are needed: one to show the open order and one to indicate that file updates are complete.

Preparing Another VTOC

John next began the design of a VTOC for program 5.0, entering factory invoices (see also chapter 7). In this design, John planned to enter the factory invoice and to check line-item charges. If these were correct, he would carry forward the sales tax sum for in-state orders and update the files to indicate that the order had been filled. If the charges were found to be incorrect, a problem flag would be set, indicating that the open order should be reviewed as a potential problem order.

CASE ASSIGNMENT 3

Prepare a VTOC for program 5.0. The names of the modules suggested for this assignment are:

   Rewrite supplier record
   Verify open order
   Display results of updates
   Update supplier file
   Get open order
   Get supplier record
   Key factory invoice
   Process line-item charges
   Update open-order file
   Rewrite open order
   Write filled-order record
   Sum sales tax
   Add sales and profit totals
   Enter factory invoice

These modules will also need to be rearranged. You may add modules to your design. Explain your reasons for any additions.

CASE ASSIGNMENT 4

Using the VTOC prepared for program 5.0 as a guide, draw the corresponding overview HIPO diagram. Inputs and outputs shown should be the same as those shown on the first system flowchart prepared for assignment 6 of chapter 7. Assume that two displays are needed: one showing the factory invoice and one indicating that file updates are completed.

REFERENCES

1. R. Pressman, *Software Engineering: A Practitioner's Approach* (New York: McGraw-Hill, 1982).
2. *HIPO—A Design Aid and Documentation Technique*, IBM Technical Publication (CC 20-1851-1), 1975.

10
Input/Output Design

INTRODUCTION

FOLLOWING the completion of the preliminary design (including its review for technical and functional accuracy), the systems design team often begins work on the design of input and output materials—forms, reports, and visual displays. This design step is one of the most creative tasks that designers are asked to do. It is also one of the most difficult. A primary design objective is to prepare input/output (I/O) materials that visually convey the formal messages of the new system. These messages can take several forms, including messages explaining how data are to be entered into processing, messages showing error and control problems, and even messages describing how to escape from a processing routine if desired.

Where to Begin?

In the design of I/O materials, the question of where to begin is often raised. Should the design team start with inputs and move to outputs, or should outputs be tackled first, followed by inputs? Many designers (including the design team at Mansfield) prefer to start with the design of computer outputs. Their reasoning is as follows: If we can specify how output documents are to be prepared, the questions of output content, media, volume, and disposition and handling can be determined. As important, we can determine what input must be entered into processing to produce this desired output.

In this chapter, we will begin with the design of system output, followed by the design of system input. In each case, we will examine the steps important to each and consider how output and input can be placed in a form that will be of greatest value to the user. The final section, on the design of interactive dialogue, considers how two-way communication (between users and the computer) might take place. In an interactive environment, the designer must determine the content of each message to be passed between the software and the user, as well as the form of dialogue best suited to handle this exchange. This chapter thus provides a methodology for understanding the essentials of I/O design together with tools and techniques important to the design process. When you complete this chapter, you should be able to

- describe the steps important to the design of system input and output;
- construct Warnier-Orr diagrams to show the hierarchy important to system input and output;
- explain the difference between logical and physical input and output data structures;
- represent system input and output visually; and
- design interactive dialogue.

DESIGN OF SYSTEM OUTPUT

The design of system output requires that the design team resolve various specific questions, including:

1. What are the output requirements for the new system?
2. What is the logical content of each output form, report, and display?
3. What is the physical makeup of each output form, report, and display?
4. How should each output form, report, and display be represented visually?
5. What are the most appropriate processing mediums (e.g., printed reports, displayed messages, plotted graphs, computer output microfilm, and so forth)?

Step 1: Defining Output Requirements

Defining output requirements should be one of the easiest tasks in the design of computer output. After all, if the systems analysis team prepares samples of terminal screen layouts and computer-printed forms and reports and places these documents in the logical specification, then the task of defining system outputs should be done, by and large. All the design team must do is trace through the logical specification to determine whether the list of system outputs is in fact complete.

Can the design team ever assume that the list of outputs provided by the analysis team is complete? Consider Mansfield, Inc., once again. John

Seevers fully expects the logical specification to be complete; even so, he is *required* to compile his own list of output requirements. John explains: "This exercise is not as counterproductive as it appears. It allows us to fully understand what outputs need to be produced and why each type of output is important to our overall design."

Step 2: Documenting the Logical Data Structure

Documenting the logical data structure is another easy task, provided the logical specification is complete. Let's review the concept of a data structure briefly. As stated in chapter 5, a *data structure* consists of a collection of data elements that describe some object or entity. The data structure important to the design of system output includes a listing of all data elements that are produced by the system and all titles and headings that explain these produced results. This distinction is important. The design team must frame all output with *titles* and *headings* to explain what the processed output signifies.

Let's consider an example to illustrate how the design team might represent a logical data structure. Suppose Pamela Abbott, the Mansfield designer, is given a mockup of a production control report for the new system (see **Figure 10–1**). The mockup suggests that the report will be produced monthly and show units produced, units accepted, and percentage accepted for each day of the month by department. In addition, it will provide totals for each department and for all departments.

Suppose also that Pamela is provided a listing of the data elements that are required in the makeup of the production control report (see **Figure 10–2**). As shown, headings and titles are capitalized and enclosed in quotes; data produced by the system are shown in lowercase form.

CONSTRUCTING A WARNIER-ORR DIAGRAM

Although the logical specification of the production control report was complete, Pamela was not satisfied with the logical output listing. She remarked, "What I need to do is to construct a Warnier-Orr diagram. With such a diagram, I can show the hierarchy to be built into the production control report."

Before continuing, let's review what we mean by a Warnier-Orr diagram. A *Warnier-Orr diagram* shows an ordered set of like objects that is wholly contained within some other set. As discussed in chapter 6, some analysts prefer Warnier-Orr diagrams over data structure diagrams in describing the logical contents of a data store. For I/O design, Warnier-Orr diagrams are generally preferred because they indicate the hierarchy within a series of sets.

Figure 10–3 illustrates a football season represented by a Warnier-Orr diagram. As shown, the season consists of a variable number (from 1 to u) of games, each game consists of two halves, each half consists of two quarters, each quarter consists of a variable number of plays (from 1 to

MANSFIELD, INC.			MAY 1, 19xx

MONTHLY PRODUCTION CONTROL REPORT

DEPARTMENT NO. 606

DATE	UNITS PRODUCED	UNITS ACCEPTED	PERCENT ACCEPTED
4-1	96	93	96.87
4-2	93	90	96.77
4-3	90	85	94.44
⋮	⋮	⋮	⋮
4-30	93	87	93.54
606 Totals	2,780	2,649	95.29

DEPARTMENT NO. 608

DATE	UNITS PRODUCED	UNITS ACCEPTED	PERCENT ACCEPTED
4-1	48	44	91.67
4-2	43	39	90.70
⋮	⋮	⋮	⋮
608 Totals	1,350	1,227	90.89

TOTALS, ALL DEPARTMENTS	14,155	13,163	92.99

FIGURE 10–1 Mockup of production control report

p), and a play is either a run or a pass or a kick, with the plus sign indicating an "exclusive or" condition. Why exclusive? Because a run will either take place or not take place (from 0 to 1). If a run takes place, neither the pass nor the kick can take place.

The brackets shown on a Warnier-Orr diagram convey a special meaning—they frame each subset and mean "is equivalent to." Thus, a football season is equivalent to a variable number of games, while a game is equivalent to two halves.

Now let's consider how Pamela used the Warnier-Orr diagram to de-

1. "MANSFIELD SUPPLY COMPANY"
2. "MONTHLY PRODUCTION CONTROL REPORT"
3. date-of-report
4. "DEPARTMENT NO.———"
5. department-number
6. "DATE"
7. "UNITS PRODUCED"
8. "UNITS ACCEPTED"
9. "PERCENT ACCEPTED"
10. date-of-production
11. quantity-produced
12. quantity-accepted
13. percent-accepted
14. "TOTALS"
15. department-quantity-produced
16. department-quantity-accepted
17. department-percent-accepted
18. "TOTALS, ALL DEPARTMENTS"
19. company-quantity-produced
20. company-quantity-accepted
21. company-percent-accepted

FIGURE 10–2 List of logical data elements

fine the hierarchy within the production control report. As a first draft, she identified the *set rankings*: **Figure 10–4** illustrates that the production control report is produced each month, for a variable number of departments, and for a variable number of days. Production totals are to be shown for each day.

As a second draft, Pamela added headings and titles and values produced by the system to the Warnier-Orr diagram. **Figure 10–5** shows this design. It now contains all of the data elements previously shown and indicates when and where the various data elements are required in processing.

WHY USE A WARNIER-ORR DIAGRAM?

Many designers prefer to use Warnier-Orr diagrams instead of HIPO or structure charts in the design of system outputs and inputs. Why? Primarily because Warnier-Orr diagrams show the nested loops to be built

Football season { Games [1, u] { Halves [2] { Quarters [2] { Plays [1, p] { Run [0,1] (+) Pass [0,1] (+) Kick [0,1]

FIGURE 10–3 Football season hierarchy

Production control report { Month { Department [1, n] { Day [1, d] { Production totals

FIGURE 10–4 Production control report hierarchy: the first draft

into computer programs. Study Figure 10–5 once again. It shows that first the title of the report is to be printed, followed by the date of the report, followed by the report headings for the first department. After these headings are printed, daily totals for each day (from 1 to d) are printed until the last day is reached. Following this, departmental totals are printed. What happens next? The process continues for the second department, and so on.

Pamela summarizes: "Warnier-Orr diagrams show logically how different sets of information are to be printed or displayed. They permit us to show the *logical execution* to be built into an output computer program."

Step 3: Defining the Physical Data Structure

Once the logical data structure is represented as a hierarchy of set rankings, the task of defining the physical data structure is greatly simplified. Consider a third Warnier-Orr diagram describing the production control report (see **Figure 10–6**). Pamela explains: "We begin by adding the words *begin* and *end* to each set within the hierarchy to indicate the beginning and ending of each program procedure. As you can tell, we have a procedure called department (1, n), which invokes or calls an-

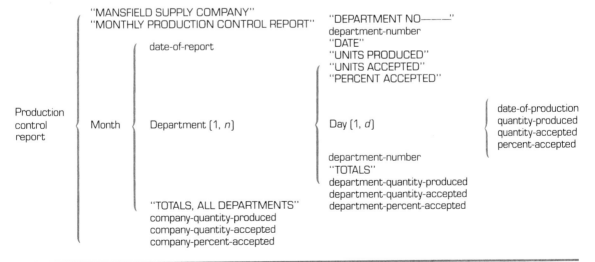

FIGURE 10–5 Production control report hierarchy: the logical design of processing

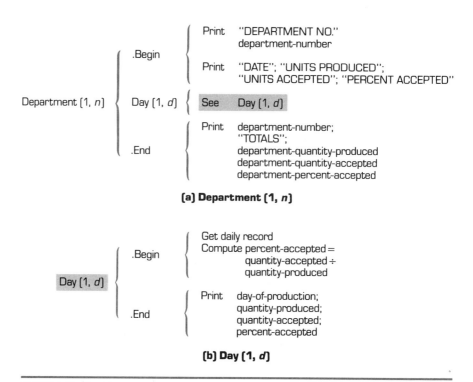

FIGURE 10–6 Physical design of department set (a) and day set (b)

other procedure, day (1, d). Confused by all of this? Let me clarify. Figure 10–6a indicates that the department heading is to be printed first, followed by the processing of subset Day (1, d), followed by the printing of departmental totals. Figure 10–6b describes the processing of subset Day (1, d). As shown, processing begins by getting a daily record for day = 1 followed by computing the percent accepted. Processing ends for day = 1 with the printing of day = 1 production totals. The next iteration of this subset sets day = 2. Processing begins and ends as before. This iterative process continues until the last daily record (day = d) is read and processed, at which time processing control is returned to the Department (1, n) subset to allow for the printing of departmental subtotals. This printing completes the first iteration of the department subset. Processing continues with the beginning of the second iteration: the printing of the next department heading (department = 2). The iteration of the department subset ends once the last department (department = n) has been processed."

Besides showing when and how the various subsets of output are to be printed, the physical data structure must indicate where, when, and how computed values are determined. For example, Figure 10–6b shows when, where, and how the percentage accepted is computed. The equation used in the computation is clearly defined as the quantity accepted divided by the quantity produced.

Step 4: Representing System Outputs Visually

Specifying how system outputs are to be represented visually consists of preparing layouts of terminal screens, along with computer-printed forms and reports. The difference between the logical and the technical specification is that system documentation is prepared to accompany each layout. Besides a Warnier-Orr diagram, system documentation provides a picture of each output document and a listing showing the characteristics of each field. The *visual picture* is usually a mockup of a report, similar to the mockup of the production control report. **Figure 10–7** illustrates *system output documentation*. As indicated, the date-of-report field is described as an alphanumeric field consisting of six numeric characters and two special characters (hyphens); the instruction column shows that a hyphen is to be placed between the second and third and between the fourth and fifth characters. The quantity produced is a numeric field. The instructions indicate that this field is to be edited as an integer. Percentage accepted is another numeric field. However, since this field is not an integer, instructions are needed to show how far to carry the decimal point and whether to round or truncate the computed value.

Besides alphanumeric and numeric fields, output consists of alphabetic fields and, in some instances, special forms such as binary or hexadecimal fields. A person's name, for example, is described as an alphanumeric field, while the number 12F1 (base 16) would be described as a hexadecimal field.

SHORT CUTS TO THE DESIGN OF COMPUTER OUTPUT

Newer computing systems frequently offer special-purpose report-writing software. With such software, a fill-in-the-blanks approach greatly simplifies the process of visually representing a report. For example, a leading software product, developed for use with microcomputers,

1. automatically plans the layout of a report to take the best advantage of the space available;
2. allows the report design to be stored for future use;
3. permits several versions of a report (up to eight) to be developed for the same output file of data;
4. prepares all reports in table form, sorting rows in either numeric or alphabetic sequence; and
5. performs calculations on numerical information stored on an output file, such as dividing units produced by units accepted to determine the percentage accepted.

System Documentation

Name of System	Designer	Date	Page 1 of 1
Quality control	Jones	6/15	
Type	Abstract		
Output analysis	Analysis of production control report		

Field	Description	Number of characters	Instruction
1. Date-of-report	Alphanumeric	8	Print: XX-XX-XX
2. Department-number	Numeric	3	

Daily Totals

Field	Description	Number of characters	Instruction
3. Date-of-production	Alphanumeric	8	Print: XX-XX-XX
4. Quantity-produced	Numeric	4	Edit as integer
5. Quantity-accepted	Numeric	4	Edit as integer
6. Percent-accepted	Numeric	6	Edit with decimal point rounded to the fourth position

Department Totals

Field	Description	Number of characters	Instruction
7. Department-quantity-produced	Numeric	5	Edit as integer with comma
8. Department-quantity-accepted	Numeric	5	Edit as integer with comma
9. Department-percent-accepted	Numeric	6	Edit with decimal point rounded to the fourth position

Company Totals

Field	Description	Number of characters	Instruction
10. Company-quantity-produced	Numeric	6	Edit as integer with comma
11. Company-quantity-accepted	Numeric	6	Edit as integer with comma
12. Company-percent-accepted	Numeric	6	Edit with decimal point rounded to the fourth position

General Comments

Headings and subtotals should be printed as per mockup of report.

Each department should begin on a new page.

Totals of all departments should be printed on a separate page.

FIGURE 10–7 System output documentation

With software products such as these, the question of how much work is left for the designer can be raised. Actually, all design work remains, with the exception of writing the portions of the computer code to produce the desired output design. The creative process of correctly representing output remains. Designing output so that it is visually appealing and conveys all necessary information continues to be a primary feature of the design process.

Step 5: Selecting the Most Appropriate Output Medium

Selecting the most appropriate output medium involves determining how system output can be distributed in a form that will be of greatest value to the user. Historically, *hard copy*, consisting of printed reports and forms, was the major form of computer output. Today's designer, however, must typically consider *soft copy* (output that is either displayed or heard), such as output written to a computer terminal or speech output.

An example will clarify which factors become important in the selection of the most appropriate output medium. Let's assume that John Seevers is required to produce as output a 400-page register showing the details of payroll processing. The register is to be produced twice each month and stored for a period of up to ten years.

The nature of this output requirement rules out any form of soft-copy or speech output. Why? Because Mansfield requires a permanent copy of computer processing for its files. Accordingly, John begins to consider output using a computer printer and computer output microfilm (COM).

ALTERNATIVE 1: PRINTING THE REGISTER USING A COMPUTER PRINTER

In studying this alternative, John commented: "Traditionally, this is the way reports are prepared. The advantages of using a line printer are several and include

1. availability of several types,
2. fast processing speed,
3. wide print width,
4. multiple-copy capability,
5. wide range of character sets and fonts,
6. offline input potential, and
7. permanent copy."

John continued: "At present Mansfield has two types of printers, a line printer and a serial printer. A *line printer* is characterized by its ability to print a line of print at a time, while a *serial printer* is limited to printing a character at a time. Unfortunately, the most powerful printer, a *page printer*, is not available. This type of printer uses either a laser beam or an electrostatic charge to produce a page of output at a time. It is suited for designs that produce massive amounts of printed output. Unfortunately, its high cost (more than $100,000) tends to rule out its use by our company."

John moves on to define the problem at hand more fully. "If a serial printer is used, it will operate at an average speed of 250 characters per second, with bidirectional printing (that is, no carriage-return time), and average a skip rate of fifty ms (milliseconds) per line. If a line printer is used, the output speed will average 150 lines per minute, with an average skip rate of ten ms per line. Finally, the print requirements for the register are as follows:

1. headings—three per page, averaging 60 characters each;
2. two-line gap;
3. fourteen entries per page, averaging 210 characters each and requiring two lines each;
4. one-line gap between entries; and
5. two-line gap to next page."

Given these requirements, John could construct a table similar to **Table 10–1**. As shown, the line printer is faster than the serial printer; however, either printer will take more than eighty-three minutes to print the 400-page register.

ALTERNATIVE 2: COM RECORDING

COM is the direct recording of computer output onto microfilm or microfiche. *Microfilm* is 16- or 35-millimeter film placed on a roll or a cartridge. *Microfiche* is 105-millimeter film, cut up into sheets, such as six by four inches.

TABLE 10–1 Worksheet comparing the time to produce a report on a serial and a line printer

	Serial Printer (in seconds)	Line Printer (in seconds)
(1) Headings (180 char)	.72	1.20
(2) 2-line gap	.10	.02
(3) 14 entries (210 char)	11.76	11.20
(4) 1-line gap per entry	.70	.14
(5) 2-line gap between pages	.10	.02
Time per page (in seconds)	13.38	12.58
Time per report (in minutes)	89.20	83.86

Notes
50 ms = .05 seconds
10 ms = .01 seconds
150 lines/minute = 2.5 lines/second = .4 seconds/line

John explains: "The advantages of COM over computer-printed output are several and include the following:

1. Lower cost—Cost savings of approximately ten to one result when COM can be used.
2. Space savings—Each roll of microfilm can store from 2,000 to 5,000 pages; each sheet of microfiche can store 270 pages of output.
3. Faster output speed—Output speed is ten to twenty times faster than either a line or serial printer; however, it is slower than a page printer, which produces from 100 to 200 pages per minute.
4. Improved page retrieval—The time it takes to find a page of output is reduced, since indexing methods can be used.
5. Greater durability—Microforms deteriorate more slowly than printed pages do."

In comparing the weight and space requirements for microfilm and microfiche against the weight and space requirements for printed output, John concluded that either COM method would generate considerable savings. His figures are shown on **Table 10–2**. As indicated, the space required to store the payroll register over a ten-year period would grow to 36 cubic feet if stored in paper form. This compares with a space of 0.04 cubic feet if these data were stored on microfiche. Likewise, the total weight difference between printed output and microfiche over a

---

TABLE 10–2 Worksheet comparing the weight and space required for computer-printed output and COM-recorded output

Computer-Printed Output

1. Page requirement: 800 pages/month = 96,000 pages in 10 years
2. Page weight = 12 lbs/1000 pages
3. Paper size = 14 7/8 in × 11 in × .004 in = .65 in^3
4. Total paper weight = 96,000 pages × 12 lbs/1000 = *1152 lbs.*
5. Total space required = .65 in^3 × 96,000 pages = 62,400 in^3 ÷ 1728 in^3/ft^3 = *36.11 ft^3*

COM (Microfilm)

1. Page requirement = 96,000 pages
2. Roll capacity = 2000 pages/roll
3. Roll weight = 4 oz.
4. Roll size = 5 in × 5 in × 1 in = 25 in^3
5. Total rolls required = 96,000 pages /2000 pages per roll = *48 rolls*
6. Total roll weight = 4 oz × 48 rolls = *12 lbs*
7. Total space required = 25 in^3 × 48 rolls = 1200 in^3 ÷ 1728 in^3/ft^3 = *.69 ft^3*

COM (Microfiche)

1. Page requirement = 96,000 pages
2. Fiche capacity = 270 pages/fiche
3. Fiche weight = 6 lb./1000 fiche
4. Fiche size = 6 in × 4 in × .01 in = .24 in^3
5. Total fiche required = 96,000 pages /270 pages per fiche = *355.55 fiche*
6. Total fiche weight = .006 lb × 355.55 fiche = *2.13 lbs*
7. Total space required = .24 in^3 × 355.55 fiches = 85.3 in^3 ÷ 1728 in^3/ft^3 = *.049 ft^3*

ten-year period approximates 1,150 pounds. John summarized: "The savings are quite remarkable when you remember that we are talking about one report. Most large data processing firms produce 500 or more reports on a monthly basis. It's no wonder that they always need additional space."

DESIGN OF SYSTEM INPUT

Much like the design of system output, the design of system input requires the design team to resolve several questions, including:

1. What are the data capture requirements for the new system?
2. What is the logical content of each source (e.g., input) document?
3. What is the physical makeup of each input processing procedure?
4. How should each type of input be visually represented?
5. Which data entry devices (e.g., key-to-disk systems, punched-card readers, optical character recognition equipment, and so forth) are most appropriate?

Step 1: Defining Data Capture Requirements

Data capture is the process of collecting and transforming source data into a medium or form that can be understood by a computer. As such, data capture is required of all input procedures. Fortunately, the work of defining each area in a system in which data capture takes place is often completed by the systems analyst in preparing the logical specification. What the systems designer must decide is how the source data at these various points in a system are to be collected and transformed. As such, *data collection* involves the readying of data for entry into the computer, while *data transformation* deals with changing source data into a form that can be understood by the computer.

Step 2: Documenting the Logical Data Structure

The design team can usually determine the types of information that need to be collected by examining the logical data structure contained in the logical specification. For the moment, however, let's assume that this specification is not complete. When that is the case, the designer might begin by listing all data elements required by each input point in a system, attempting to define the minimum set. Why minimum? Because data entry is a labor-intensive activity and is very expensive. John explained: "We estimate that we have to pay our data entry personnel more than $10 per hour when we include fringe benefits. In return, the data entry operator is able to average 5 to 10,000 keystrokes per hour. What is a keystroke? A *keystroke* occurs when the operator depresses a key on the terminal keyboard. Thus, we can minimize data entry costs only if we specify the minimum set of data to be keyed into processing."

John continued: "Another reason for a minimum set is that all data that are entered into the computer must be edited for correctness. These editing instructions require considerable computer code. Our objective in writing editing instructions is to forbid the entry of incorrect data into processing."

A MINIMUM SET EXAMPLE

Suppose that a design team such as Mansfield's is required to design a minimum set of input data. Initially, the team might be given a sample source document, such as the sample production control slip shown on **Figure 10–8.** This particular document is limited to ten possible entries, beginning with the name of the department and ending with the name of the person who approved the slip.

Suppose next that the design team finds fault with this source docu-

```
PRODUCTION CONTROL SLIP

Department                              606
Date                                  4-13-XX

Units Produced:
    Station              1              500
    Station             16              124
    Station             34              342
    Station             22              185
    Station              5              445

Units Accepted:
    Station              1              432
    Station             16              116
    Station             34              312
    Station             22              174
    Station              5              413

Prepared by:   Roger Mathews
Approved by:   Sy Holdberg
```

FIGURE 10–8 Source document for the production control system

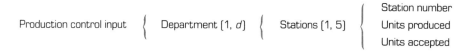

FIGURE 10–9 Production control slip input hierarchy

ment. After constructing a Warnier-Orr diagram (see **Figure 10–9**), the team concludes that the station number, units produced, and units accepted should be keyed in sequence to complete a station subset.

The team continues their analysis by beginning to fill in the Warnier-Orr diagram to describe, logically, how the data entry procedure should work. As shown in **Figure 10–10,** the team assumes that data entry will be accomplished through the use of a computer terminal. The data entry operator will initially be asked to enter the terminal number, the department number, and the date. Following this, the first department number is entered and a second date, if different from the current date. Next, the first station number is entered, together with the units produced and the units accepted for that station.

The value of beginning with a preliminary Warnier-Orr diagram, followed by adding details to the hierarchy once it is known, is twofold:

1. the designer can clarify the way in which data are best sequenced and

2. the designer can communicate how and when these data are to be entered into processing.

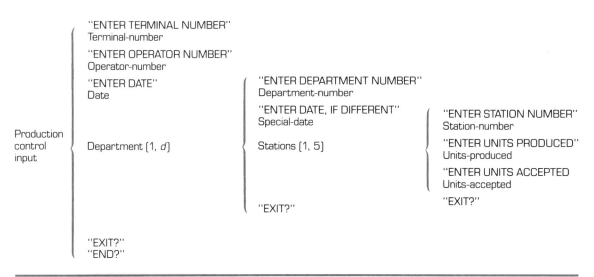

FIGURE 10–10 Production control slip input hierarchy: the logical design of processing

Consider Figure 10–10 once again. The Warnier-Orr diagram indicates that the date needs to be entered only once for all departments, unless special conditions exist. Likewise, the department number must be entered once, regardless of the number of stations reporting production totals for that department.

The word *exit* has been added to each set shown in Figure 10–10. What does exit mean? Actually, the words *escape* or *abort* could have been used instead. *Exit* is the designer's way of indicating the point at which a user can leave one subset to return to an earlier subset of a program or set of programs. Suppose that a user enters a station number only to realize that units accepted are missing from the production control slip. Instead of continuing by inserting a null amount for units accepted, the user decides to exit. This returns control to the earlier set (which, as shown, is the beginning of the department subset). Then, the user can enter a department number or decide to exit once again.

Step 3: Defining the Physical Data Structure

Once the logical data structure is fully documented, the details of the physical design remain. The designer might start, as before, by adding the words *begin* and *end* to each set within the hierarchy in order to set the boundaries of each program procedure. Following this, the steps within a procedure can be written.

Let's suppose that our design team is asked to transform the logical Warnier-Orr diagram into a physical design of processing. **Figure 10–11** illustrates the revised diagram. As indicated, the designers initially plan to display station number, units produced, units accepted, and control-E (the control key and the letter *E* on the terminal keyboard) to exit.

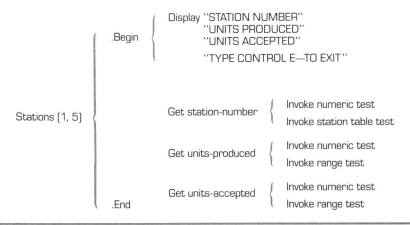

FIGURE 10–11 Production control slip input: the physical design of the stations subset

Following this, the interactive portion of processing occurs, beginning with the instruction to get the station number from the data entry person. Once the station number is entered, it is subjected to two data entry edit tests: a numeric test and a table lookup test

1. A *numeric test* determines whether the station number consists of numeric characters (instead of alphabetic or alphanumeric characters). If a nonnumeric character is found, an error condition must be reported, such as ALPHABETIC CHARACTER RECEIVED, PLEASE REENTER THE STATION NUMBER.

2. A *table lookup test* determines whether the number keyed to processing matches a similar number stored on a station number table. If a number is entered that does not match, an error condition is reported, such as THE STATION NUMBER DOES NOT MATCH. PLEASE REENTER.

Once the station number is accepted, the interactive portion of the program gets the number of units produced and tests that number with a numeric test and a range test (i.e., a test to determine whether the number entered appears to be either too high or too low).

Step 4: Representing System Inputs Visually

Defining the physical data structure helps the design team partially explain how the layout of a terminal screen might appear. By carefully studying the Warnier-Orr diagram, for example, the designer can determine which types of information to display and which fields are variable. Even so, the layout process cannot be viewed as mechanical. It requires considerable skill to properly lay out an input screen.

Consider **Figure 10-12.** This figure illustrates how a designer might begin the layout for the production control slip. Instead of beginning with the first set shown on the Warnier-Orr diagram, the designer begins with the last. The conventions used are uppercase letters to indicate where data have been carried over from a previous set and lowercase letters to show where data must be entered to complete the iterations of the current set. Thus, the department number and date would be entered and edited before the station number, units produced, or units accepted were entered into processing.

In the layout of an input screen, the designer not only identifies headings and fields but also specifies field lengths and whether each field is alphabetic, numeric, or alphanumeric. A documentation convention is used here as well. If COBOL is to be used in implementing the design, the designer might select X to describe all alphabetic fields, 9 to mark numeric fields, and V to show where the decimal point will be placed.

The designer usually adds notes to each layout sheet to help explain the operational aspects of processing. The notes shown on Figure 10-12 indicate how data are to be entered (type in numbers using the keyboard), what to do if an input error is made (edit using back arrow), how

```
                    PRODUCTION CONTROL SLIP

    DEPARTMENT:  999       NAME:   XXXXXXXXXXXXXXXXXXX
    DATE:        99-99-99

    STATION              UNITS              UNITS
    NUMBER               PRODUCED           ACCEPTED
      999                99999              99999
      999                99999              99999
      999                99999              99999
      999                99999              99999
      999                99999              99999

    TYPE CONTROL-E TO EXIT
```

Notes:
1. Type in numbers using keyboard.
2. Edit using back arrow.
3. Hit return key to move cursor to next column or beginning of next row.
4. Type "HELP" for instructions.
5. Type "CONTROL-E" to exit.

FIGURE 10–12 Production control slip input: the terminal screen layout

cursor movement is to be controlled (hit return key to move cursor to the next column or the beginning of the next row), what operators can do if they need assistance (type "HELP" for instructions), and, finally, how to exit (type "CONTROL-E" to exit). Are these notes important? Definitely. The key to a successful input design lies in developing a visually pleasing display combined with an easy-to-follow operational procedure.

SHORTCUTS TO THE DESIGN OF COMPUTER INPUT

The newer computers also offer special-purpose screen management and forms design software. Equipped with this software, the designer approaches the design of input forms as follows (see **Figure 10–13**):

1. A form is created, showing the titles of the various fields of data to be keyed to processing.

2. Once designed, the form is stored; it is later retrieved and used, modified, or deleted.

3. Data entry personnel use the form to enter input data into processing.

4. Entered data are edited for correctness by the computer; once edited, approved data are stored on an approved input file.

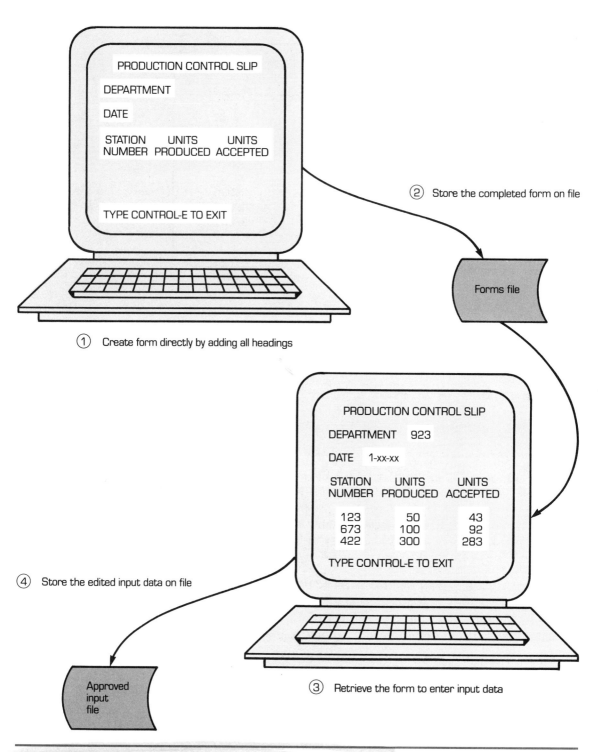

FIGURE 10–13 Forms design software and the design of data entry forms

With forms design software, the designer can quickly modify the way in which input is fed to processing. Does this improved efficiency destroy the creative process of input design? Most practicing designers believe not. To the contrary, they seek out screen design and forms handling software to simplify the mundane tasks of manually preparing input layout mats and writing computer code to describe the layout of a mat on a terminal screen.

Step 5: Selecting the Most Appropriate Input Medium

Selecting the most appropriate input medium involves a process much like the process followed in the selection of an output medium. With input, the designer generally considers some form of soft copy, such as input instructions written to a computer terminal or transmitted by means of a light pen. However, this was not always the case. Data processing depended on punched cards as input hard copy for more than one hundred years.

Various data capture alternatives currently exist for the systems designer to consider. Several of these alternatives are

1. punched card,
2. punched paper tape,
3. key to disk,
4. key to diskette,
5. key to cassette,
6. magnetic ink character recognition (MICR),
7. optical mark recognition (OMR),
8. optical character recognition (OCR),
9. voice data entry,
10. writing tablets, and
11. touch-sensitive pads.

The most popular of these is some form of key entry system, such as key to disk, key to diskette, or key to cassette. With these systems, the data entry operator uses a keyboard and a computer terminal to enter data; these data are then stored on a magnetic disk, a diskette (a floppy disk), or a cassette. Key entry systems are not always used, however. Banking institutions depend heavily on MICR equipment in the processing of bank checks, universities use OMR equipment extensively in processing student grades, and supermarkets are rapidly converting over to OCR equipment and to *bar coding*—the coding of products using optical bars to identify products.

An example will clarify the factors that influence the selection of the most important input medium. We will let John Seevers explain the process involved in deciding between two data capture alternatives.

ALTERNATIVE 1: KEY-TO-DISK DATA CAPTURE

"A *key-to-disk system*," John explained, "uses computer terminals to enter and review keyed data. Once keyed data are verified as correct (by the data entry operator and the computer), the data are transmitted and stored on a magnetic disk. Actually, there are several advantages associated with using a key-to-disk system:

1. The operator can visually inspect keyed data before storing it.

2. The computer can be used to edit data before storing it.

3. Data capture can take place at locations apart from the physical location of the computer.

4. Data entry stations, or workstations, are often designed to be self-contained, thus protecting against a complete system breakdown.

5. Data transmitted to a large computer from a workstation can be stored locally, thus permitting retransmission should data be lost or destroyed.

"Unfortunately, two major disadvantages are associated with a key-to-disk system, or with any key entry system: high expense and a slow rate of data capture. Its expense is high because data entry remains a labor-intensive activity. And data capture is slow because even highly skilled data entry personnel average less than 10,000 keystrokes per hour."

ALTERNATIVE 2: KEY-TO-DISK COMBINED WITH
OPTICAL CHARACTER RECOGNITION

"With OCR, we can scan hand printings or markings directly," John explained further. "The applications that seem to be best suited for OCR include hand-printed sales orders and journal vouchers and turnaround documents that can be printed in an OCR type font. Our analysis shows the advantages of OCR equipment to be several, including:

1. A person other than a trained data entry operator can prepare computer input materials.

2. Data are captured directly, at their source, by people who know what the data should look like and what they mean.

3. *Turnaround documents*, namely those that are initially printed by one computer application and later returned as input to another computer application, can be printed in an OCR type font.

4. OCR equipment is reliable and fast, processing up to 1,000 forms per hour.

"Unfortunately, our analysis also indicates several disadvantages of OCR equipment. These include:

1. The equipment is expensive.

2. Hand printings or markings are required.

3. Error rates are high if hand printings or markings are not done carefully.

4. Some OCR type fonts do not allow full use of the alphabet."

John began to explain Mansfield's problem. "At present we have a key-to-disk system with four workstations. The cost of this system is $600 per month and, frankly, the system is not adequate. We must either expand the key-to-disk system, by adding two or four additional workstations, or acquire an optical reader to handle some of our input requirements. We estimate that 50 percent of our data entry work load could be handled by OCR equipment."

In choosing between the two alternatives, John prepared a worksheet showing the costs for the two alternatives, annual keystroke requirements, capacity ratings for data entry operators, and use estimates for the OCR equipment (see **Table 10-3**). John commented: "Data entry operator wages run $700 per month; however, we can hire people part-time. If we expand the number of workstations on the key-to-disk system, we will need an additional magnetic tape drive. If we acquire OCR

TABLE 10-3 Worksheet comparing the cost requirements and ratings of key-to-disk systems versus OCR equipment

Cost Comparisons:

Item	Cost
Key-to-disk with 4 workstations	$600/month
Key-to-disk with 6 workstations	750/month
Key-to-disk with 8 workstations	900/month
Magnetic tape (with 6 or 8 workstations)	300/month
Data entry operator	$700/month
Optical scanner	$2,000/month
OCR forms	$5/thousand

Keystroke Requirements

Annual number of transactions	2 million
Average number of keystrokes per transaction	30
Annual keystrokes (without verification)	60 million
Annual keystrokes (with verification)	120 million

Data Entry Operator Ratings

Keystrokes per hour per operator	10,000
Annual working hours per operator	2,000
Annual keystrokes per operator	20 million

OCR Requirements

Annual number of transactions	1 million
Average transactions per OCR form	5
Annual forms processed	200,000

equipment, we will need to purchase special OCR paper, at a cost of $5 per 1,000 pages."

"Let me explain the keystroke requirements and capacity ratings briefly," he added. "We process two million transactions yearly, which translates into sixty million keystrokes. We also key verify each transaction. *Key verification* means that we double key to verify the accuracy of the data entered into processing. This step in our operation adds another sixty million keystrokes to our data entry load."

John continued: "We estimate that each data entry operator can average twenty million keystrokes a year. That's impressive, isn't it? This means that if we only hired data entry operators, six people could handle the input work load."

Table 10–4 compares the cost of moving to a key-to-disk system with six workstations (one for each data entry operator) with the cost of a combined system. John determined that a six-workstation system would cost $63,000. This cost was compared with the combined key-to-disk and OCR system cost of $57,400. Even then, John was undecided.

Eventually, he reached the following conclusion: "Alternative 2 is better because it offers us more flexibility. If we acquire the OCR equip-

---

TABLE 10–4 Worksheet comparing a key-to-disk system with six workstations to a combined optical scanner, key-to-disk system with four workstations

Alternative 1: Key-to-disk only

1. Required number of operators:

    120 million keystrokes/20 million keystrokes per operator = 6 operators

2. Projected costs:

    (a) Wages—6 operators @ 700/mo × 12 = $50,400
    (b) Key-to-disk with 6 workstations = $750 × 12 = 9,000
    (c) Magnetic tape unit = $300 × 12 = 3,600

    Total cost $63,000

Alternative 2: Key-to-disk and optical scanner

1. Required number of operators:

    60 million keystrokes/20 million keystrokes per operator = 3 operators

2. Projected costs:

    (a) Wages—3 operators @ $700/mo × 12 = $25,200
    (b) Key-to-disk with 4 workstations = $600 × 12 = 7,200
    (c) Optical scanner unit = $2,000 × 12 = 24,000
    (d) OCR forms (200,000/1,000) × $5 = 1,000

    Total cost $57,400

ment, we will have at our disposal considerable capacity to process material for new systems. Likewise, if key-to-disk activity increases, we can add another operator to our existing equipment to gain an additional twenty million keystrokes per year."

DESIGN OF INTERACTIVE DIALOGUE

One of the most creative and challenging aspects of I/O design is the design of *interactive dialogue*—the communication that takes place between the user and the computer system. What the designer must anticipate is how the user will want to exchange formal messages with the software design. This exchange often takes place over a visual display terminal, equipped with a keyboard, a "mouse," or a light pen.

The designer must answer the following questions in the design of interactive dialogue:

1. What message content must be exchanged between the user and the software design?

2. What form of interactive dialogue is best suited to handle this exchange of messages?

In addressing these questions, the designer must consider the various types of formal messages, the visual representation of message content, and methods of conversing with the user.

Four Types of Formal Messages

In an interactive processing environment, four types of formal messages are often displayed for the user's benefit. These types are

1. *processing commands*—messages that specify which computer program is to be called into processing. These types of instructions are also referred to as *program processing menu instructions* or *main menu instructions*.

2. *file commands*—messages that specify how information on a data file or database is to be modified. These types of instructions are often called *file processing menu instructions*.

3. *static user instructions*—messages that inform the user of some condition in processing and that do not require the user to take action.

4. *dynamic user instructions*—messages that inform the user of some condition in processing and that require the user to take action.

Walter Ruckman, the Mansfield designer, explained the relationships between these four types of formal messages in the design of the menu-based payroll system. "With this system," he explained, "the user is initially presented with a program processing menu and a message that asks the user to indicate the program to be executed."

```
            1.  UPDATE HOURS WORKED
            2.  UPDATE EMPLOYEE MASTER FILE
            3.  DISPLAY HOURS WORKED
            4.  DISPLAY EMPLOYEE PAYROLL ACCOUNT

            5.  PRINT PAYROLL REGISTER
            6.  PRINT HOURS WORKED REGISTER
            7.  PRINT EMPLOYEE MASTER REGISTER
            8.  PRINT EMPLOYEE PAYCHECKS
            9.  PRINT PAYROLL JOURNALS
           10.  PRINT QUARTERLY REPORTS
           11.  PRINT YEAR-TO-DATE REPORTS
           12.  BACKUP FILES
           13.  RESET FILES
           14.  EXIT


                    ■ ←                      Message
                       Prompt               ╱ window

           WHICH PROGRAM DO YOU WISH TO EXECUTE?
```

FIGURE 10–14 Program-processing menu with message window

Figure 10–14 illustrates this design. As indicated, the user must depress one or two keys to select a number from one to fourteen. A selection in effect commands the program to take a particular action.

Walter continued: "Let's assume that the user selects menu choice 2, UPDATE EMPLOYEE MASTER FILE. This selection leads to the display of a file processing menu (see **Figure 10–15**). Once again the user is asked to select a single item on the menu. As illustrated, the user can decide to add, delete, or change a record, or return to the main menu (the program processing menu).

"A static message follows the decision to add a new record," he added. "The software is designed to respond with the message:

```
YOU ARE ABOUT TO ADD NEW EMPLOYEES TO THE EMPLOYEE
   MASTER FILE.
```

This is followed by a dynamic message,

```
          ENTER NEW EMPLOYEE I.D.: ■ _ _ _ _
```

where the message is followed by a prompt and a flashing underscore. A *prompt* is an indicator that the computer is ready to accept input. The

```
1. ADD A NEW EMPLOYEE RECORD TO THE FILE
2. DELETE AN EMPLOYEE RECORD FROM THE FILE
3. CHANGE AN EMPLOYEE RECORD STORED ON THE FILE
4. RETURN TO THE MAIN PROCESSING MENU



    ENTER THE TYPE OF PROCESSING YOU REQUIRE.
```

FIGURE 10–15 File-processing menu with message window

flashing underscore shows the location of the cursor. Finally, this message indicates that a five-digit employee identification number is to be entered into processing."

Constructing a Dialogue Tree

Besides showing processing and file menus and listing static and dynamic messages, the designer must illustrate the sequence in which messages are to be transmitted. A *dialogue tree* is one method of showing such sequences. As shown in **Figure 10–16,** a dialogue tree has multiple branch points when menus are used, and forks at yes or no points. Trace the steps shown in Figure 10–16. The dialogue tree should lead you to conclude the following:

1. When an initial response of 2 is received, the program branches to a procedure to DELETE A RECORD from the employee master file.

2. Before a record is deleted, the user is asked to verify that the employee name is correct. The message reads: EMPLOYEE NAME OK?

3. If the name is correct, the tree forks and asks: SURE YOU WISH TO DELETE?

4. If the name is incorrect, the tree forks and asks: NAME OK NOW?

5. If the name is correct (is OK now) and the user responds yes to the question SURE YOU WISH TO DELETE, the record is removed from the file.

6. If the name is not correct, or if the name is correct but the user decides not to delete, control is shown as "return to start"—namely, a loop back to the start of the tree.

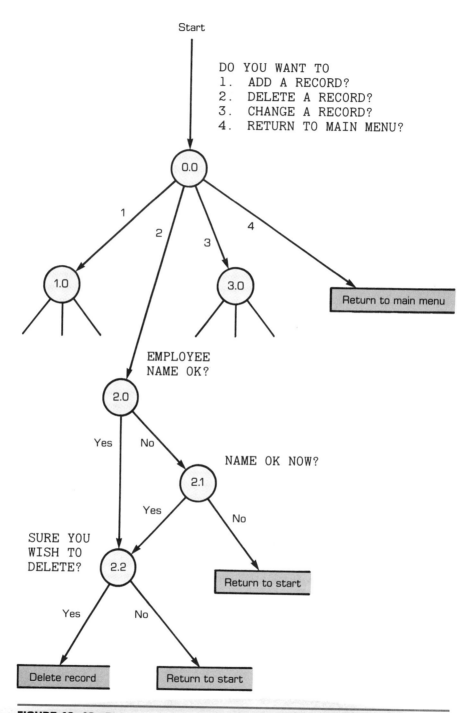

FIGURE 10-16 Dialogue tree showing branches from prompting menus

Isn't this tree incomplete? If the employee name is not correct at node 2.0, how could the name be correct at node 2.1? An expanded dialogue tree, such as the one shown in **Figure 10–17,** helps fill in the missing messages. The more detailed tree shows a node with an X. This is a *nonrestricted node* (i.e., a node that is not restricted to a prescribed number of choices). The first nonrestricted node indicates that it is necessary to find an employee record before testing to determine whether the name is correct. Moreover, if a record is found but the name is incorrect, a second attempt (as noted by a second unrestricted node) is made to find the correct employee record. If this second search is successful, the user is asked: NAME OK NOW?

Finally, a dialogue tree can be extended to show the processing requirements associated with such items as determining the correctness of keyed data values. Consider **Figure 10–18.** As shown, the processing requirements that precede the questions of EMPLOYEE NAME OK? and NAME OK NOW? can be clarified. A tracing of this expanded design leads to the following conclusions:

1. The message ENTER EMPLOYEE ID is displayed (see D.1).

2. After the identification number is entered, processing determines whether the record can (RECORD FOUND) or cannot (RECORD CANNOT BE FOUND) be found (see D.11 or D.12)

3. If the record cannot be found, the user can enter a second employee identification number (loop ≤3).

4. If the second attempt is not successful, a third attempt is possible; however, this is the last attempt. After three unsuccessful trials, control is returned to start (loop >3).

5. If a record is found, the user is asked to verify the correctness of the employee name (see 2.0).

6. If a record is found, but for the wrong person, the user responds with no, at which point control forks to node D.1, which asks the user to enter the employee identification number.

Picture-Frame Analysis

The designer must often supplement the information shown on the dialogue tree with pictures of the proposed dialogue. A *picture-frame analysis* illustrates frame by frame how the dialogue will be presented to the user. **Figure 10–19** shows how such an analysis works in practice. Walter explained this design.

"What we are attempting to show," he said, "is how a user would construct a *query:* a question to be responded to by the computer. As indicated by the first frame, the user must specify whether to find the truck with mileage greater than, less than, or equal to some value. The user specifies less than. In frame two, the user is asked the question: Less than what? The user responds by entering the number 15,000. In frame

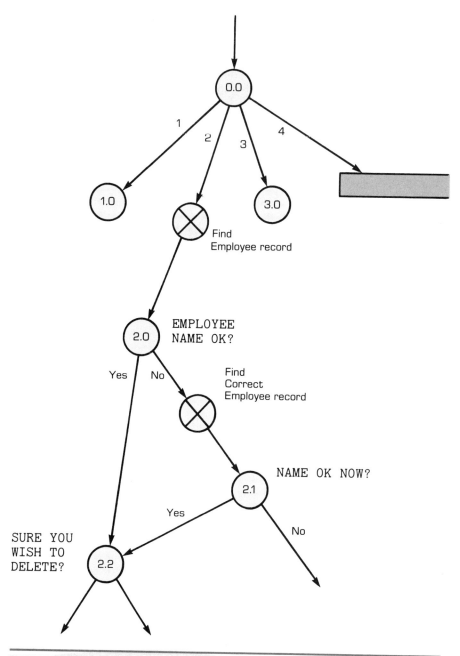

FIGURE 10-17 Dialogue tree showing location of entry of all unrestricted data sources

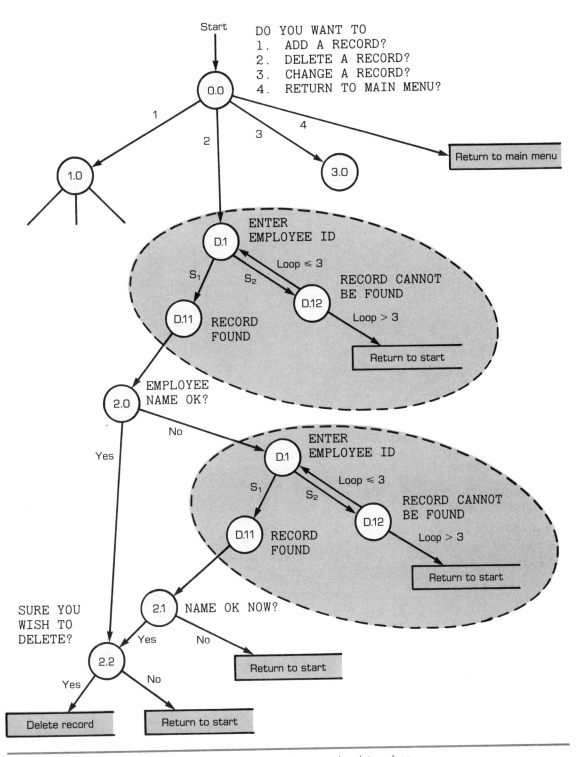

FIGURE 10–18 Dialogue tree showing requirements with processing data values

```
                    Decision Support Transform


   new query:
    Find the truck with mileage

                            greater than
                            less than
                            equal to



    use arrow keys to move selection, return to choose
         use backspace key to move back one level




                    Decision Support Transform


   new query:
    Find the truck with mileage less than



                   less than what? 15000



    type in number using keyboard—edit using back arrow
    return to accept—use backspace key to move back one
                              level
```

continued

FIGURE 10–19 Picture-frame analysis showing frame-by-frame construction of a user query

three, the user must indicate the time frame of interest. By selecting no date, the user chooses not to consider the question of time. Finally, in frame four, the user must specify that the query is complete. The query reads: FIND THE TRUCK WITH MILEAGE LESS THAN 15000."

```
                    Decision Support Transform


new query:
  Find the truck with mileage less than 15000 in (the)

                         no date
                         past week
                         past month
                         past year



  use arrow keys to move selection, return to choose
       use backspace key to move back one level




                    Decision Support Transform


new query:
  Find the truck with mileage less than 15000.

                         done
                         and
                         or



  use arrow keys to move selection, return to choose
use backspace key to move back one level
```

FIGURE 10–19 continued

Picture-frame analysis provides information unlike that from any other I/O methodology. It provides a walkthrough of a sample dialogue designed to take place between the user and the software design. In complicated design environments, such as the query building design, picture-frame analysis greatly aids all members of the design team. The frame-by-frame dialogue makes it possible to visualize how information will be entered into processing and processed following its entry.

Screen Design

Screen design deals with the layout of formal messages on a screen for presentation to the user. For example, when the user is asked to make a selection from a menu, the designer must be able to display the menu as well as provide space for data to be entered and messages to be displayed. A *data entry window* provides the space needed to enter data into processing, while a *message window* provides space to display messages, such as WHICH PROGRAM DO YOU WISH TO EXECUTE.

How much data can or should be placed on a single terminal screen? In part, the answer depends on the size of the screen. Small screens are often limited to sixteen lines of output and sixty-four characters per line. Somewhat larger screens are limited to twenty-five lines of output and eighty characters per line.

While no standards exist for screen design, the following guidelines are useful:

1. Design messages to read from left to right and from the top to the bottom of the screen.
2. Use reverse video (black on white or white on black) to show where data are to be keyed to the screen.
3. Use a flashing underscore to show the location of the prompt (that is, the location to enter data into processing).
4. Design message windows to help the user know what data to enter or to indicate what type of error was made.
5. Add "white space" as needed to all designs to improve overall readability.
6. Avoid distortion by refraining from the use of screen edges.
7. Design a set of commands to avoid requiring the user to work through page after page of menus. A CONTROL-H command might allow the user to proceed directly to updating hours worked, for example.

Screen design must also consider cursor movement. In this instance, the objective is to minimize for the user the work required to position the cursor on a screen. Fortunately, there are several ways of achieving this objective, including the following:

1. Design input procedures so that following the entry of data and depression of the return key, the cursor jumps to the next field of input.
2. Design input procedures so that the cursor jumps to the next field of input following the depression of the return key, even in the absence of keyed data.

3. Use a mouse to propel the cursor across the screen either diagonally or in a circle. A *mouse* is a hand-held cursor controller that signals the computer to reposition the cursor.
4. Use a touch-sensitive screen when it becomes important for the user to select a single item from a list of several items.

The Mansfield, Inc., Case Study

Satisfied with the preliminary design work done by his staff, John Seevers decided to have Carolyn and Pamela work on the design of system output and input, while he worked with Walter on the design of system files. "By dividing our effort," John commented, "we can move along faster."

Defining Output and Input Requirements

Before starting their design effort, Carolyn and Pamela defined all output and input requirements. Of the eighteen programs shown on the program processing menu, Pamela determined that ten printed reports and twenty-six display screens, including ten to show the status of reports being printed, would need to be designed.

Carolyn examined source documents. She remarked, "The source documents important to the new system are the packing slip (which contains customer shipping information), the invoice remittance, the monthly statement remittance, the accounts receivable (adjustment) journal voucher, the new customer slip, the customer change form, the new product slip, and the product change form."

Warnier-Orr Diagrams

Following the definition of all I/O requirements, Carolyn and Pamela developed Warnier-Orr diagrams to document the logical structure of each major system output and input. **Figure 10–20** shows their design of the cash payments register. Carolyn explained their design.

"Initially, we print the name of the register followed by the date. Then we print the cash remittances processed during the month. If the remittance follows from an invoice, a single invoice number, cash payment, and general ledger (G/L) code are printed. However, if the remittance follows from a monthly statement, several invoice payments typically are processed. For each payment (from 1 to p), the payment number (from 1 to p), invoice number, cash payment and G/L code are printed. After all remittances are printed, total individual invoice payments and total monthly statement invoice payments are printed."

Representing System Input Visually

Pamela explained how she used Carolyn's description of system output in designing system input display screens.

"As shown in **Figure 10–21,** the monthly statement remittance display contains all the variables shown on the Warnier-Orr diagram. The cursor is initially positioned to require input of the remittance number, followed by the customer

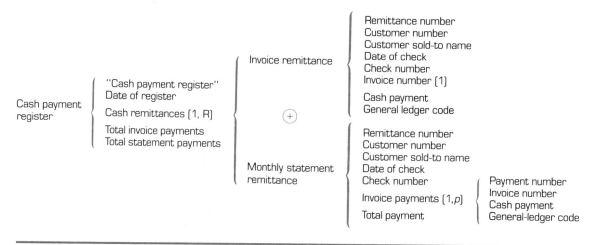

FIGURE 10–20 Warnier-Orr diagram showing the structure of the cash payment register

number. The remittance number tells us whether an invoice remittance or a monthly statement remittance is to be processed. The customer number allows us to search the customer database to determine whether the customer is on file. If no customer is found, the computer responds with

CUSTOMER RECORD NOT FOUND. PLEASE REENTER.

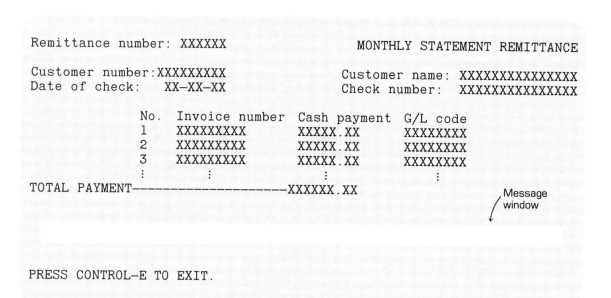

FIGURE 10–21 Monthly statement remittance display, with message window

"After the customer number is entered, we enter the date of the check, the check number, the first invoice number, the cash payment for this invoice, the G/L, the second invoice number, and so forth, until all invoice payments are entered. Finally, we enter the total payment. At this point, the computer sums the individual cash payments and compares this sum with the total payment entered into processing. If a difference is indicated, the computer responds with:

PAYMENT TOTALS DO NOT AGREE. PLEASE REENTER CASH PAY-
MENT AND TOTAL PAYMENT TOTALS.

SUMMARY

The design of computer output and computer input follows directly from the design of the logical specification. Of the work to be done, the task of visually representing system outputs and inputs is most challenging. Some designers view this aspect of design to be the most difficult portion of the design process.

Warnier-Orr diagrams are especially helpful in the design of system outputs and inputs. These diagrams show the structure within each type of output or input; they view outputs and inputs as a series of nested loops, in which the left-hand sets can invoke the right-hand sets.

Modern technology continues to simplify the design of both input and output. Current software products permit forms to be created and stored in coded form, thus avoiding the need to prepare forms by hand and prepare coded instructions for each form.

Selecting the most appropriate output or input medium is complicated by various alternatives. Historically, the printed report was the main form of output, while the punched card was the main form of input. Today, soft-copy displays are equal in importance to printed output; direct key entry of data has all but eliminated keypunched forms of input.

The design of interactive dialogue is a special feature of both input and output soft-copy design. The design team must determine the message content to be exchanged between the users and the software design and decide on the interactive dialogue best suited to handle this exchange of messages. Dialogue trees and picture-frame analysis are both used to illustrate the sequence in which messages are to be transmitted. Screen design deals with the way in which messages are to be formally transmitted to or received from the user.

REVIEW QUESTIONS

10-1. Why should the design team compile its own list of outputs, thus duplicating work done by the systems analysis team?

10-2. Why do some designers prefer to use Warnier-Orr diagrams in place of HIPO diagrams or structure charts when designing system outputs and inputs?

10–3. How is each field described by system output documentation?

10–4. Of what value is special-purpose report-writing software?

10–5. What is COM recording and what are its advantages over computer-printed output?

10–6. How does data capture differ from data collection and data transformation?

10–7. Why should the minimum set of input data always be key entered?

10–8. What are some advantages of a key-to-disk system?

10–9. What processing steps are followed with forms design software?

10–10. What questions must the systems designer answer in the design of an interactive dialogue?

10–11. How do processing commands differ from file commands? How do static instructions differ from dynamic instructions?

10–12. Compared with system input and output documentation, of what value is a dialogue tree?

10–13. What advantage is offered by picture-frame analysis?

10–14. How does screen design differ from picture-frame analysis?

EXERCISES

10–1. What is wrong with the Warnier-Orr diagram shown below?

10–2. Suppose that we are shown the following Warnier-Orr diagram. Suppose next that we need to print a three-month report, by department and by day, showing the daily totals of employees present and absent. How many of these daily totals will be printed?

Month (3) { Department (10) { Days (30) { Number of employees present
Number of employees absent

10–3. Design a Warnier-Orr diagram to describe how to keep a detailed record of balls and strikes for each batter on a baseball team for the entire season. Assume that each game in the season (the exact number is unknown) lasts nine innings.

10–4. Construct a Warnier-Orr diagram showing the structure of the budget appropriation display shown below. (Hint: Is this display based on subsets of department numbers or subsets of item numbers?)

```
DEPARTMENT: 0934          —   PERSONNEL
BUDGET ITEM: 637-001      —   TRAVEL EXPENSE
DATE OF REQUEST: 11-15-8X     OF APPROPRIATION: 12-14-8X

                REQUEST         APPROPRIATION        ALLOCATION

TOTAL           $10,000         $8,000               $7,000

1ST PERIOD      2,000           1,500                1,500
2ND PERIOD      2,500           2,000                2,000
3RD PERIOD      1,500           1,500                1,500
4TH PERIOD      4,000           3,000                2,000

LEVEL APPROPRIATED: 80%    LEVEL ALLOCATED: 88%
```

10–5. Suppose that the fourteen entries shown on Table 10–1 average 190 characters instead of earlier-estimated 210. Which printer would you select if your objective was to minimize total print time?

10–6. Construct a dialogue tree using the four picture frames shown on Figure 10–19. Include in your drawing the location of the nonrestricted node and show the message at this location.

10–7. Construct a second dialogue tree showing the details of frame 2 on Figure 10–19. Assume that edit tests are designed to determine whether the mileage figure entered by the user contains a nonnumeric character, is less than zero, or is greater than 99,999. The following design rules hold in building this dialogue:

1. If a nonnumeric character is found, the message NONNUMERIC CHARACTER, PLEASE REENTER will be displayed. The user will be given one more chance to enter an all-numeric mileage figure.
2. If a negative number is found, the message MILEAGE IS NEGATIVE, PLEASE REENTER will be displayed. The user will be given one more chance to enter a positive number.
3. If a number greater than 99,999 is found, the message MILEAGE IS TOO HIGH, PLEASE REENTER will be displayed. The user will be given one more chance to enter a number within the proper range.

WORLD INTERIORS, INC.—CASE STUDY 10
DESIGNING DATA INPUT SCREENS AND WARNIER-ORR DIAGRAMS

Introduction

While working on the HIPO charts for the new system, John Welby instructed his newly hired assistant, Jane Strothers, to begin the design of system inputs and outputs. He specifically asked her to develop the input display screens. When she accomplished that part of the design, she was to begin work on the design of Warnier-Orr diagrams, showing how stored and processed data were to be printed.

Jane initially decided to limit her design work to the main types of system inputs and outputs. First, she would undertake an I/O analysis, using as guides the DFDs and HIPO charts prepared by John. Next, she would design data entry screens for information shown on the customer order and the factory invoice. After this, she would enlarge her task to include the design of Warnier-Orr diagrams to be used in producing three system registers: the daily order register, the filled-orders register, and the problem-orders register.

John liked her approach. He commented: "After we get these preliminary versions of system inputs and outputs, we can sit down and finalize all layouts, including those for computer files. We can also come to some agreement on the types of processing controls to be built into our system design."

The Customer Order and the Customer Invoice

Jane found little wrong with WI's customer order or invoice forms. The customer order (see Figure 5–19) was easy to use and could be completed with little difficulty. She also liked the customer invoice (see Figure 5–20), though she decided to make the following change: she would add a customer key to the form, with the key consisting of the first five digits of the postal code plus the first three letters in the customer's last name. "Adding the key should help," remarked Jane. "When people refer to their account, they'll probably give us their customer key number."

Customer Order Input Display

While she was satisfied with customer order and invoice forms, Jane was quite dissatisfied with the current data entry displays designed for entering customer order information into processing. "The problem," she thought, "is the small screen (sixteen lines by sixty-four characters). Currently, three display pages are required to enter the details of a single customer order into processing [see **Figure 10–22**]. Besides needing to skip from one display page to another, we are forced to key in too much data. With John's new design, we should be able to avoid entering a substantial amount of line-item information."

Factory Invoice Input Display

In researching WI's existing system, Jane could not find any notes describing what data entry requirements were needed, if factory invoice data were entered into processing. All she could find was the display screen used to enter supplier name and address information (see **Figure 10–23**) and copies of typical factory invoices (for a sample, see Figure 7–18).

Jane decided that it would be necessary to enter only a limited amount of information into processing for each factory invoice. Besides the supplier number and the keys to access records stored on the open-order file, the factory invoice identification and terms data would need to be key entered. In addition, some line-item information would have to be key entered to check the accuracy of the invoice and to determine the acceptability of the profit margin.

Jane decided to add a message window to this input display. The window would allow for various messages to be passed on to the data entry operator. For example, the operator could be told that an error had been made. Likewise, the operator could be asked whether the open order should be treated as a problem order.

Finally, Jane realized that it was critical to key factory invoice information into processing as quickly as possible. "I wonder how much money the company has lost by its failure to take advantage of cash discounts," she commented. "For most suppliers, a cash discount of 2 percent is available if an invoice is paid within ten days past a specified cutoff date. Past-due balances or

```
CUSTOMER INFORMATION

CUSTOMER FIRST NAME : XXXXXXXXXX
 CUSTOMER LAST NAME : XXXXXXXXXXXXXX
       PHONE NUMBER : (XXX) XXX-XXXX

     STREET ADDRESS : XXXXXXXXXXXXXXXXXXXX
       CITY ADDRESS : XXXXXXXXXXXXXX
      STATE ADDRESS : XX
        POSTAL CODE : XXXXX-XXXX
```

(a)

```
CUSTOMER ORDER INFORMATION

       ORDER NUMBER : XXXXX
        TERMS CODE  : X

    SUPPLIER NUMBER : XXX
       SHIP TO CODE : X
      SHIP VIA CODE : X

 SPECIAL INSTRUCTIONS : XXXXXXXXXXXXXXXXXXX
                       XXXXXXXXXXXXXXXXXXX
                       XXXXXXXXXXXXXXXXXXX
```

(b)

```
LINE-ITEM INFORMATION

ITEM NO.  DESCRIPTION              QUAN.   FINISH    PRICE

XXXXXXX   XXXXXXXXXXXXXXXXXXX      XXX     XXX       XXX.XX
XXXXXXX   XXXXXXXXXXXXXXXXXXX      XXX     XXX       XXX.XX
XXXXXXX   XXXXXXXXXXXXXXXXXXX      XXX     XXX       XXX.XX
XXXXXXX   XXXXXXXXXXXXXXXXXXX      XXX     XXX       XXX.XX
XXXXXXX   XXXXXXXXXXXXXXXXXXX      XXX     XXX       XXX.XX
XXXXXXX   XXXXXXXXXXXXXXXXXXX      XXX     XXX       XXX.XX
XXXXXXX   XXXXXXXXXXXXXXXXXXX      XXX     XXX       XXX.XX
XXXXXXX   XXXXXXXXXXXXXXXXXXX      XXX     XXX       XXX.XX

SALES SUBTOTAL  XXXXX.XX
SALES TAX         XXX.XX
SALES TOTAL     XXXXX.XX         CUSTOMER PAYMENT XXXXX.XX
```

(c)

FIGURE 10-22 Display screens used in entering customer order information into processing

```
SUPPLIER INFORMATION:

            SUPPLIER NAME   :  XXXXXXXXXXXXXXXXXXXX

           STREET ADDRESS   :  XXXXXXXXXXXXXXXXXXXX
             CITY ADDRESS   :  XXXXXXXXXXXXXXX
            STATE ADDRESS   :  XX
              POSTAL CODE   :  XXXXX-XXXX

          SUPPLIER NUMBER   :  XXX

    ENTER ITEM NUMBERS AND ITEM NAMES ON NEXT PAGE.
```

FIGURE 10–23 Display screens used in entering factory invoice information into processing

DAILY REGISTER FOR 05/04/XX

SUPPLIER H87	ORDER TO: XXX SUPPLIER XX NAME XXX				
INVOICE NO. 22345	ORDER NO. 21050		ORDER DATE 05/03/XX		
TERMS—CASH	SHIP TO CUSTOMER		SHIP VIA UPS		
ITEM NO.	ITEM DESCRIPTION	QUANTITY	FINISH	PRICE EACH	PRICE EXTENDED
H37 F352	VERTICAL BLIND	1	100	74.70	74.70
H37 V352	VALANCE	1	100	8.10	8.10
	SUBTOTAL 82.80		SALES TAX NONE		SALES TOTAL 82.80
SPECIAL INSTRUCTIONS—NONE					

INVOICE NO. 22352					

SUPPLIER J50					

FIGURE 10–24 Daily order register

charges outstanding by more than thirty days are assessed a finance charge. This monthly charge is usually 1.5 percent of the past-due balance."

Daily Order Register

Figure 10–24 shows a sample from WI's current daily order register. Jane liked the format of this register and thought it would need little change. "What I must do, however, is draw a Warnier-Orr diagram showing the design of this printed output," she indicated. "This diagram will provide useful documentation for the new system."

CASE ASSIGNMENT 1

Using Figures 5–19, 5–20, and 7–18, along with the DFDs, VTOCs, and HIPO diagrams as guides, analyze the data flow into the daily order file. Organize your data items with three main headings: customer information, customer order information, and line-item information. Specify whether data are to be key entered once for all orders, key entered for each order, calculated, or transferred from a file.

CASE ASSIGNMENT 2

Design a data input screen showing how data recorded on a customer order are to be entered into processing. Use a twenty-four-line-by-eighty-character display screen layout sheet in preparing your design. Use existing field lengths, estimating new ones as required. Explain your design once it is completed.

CASE ASSIGNMENT 3

Design a data entry screen showing how data recorded on a factory invoice are to be entered into processing, using a twenty-four-line-by-eighty-character display screen layout sheet. Include in your layout a window for displaying messages. Explain the various messages to be displayed within the message window. Explain field lengths and window size as necessary.

CASE ASSIGNMENT 4

Design a Warnier-Orr diagram to illustrate the data structure of the new system's daily order register.

11
Data File and Database Design

INTRODUCTION

THE design of computer files and the database is often undertaken in parallel with the design of input and output materials. Once again we will let John Seevers, Mansfield's lead systems designer, comment about this design step: "With file design we must also be creative. An important objective is to minimize the use of external storage space, while maintaining efficiency in computer program processing. We must be especially careful in structuring all files. Why? We discovered the hard way that it is easier to change the logic of one of our computer programs than to change the structure of data stored on external files."

Let's review the terms used in the logical description of a data store before continuing this discussion. As stated in chapter 6, a data store describes an object or entity and is made up of logical records. Each record can be described by its attributes and identified by one or more key attributes. In addition, each record can contain one or more internal or external pointer attributes.

These terms are also helpful in describing the contents of physical files and in examining the contents of a database. We know, for example, that a physical file physically exists and contains records. Moreover, each record is made up of attributes and usually contains key and pointer attributes.

In this chapter, we will examine the difference between physical and logical files before beginning our discussion of record and file structures, the design of relational files, and the design of system files. This discussion is required to show that physical files may be quite different from their logical counterparts. The objectives of this chapter are similar to those of the previous one. In addition to a methodology for understanding the essentials of file and database design, tools and techniques are provided as well. When you complete this chapter, you should be able to

- describe the different types of record and file structures;
- prepare entity-relationship diagrams to explain the contents of database relations;
- understand the rules important to designing a relational database;
- know how to determine file processing requirements; and
- prepare physical file designs.

THE DESIGN OF PHYSICAL FILES

The design of computer files typically follows several basic rules, some of which are simply common sense. For example:

1. Do not place records in a computer file unless record attributes can be easily maintained. For example, a record should not contain the number of children unless that number can be revised easily as family membership changes.
2. Do not place records in a computer file unless the cost of storage can be justified. Do not store an employee's nearest relative or home phone number, for instance, if these data are never used, or used infrequently.
3. Do not place logical records in a file if records can be blocked to save file space and improve the efficiency of processing. *Blocking* is the grouping of logical records to form a single physical record, while the *blocking factor* is the number of logical records contained in a single physical record.
4. Do not place the names of events, objects, or persons in a file if codes and key attributes would be more appropriate. Instead of storing the name of a product, for instance, it might be better to store the number assigned to a product.

Physical Versus Logical Records

Let's modify our set of terms to give additional meaning to these common-sense rules. A *file* refers to a collection of records, while a *file storage device* specifies the hardware used as a data recording and storage medium. Thus, records in a file are blocked to save space on a file storage device. Next, let's compare terms. A *physical record* is the unit of data transferred between a file storage device (e.g., a magnetic tape or magnetic disk storage device) and the main memory of a computer. A *logical record*, in contrast, is the unit of data called for by an instruction contained in a computer program. Consider the following example.

Suppose that we have 1,000 10-byte logical records, in which a *byte* is required to represent a character. Given this information, we can determine that 10,000 bytes of record storage will be needed. Suppose next that the input/output (I/O) buffer size permits records of up to 1,000 bytes in size to be passed between the internal memory of the computer and a file storage device. This buffer size permits 10-byte records to be blocked by a factor of 100 (10 bytes/record × 100 = 1,000 bytes). Or one physical record can be defined as

```
Physical record = {logical record} 100
```

In other words, we need store only ten physical records of 1,000 bytes each. How is processing changed by blocking? During processing, a physical record would be read from storage into memory. A search of the physical record would then be made to locate the logical record called for by a computer program.

Let's consider next the difference between a physical file and a logical file. A *physical file* consists of physical records written to a file storage device; a *logical file* is the collection of logical records called for by the system design. A design might require a collection of project records (e.g., a logical collection); however, in order for this collection to be read from storage, it is necessary first to read data from two physical files—one containing identification (nonvariable) information about a project, such as the project name, and another containing variable information, such as hours worked by employee. We can define this situation as follows:

```
Logical project file = Physical-project-identification-
                       file + physical-project-hours-
                       worked-file.
```

And since we also know that files consist of collections of records, we can write

```
Logical project record = Physical-project-identifica-
                         tion-record + 1{physical-project
                         hours-worked record}n
```

where one or more hours-worked records exist for each project.

Because the difference between physical and logical records and files is so important to file design, we will refer to these concepts throughout this chapter, and especially during our discussion of organizing physical files. Before this material can be understood, however, we must consider record structure and methods of file structure.

Record Structure

There are three types of record structures that a file designer must consider: fixed-length structure, variable-length structure, and multidimensional (or mixed) structure.

FIXED-LENGTH STRUCTURE

The simplest type of record structure is the fixed-length structure. Each record contains the same number of attributes or *physical fields;* each record is allocated the same number of bytes. **Figure 11-1** illustrates a fixed-length record, which as shown is an inventory record. Given the byte requirements for each record and the number of products to be stored, it becomes an easy matter to determine total file storage requirements. Assume, for instance, that 10,000 inventory items must be stored. Since each physical record requires 52 bytes, 520,000 bytes of storage space must be made available.

VARIABLE-LENGTH STRUCTURE

A variable-length structure holds when either records do not contain the same number of fields or the number of fields is constant but field length is variable. Imagine that a product can have more than one name. As shown by **Figure 11-2a,** a product record requires either 52 bytes for one name or 72 bytes for two names. Imagine next that the record length is determined by the physical requirements of each field. As shown by **Figure 11-2b,** the record with a product name of SPROKETS-PLUS is 7 bytes longer than the record with a product name of T-TOYS.

MULTIDIMENSIONAL STRUCTURE

A multidimensional structure consists of several record segments, in which each segment consists of a fixed or variable portion, or some combination thereof. **Figure 11-3** illustrates such a structure. The first part of this two-record set is the *mandatory,* or required, portion, while the second part is an *appended,* or optional, portion. Suppose, for example, that this record structure were selected to store student records. In design, each student would be assigned a unique number, which would be stored as the primary key. Student information, such as the date started school and date of birth, would also be collected and placed in the fixed portion of the mandatory record, while other student information, such as student name and address, would be collected and placed in the variable portion of the mandatory record. Finally, student grades would be

Field Name	Field Length (in bytes)
Product-number	8
Product-name	20
Beginning-quantity	8
Ending-quantity	8
Current-quantity	8

FIGURE 11-1 Fixed-length record structure

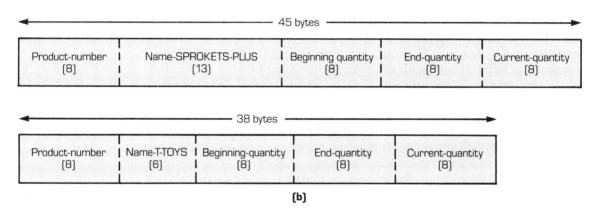

FIGURE 11–2 Variable-length record structure in which the number of fields varies (a) or the field length (in bytes) is permitted to vary (b)

appended to the student record, in which each record consisted of student number, the date the course was taken, the course number, and the course grade. Given these facts, we could define a student file as follows:

```
Student-file = Mandatory-student-record + 1{appended
               course-grade-records}n,
```

where n is the number of courses taken by the student for which a grade was recorded, and where

```
Course-grade-record = Student-number + date-taken +
                      course-number + course-grade
```

In this instance, the student number and the date the course was taken are key attributes (primary and secondary keys, respectively). They per-

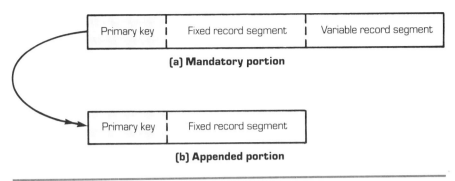

FIGURE 11-3 Multidimensional record structure

mit the retrieval of all courses taken by a single student, or of the students enrolled in a single course.

File Structure

Along with different types of record structures there are also different types of file structures. The three most common structures are sequential, direct, and indexed-sequential.

SEQUENTIAL ORGANIZATION

The simplest type of file structure is sequential organization, in which records are stored in a file in sequence, generally in serial order. Both magnetic tape and disk file storage devices permit the sequential storing of data. When magnetic tape is used, one physical record follows another, with each record kept separate by an *interrecord gap* (IRG) (see **Figure 11-4**). When a magnetic disk is used, one physical record also follows another; however, bits rather than bytes are written serially. Here, too, each physical record is kept separate by an IRG.

Several advantages are associated with sequential file structure. As stated by Ellzey[1] and others, a sequential file structure is easy to create, is easy to search, uses storage space efficiently, and permits records to be blocked and stored on either magnetic tape or disk file storage devices. Moreover, if data must be printed in serial order, such as by student number, date of entry, or customer name, the sequential structure makes sense.

The disadvantages of sequential structure are also several, including:

1. Lengthy search time—If a file contains n records, then on the average $(n + 1)/2$ records must be examined before the record of interest can be found.

2. Record addition—If the next record is assigned a primary number of the last primary number plus one, then the full file of records must be read to append a new record to the file.

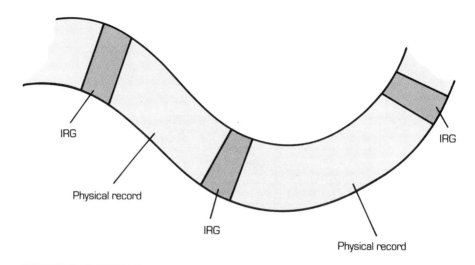

FIGURE 11-4 Physical record storage on magnetic tape

3. Record deletion—If a record must be deleted, the file must first be searched to find the record, and then null values must be inserted to indicate that the record has been deleted. The alternative is to rewrite the entire file to physically remove the deleted record.

DIRECT ORGANIZATION

Direct or random organization stores records in locations determined from a record's key attribute value, by means of an algorithm or *hashing scheme*. For an example of such a scheme, consider the following design problem.

Suppose we know that a business has 500,000 records, numbered from 1 to 500,000, and that 250,000 (or 50 percent) of these records are active at any given time. Suppose further that a single magnetic disk storage device is available and that this device stores up to 300,000 accounts. How should these records be stored?

One solution to this problem would be to divide the record key of the customer by 1.66667 (500,000 accounts divided by 300,000 storage locations) and use the computed value to indicate the storage location. Record 500,000 would thus be placed in storage location 300,000, record 250,000 would be placed in storage location 150,000, and so forth. This particular hashing schedule is called the division/remainder method and is one form of *indirect addressing*.

When is direct organization advised? When response time (i.e., the time between a request for data and its output) is critical, such as in immediate-response systems; when a sequential listing of records is not required; and when magnetic disk storage (or some other online storage medium) is available. Direct organization cannot be used with magnetic tape storage, since magnetic tape is limited to sequential organization. Likewise, direct organization does not permit records to be blocked without difficulty.

INDEXED-SEQUENTIAL ORGANIZATION

A popular form of file organization, one that offers the advantages of sequential and direct organization, is indexed-sequential organization. This method stores records in sequence, such as in numeric or alphabetic order. It also permits records to be retrieved directly, although in a manner different from that of direct organization. With indexed-sequential organization, indexes are maintained, showing relative locations of records stored on file.

Let's consider an example in which the storage medium is a magnetic disk. As shown in **Figure 11-5,** one way to determine where records are stored is to create two indexes: a cylinder index, showing the highest primary key value stored on a disk cylinder, and a track index, showing the corresponding primary key value on each track within each cylinder. Suppose we know that 320 fixed-length records are stored on cylinders 21 and 22 and we wish to retrieve a record whose primary key value is 240. A search of the cylinder index tells us that the key value is greater than 160 but less than 320. Thus, the record is stored on cylinder 22. A secondary search of the track index for that cylinder tells us that the record of interest can be found on surface 10 of cylinder 22. With this information, the computer can go to a specific recording surface, read the data contained on the track, and find the specific record of interest.

As this example demonstrates, indexed-sequential organization is designed to support sequential, and relatively fast, direct access to records stored on an online storage device. Similarly to direct organization, however, indexed-sequential organization makes it difficult to store variable-length records; in addition, indexed-sequential organization cannot be used with magnetic tape. Finally, additional storage space is required with this method to hold indexes; additional processing time is required to modify indexes when records are added or deleted from a file.

DESIGN OF RELATIONAL FILES

This overview of record and file structures suggests that it would be nice if all information systems stored nothing but basic facts about objects and used the same number of attributes to describe each object. For example, if we only wanted to store information about product records and each record contained five fields, as in Figure 11-1, then we could create either a sequential, direct, or indexed-sequential file containing 52-byte fixed-length product records. Unfortunately, this would not be realistic. Most file design problems store *relationships* or associations between different objects. Suppose that one object is a group of students, while another is a set of classes. In processing, the interesting information is

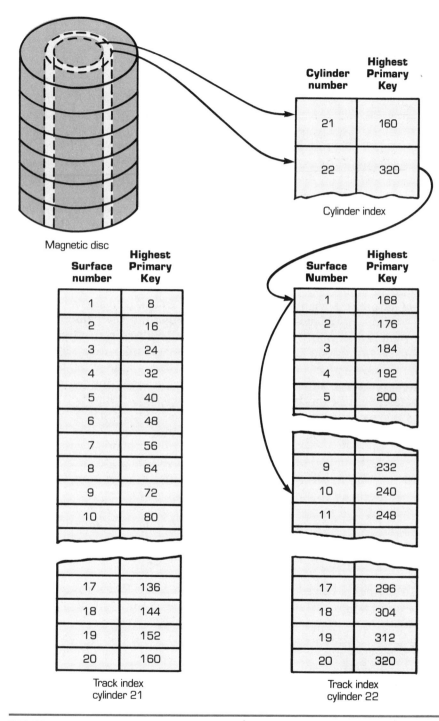

FIGURE 11–5 An example of indexed-sequential organization

to find out the relationship between each student and his or her classes (e.g., which classes a student is enrolled in) and the relationship between each class and its students (e.g., which students are enrolled in a class). How do we design files for situations such as these? Let's consider the construction of entity-relationship diagrams, which provide a valuable tool in separating objects from their relationships.

Entity-Relationship Diagrams

Many systems designers rely heavily on entity-relationship diagrams to model the file processing specifics of a system. Introduced by Chen[2,3] in 1976, an *entity-relationship (E-R) diagram* shows individual entities (object sets) and occurrences and their relationships. **Figure 11–6** illustrates an E-R diagram. As shown, each entity set is shown as a rectangle, each relationship as a diamond-shaped box. An *entity set* represents a distinct set of objects such as customers, projects, books, ships or soldiers. A *relationship* describes a meaningful interaction between objects. For example, suppose that a borrower (the first object) reserves a book (the second object). In this instance, the word *reserves* describes the relationship between the borrower and a book.

E-R diagrams are drawn to show the number of *entity occurrences* between entity sets. If there is no maximum, the number is shown as "many." **Figure 11–7a** illustrates a one-to-many (1 to N) E-R diagram. The interpretation of this diagram is as follows: one department is as-

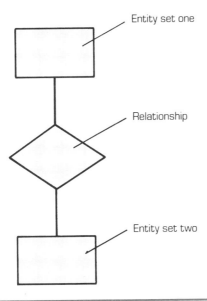

FIGURE 11–6 An E-R diagram

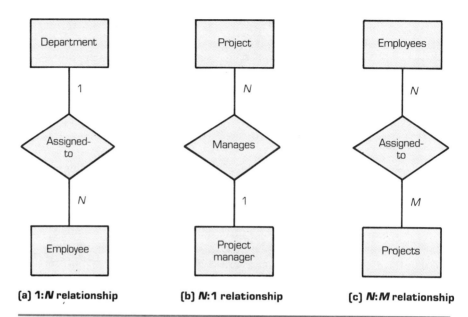

FIGURE 11-7 1 to N, N to 1, and N to M E-R diagrams

signed to each employee; however, N employees are generally assigned to a department. Likewise, a many-to-one (N to 1) E-R diagram can be drawn. **Figure 11-7b** reads as follows: N projects can be managed by one project manager; however, one project manager manages each project. Finally, a many-to-many (N to M) E-R diagram can be drawn. **Figure 11-7c** shows: N employees can be assigned to work on each project, and M projects can be assigned to each employee.

Besides entity occurrences, the attributes associated with each entity set are usually added to each E-R diagram. **Figure 11-8** illustrates that

1. The entity set "borrower" consists of three attributes: the borrower number, the borrower name, and the borrower telephone. The key attribute is the borrower number.

2. The entity set "book" consists of five attributes: the ISBN (the international standard book number), the author, the title, the publisher, and the date of publication. The key attribute is the ISBN.

3. The relationship "checks-out" consists of three attributes, the borrower number, the ISBN, and the return date; however, two of these are key attributes (hence they are underlined) and are, in fact, the same key attributes as those shown for the entity sets. The return date is added to the relationship checks-out instead of to the entity set called book, because it describes a variable relationship between a borrower and a book. Since many books may have the same return date, the return date acts as a secondary key attribute.

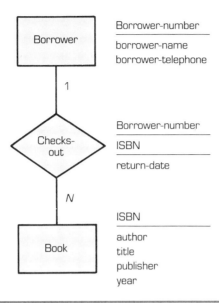

FIGURE 11–8 E-R diagram complete with attributes

This replication of key attributes in describing the relationship does not occur by accident. Rather, by repeating primary and secondary keys, the file designer can eliminate all repeating nonkey groups. For example, suppose that a borrower checks out all three books. We might organize this information as one large record, such as the record shown on **Figure 11–9.** However, as indicated by the shaded areas, this record contains repeating key and nonkey groups, with the nonkey groups indicating the borrower name and telephone number, respectively.

Now let's suppose we create two physical records to store the relationships indicated by the E-R diagram. **Figure 11–10** illustrates this new method of organization. In this case, all repeating nonkey groups are eliminated.

The process of eliminating all repeating nonkey groups is called the *normalization of relations*. Through normalization, the quantity of data to store is generally minimized. Following the process of normalization, all records will be of fixed length. Likewise, normalization typically leads to the creation of several physical files, called *relations*. As shown by the library example (see Figure 11–10), two relations are created: one to store borrower records and one to store book records. While a checks-out relation might also be created, in practice it is not. Why? Because in a 1-to-N relationship, the primary object (in this instance the borrower) is made up of facts needed to describe a set of associated objects (in this instance books).

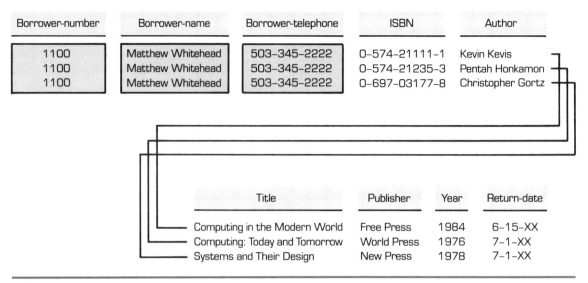

FIGURE 11-9 Single record with repeating groups

Relational Design Rules

To gain a better appreciation of relational file design, let's summarize relational design rules:

1. A relational file is limited to fixed-length records.
2. All repeating nonkey groups must be eliminated from a record.
3. A nonkey field must provide a fact about the key.
4. Storage is restricted to items that cannot be derived, (i.e., computed).

Borrower-number	Borrower-name	Borrower-telephone
1110	Matthew Whitehead	503-345-2222

(a) Borrower record

ISBN	Borrower-number	Return-date	Author	Title	Publisher	Year
0-574-21111-1	1110	6-15-XX	Kevin Kevis	Computing in the Modern World	Free Press	1984
0-574-21235-3	1110	7-1-XX	Pentah Honkamon	Computing: Today and Tomorrow	World Press	1976
0-697-08177-8	1110	7-1-XX	Christopher Gortz	Systems and Their Design	New Press	1978

(b) Book record

FIGURE 11-10 Normalized relations

Other rules govern the number of physical files to create:

1. In a one-to-one relationship with membership required, only one physical file is required.
2. In a one-to-many relationship with membership required, two physical files are required.
3. In a many-to-many relationship, three physical files are created.

Let's return to our student and class example to illustrate how these rules work in practice. As shown by the E-R diagram (see **Figure 11–11**), the entity sets student and class are linked together by the relationships "is-enrolled-in" and "contains." We can read this diagram as:

Each student is enrolled in <u>N</u> classes (and)
Each class contains <u>M</u> students.

In the construction of relational files for this design problem, one file would contain student information, which we can now define as

Student (<u>Student-number</u>, student-name)

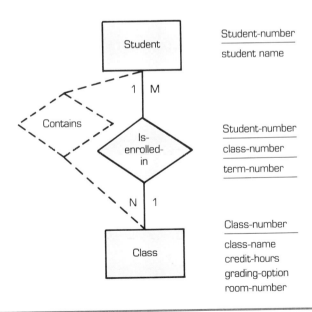

FIGURE 11–11 Student-to-class E-R diagram

Likewise, we can define class and "is enrolled in" as:

Class (<u>Class-number</u>, class-name, credit-hours, grading-option, room-number)
Is-enrolled-in (<u>Student-number</u>, <u>class-number</u>, <u>term-number</u>)

Note that we might also define the file named "contains"; however, this file would store the same data as the is-enrolled-in file. (Howe calls this a preposted condition.[4])

To complete this problem, let's suppose that we wish to store each student's grade in a course and compute the student's total credit hours and grade point average (GPA). Would we need to create a new file? The answer would be yes if a grade were not related to a student and a class. However, since it is, we can simply append it to the relationship file, so that

Is-enrolled-in(<u>Student-number</u>, <u>class-number</u>, <u>term-number</u>, grade)

Viewed another way, the layout of this new file would look as follows:

Student-number	Course-number	Term-number	Grade
111	234	F8XX	B
111	324	W8XX	A

What about total credit hours and GPA? Would these be stored as well? In this instance, the answer would be no, since both can be computed. By knowing the course number, we can retrieve the credit hours for each course and sum total credit hours in this way, or calculate the GPA directly (i.e., by multiplying the grade points earned and dividing that figure by the total credit hours earned).

The rules of relational design, while seemingly simple enough, are more difficult to implement than they appear. Suppose that Mansfield, Inc., decides to expand its payroll system to account for project times. We will let Carolyn Liddy, the systems specialist, explain the logical steps in processing.

"We needed to create a project file to store the number, name, scheduled hours, and hours consumed for each project," she said. "Then, in addition, we were required to modify the data entry portion of the payroll system to collect for each employee the project number they worked on, the date, and the number of hours worked on the project.

"**Figure 11-12** illustrates the E-R diagram describing this modification," Carolyn continued. "As indicated, the employee entity set is the same as before, with each employee record consisting of an employee number and record subgroups. These subgroups were documented using a Warnier-Orr diagram (see Figure 6-12). The project entity set contains

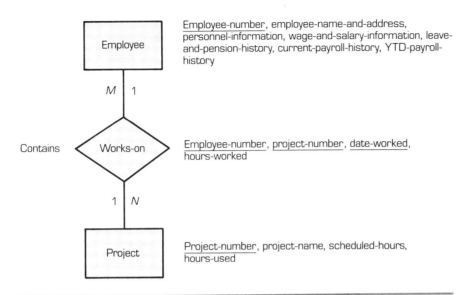

FIGURE 11–12 Employee-to-project E-R diagram

much smaller records. Each project record was limited to a project number, project descriptors (project name), and project planning and control variables (scheduled project hours). Finally, the works-on file consists of information describing the relationship between employees and projects. Besides the employee number and project number, each works-on record includes the date (a third key) and the number of hours the employee worked on a project (hours worked). The value of this relational file is that it permits us to create a variety of project-reporting lists (see **Figure 11–13**). For example, we can show the employees who work on a project, the projects an employee has worked on, or the projects worked on during a specified day of the week."

Project number		Employee number		Date worked	
employee-number	1	project-number	1	project-number	1
employee-number	2	project-number	2	project-number	2
employee-number	3	project-number	3	project-number	3
.		.		.	
.		.		.	
employee number	X	project-number	Y	project-number	Z

FIGURE 11–13 Lists produced from the works-on file

DESIGN OF SYSTEM FILES

Following the modeling of file relationships, the systems design team is ready to design system files. The objectives of system file design are to provide for efficient use of auxiliary file storage devices, such as magnetic tape or disk devices, and contribute to efficiency in computer program processing. Two design steps are required in realizing these objectives: determining file processing requirements and preparing the physical file design.

Determining File Processing Requirements

Systems design requires the development of different types of files. These include master files, transaction files, summary files, suspense files, backup files, and change files (see Eliason[5] regarding the choice of each based on the system to be implemented).

MASTER FILES

Master files contain relatively permanent records describing the details of an event, an object, or a person. They typically represent the entity sets shown on E-R diagrams and are labeled as master files on system flowcharts. Since most master files must be updated online, either direct, indexed-sequential, or relational file organization is required. With sequential organization, inefficiency in computer program processing results.

TRANSACTION FILES

Transaction files contain relatively temporary records describing the details of an action or event. They typically represent the relational files shown on E-R diagrams and are created by a specific computer processing run. Transaction files are often organized as sequential files when a single listing of transaction records is required. When several lists are needed, such as the listing possible from the employee-to-project relationship (see Figure 11–13), however, a relational method of file organization is advised.

SUMMARY FILES

Summary files contain condensed versions of master or transaction file records. These files are most often used in passing data from one computer application to another or for storing historical information when it is not necessary to preserve all the details of processing. Suppose that a historical summary of total hours worked by year and month and by project is required by Mansfield. The file could be called project summary and defined as follows:

```
Project-summary (Project-number, year, month, total-
                 hours-worked, total-employees-involved)
```

How would this file be created? At month end, a small computer program would be required to sum hours worked by project and to count the number of employees involved. Historical summary files are usually organized as sequential files and stored on magnetic tape files, while summary files passed from one computer application to another are often organized as indexed-sequential or as relational files. The choice of file organization depends on how often individual records need to be retrieved and brought into processing. With historical records, most often the entire group of records needs to be analyzed, rather than an individual record. Hence, it makes more sense to load and read blocked historical records from a magnetic tape than to keep these records on-line.

SUSPENSE FILES

Suspense files store records in error. These files are usually created to store keyed data that, for some reason, contain errors as identified by an input error-checking computer program. One alternative to this problem is to require data entry operators to rekey all data in error, thus avoiding the creation of any suspense files. Another alternative is to store data in error online. Once the problem with the record has been identified, the data in error can be retrieved, edited, and tested once again. If all tests are successful, the record is entered into processing as a normal record. If a test detects still another error, the record is returned to the suspense file to await further editing. Because individual records must be brought in and out of processing, and because a sequential listing of errors is often not required, suspense files can be organized as direct files. Indexed-sequential and relational organization are other design options, however.

BACKUP FILES

A backup file is a copy of a master, transaction, summary, or suspense file that is made for security reasons. Most often backup files are organized sequentially to show how records are physically stored on tape or disk files. File designers must specify the backup intervals for all files. They may decide to backup all transaction and suspense files daily and all master files every three days, for example.

CHANGE FILES

A change file is a special type of backup file and contains adjustments made to master and suspense files. Most change files show before- and after-change versions of a record to clearly document the effects of a processing change. Similar to backup files, change files are organized as sequential files.

Database Considerations

Besides determining which types of files are needed by a system, the design team must ascertain whether to store data on a database (presupposing for the moment that this option is feasible) or on files separate from the database. As stated in chapter 6, a *database* is a collection of interrelated data, with minimum redundancy, that serves one or more computer applications. An employee database, for example, might be shared by a payroll processing computer application, a personnel management computer application, a project management computer application, a job costing computer application, and a wage and salary computer application. Instead of storing the same employee data for each application, a single pool of data is maintained for all applications that require employee information (see **Figure 11–14**).

A *database management system* (DBMS) is required by an organization to establish files that can be shared among applications. Such a system usually contains four software components: a data description component, a data manipulation component, a query-language component, and a database utility component. While an in-depth discussion of any one of these components is beyond the scope of this text, an overview of each can be provided.

The *data description component* includes a *data description language* (DDL) to define the files (relations), records, attributes, and key

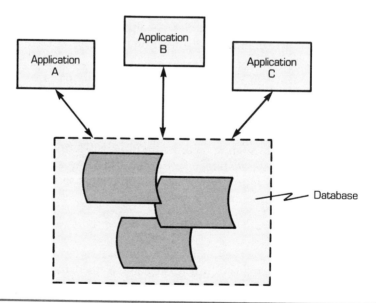

FIGURE 11–14 Database permitting several applications to share stored data

attributes to be stored within the database. It is also used to specify data representation (e.g., binary or byte, attribute length, maximum or minimum values, and other facts).

The *data manipulation component* includes a *data manipulation language* (DML) that is to be embedded within an application program. A program written in COBOL, for example, would contain DML instructions if the program required data from the database. DML instructions permit the application program to send and receive messages and data from the database.

The *query-language component* includes a *query language* that permits users to access data from the database independent of any application program. Consider the following query written in the language *structured query language* (SQL):

```
SELECT EMPLOYEE-NAME
FROM EMPLOYEE
WHERE DEPARTMENT-CODE = 500
```

In this example, users specify that they wish to select from the pool of employee data the names of the employees assigned to department 500.

The *database utility component* consists of a series of utility programs that allow the person responsible for managing the database (generally a *database administrator*) to create, restore, back up, and modify the database, independent of application programs.

Database Types

There are several types of commercial DBMSs, including those based on linked-list data structures, hierarchical or tree data structures, plex or network data structures, and the relational model. Once again, we will provide an overview of each.

A *linked-list* approach features internal and external pointers to permit records in common to be chained together. As discussed in chapter 6, linked-list systems permit inverted files to be created in processing; as shown in **Figure 11–15,** they permit both the internal and external chaining of records.

A *hierarchical* or *tree* approach features associations between entity sets. As illustrated in **Figure 11–16,** a single department is described as a collection of employees. In the drawing of a schematic of this relationship, the double arrowhead implies many, while the single arrowhead indicates one.

With a hierarchical design, the top of the tree is called the root, the middle-level nodes are referred to as branches, and the lowest-level nodes are called leaves. A parent-child relationship is also used in describing

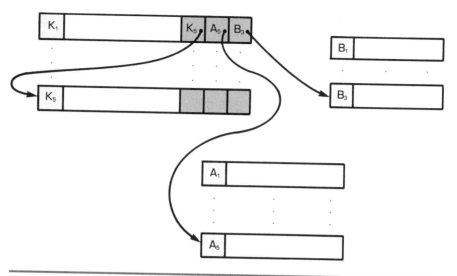

FIGURE 11–15 Linked-list approach featuring internal and external pointers

this type of design. A superior node, such as an employee's department, is called the parent; a node subordinate to another is called the child. Thus, an employee is a child of a department.

A *plex* or *network* approach is a structure in which a child has more than one parent (see **Figure 11–17**). DBMSs vary in the ease with which they can represent these types of data structures. Some DBMSs allow a child to have a maximum of two parents (a simple network). A child with more than two parents must be redefined.

A *relational model* approach, as discussed earlier in this chapter, portrays data in a two-dimensional array or table form. As such, it is easy to understand, can handle both simple and complex networks, and

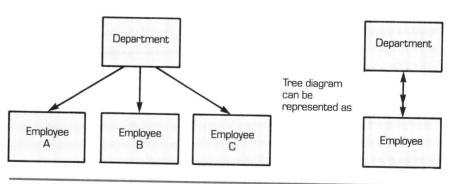

FIGURE 11–16 Hierarchical or tree schematic relationship

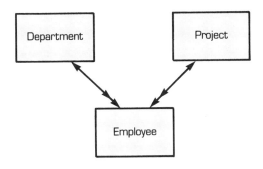

FIGURE 11–17 Network schematic relationship

requires no special indexes or pointers. How, then, can data be linked together? The answer lies within the data itself. Two records are related if they contain identical key attributes. For example, by matching borrower numbers, it can be determined whether a borrower has checked out a book (see Figure 11–10). By matching ISBNs together with a borrower number, it can be determined whether a specific book has been checked out.

Preparing the Physical File Design

Preparing the physical file design is a multistep process. The design team must

1. determine which types of files are required by the system, including those to be housed within the database;

2. specify file relations and records; and

3. make file-size projections and determine the impact of the design on file storage devices.

In determining which types of files are required, the design team might begin by altering system flowcharts to specify file types and database requirements. **Figure 11–18** provides an illustration. This flowchart shows that the program update will involve two relations maintained by a relational DBMS and a sequentially organized product change file. All relations and files, as indicated, will be magnetic disk storage devices; for magnetic tape storage, a magnetic tape flowcharting symbol is required.

Once the types of files are known, file relations and records can be specified. In practice, this specification is straightforward, provided E-R diagrams have been prepared in advance.

Record layout forms are part of the file relation and record specifica-

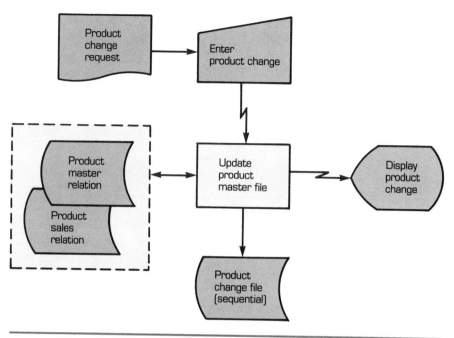

FIGURE 11–18 Altering system flowchart to specify physical files required in processing

tion. Similar to system output documentation shown in chapter 10 (see Figure 10–7), these forms name each field; show whether the field is alphabetic, alphanumeric, binary, numeric-packed, or numeric; indicate the source of the data (source document, computed by computer program, or transferred from another file); specify the field length in characters and in bytes; and provide special instructions, such as maximum and minimum field values.

Blocking and making file-size projections are other factors important to the physical file design specification. Suppose that the design team plans to block logical records written to the product change file shown on Figure 11–18. Suppose further that the designers determine that a maximum of 3,000 records will need to be stored, that each record will require 53 bytes of storage, and that the maximum physical record size is 1,000 bytes. Given this information, the design team can calculate the blocking factor as 18 (1,000 divided by 53), determine that each physical record will require 954 bytes (18 times 53 bytes divided by the record), and estimate that 167 physical records (3,000 divided by 18) will be stored at any one time.

Next, suppose the data in question are to be stored on magnetic disk and that eight physical records can be stored on a single track. Knowing this permits the design team to calculate the following:

$$\text{Logical records per track} = \text{Blocking factor} \times \frac{\text{physical records}}{\text{track}}$$
$$= 18 \times 8 = 144$$
$$\text{File size (in tracks)} = \frac{\text{Maximum logical records}}{\text{Logical records per track}}$$
$$= \frac{3{,}000}{144} = 20.83 = 21$$
$$\text{File size (in cylinders)} = \frac{\text{File size (in tracks)}}{\text{Tracks per cylinder}}$$
$$= \frac{21}{9} = 2.3 = 3$$

Of what significance are estimates such as these? By estimating the total cylinders required for all new system files, the design team can determine the impact of the system on existing file storage space and indicate whether existing space will be sufficient. If it is not sufficient, the design team must acquire additional disk capacity or redesign the system to reduce overall disk storage requirements. Writing the product change file to magnetic tape storage instead of magnetic disk storage, for example, will release three cylinders of disk storage for other system uses.

The Mansfield, Inc., Case Study

With Carolyn and Pamela working on the design of system output and input, John and Walter turned their attention to the design of computer files. "We will be working with a relational DBMS," remarked John. "Perhaps the best way to approach the design effort is to select the files to be managed by the DBMS, develop the E-R diagrams for these files, and prepare a record format analysis of each required relation."

Defining DBMS Requirements

Following careful study of the logical design for the billing and receivables system (see Figure 7–15), John and Walter concluded that customer and product relations, accounts receivable relations, and sales summary relations would need to be designed. They also concluded that the invoice transaction file would not be part of the DBMS. Walter explained: "We need to be able to retrieve individual customer, product, receivables, and sales summary records. In addition, we will need a sequential listing of all records stored on these four files. Thus, our file organization choices were limited to either using the DBMS or creating indexed-sequential files. We selected the DBMS to store all four files."

He added: "The invoice transaction file was kept separate from the DBMS because we did not need to retrieve individual records in this instance, nor did we consider it necessary to store all invoice details in the database, especially when the important items, such as receivables and sales totals, were stored elsewhere. We concluded that the invoice transaction file would be easier to work with if we left it as a sequentially organized collection of records."

Entity-Relationship Diagrams

The entity-relationship diagrams for the customer, product, accounts receivable, and sales summary files clarified how all records were to be stored. "The product file, for example, is made up of four relations," John explained. "Besides a main product description relation, a product price, product cost, and product location relation are required (see **Figure 11–19**). The product price and cost relations are necessary because our costs and prices change periodically. If a customer returns a product for a credit, we must be able to supply the correct price and cost figures. The product location relation is necessary

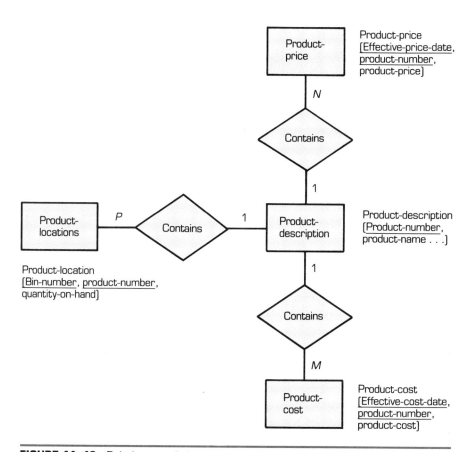

FIGURE 11–19 Relations needed to store product records

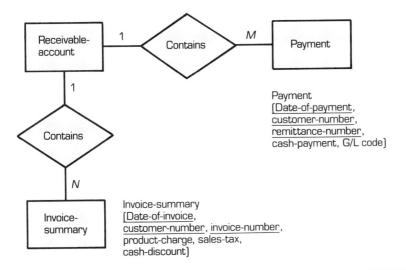

FIGURE 11–20 Relations describing the accounts receivable file

because a single product can be stored in several bin (warehouse) locations. We need to know the quantity on hand at each location.

"The accounts receivable file is subdivided into three relations," John continued, "a receivable account relation, a payment relation, and an invoice summary relation (see **Figure 11–20**). The receivable account relation stores basic information about each of our customers. The payment relation is mandated since customers can make several payments during a month. Likewise, the invoice summary relation is mandated in that customers can receive several invoices during the month."

Record Format Analysis

The final step of the file design process was to describe in detail each field within each relation. "This task is easier than preparing E-R diagrams," explained Walter. "For example, we defined the product price relation as follows:

```
Product-price(Product-number, effective-price-date,
              product-price)
```

A record format analysis helps complete this definition. We begin by building a table for each relation, such as the table shown below. This particular table tells us that the product price relation is limited to three numeric fields, that the total characters per record is eighteen, and if all numeric characters are stored in packed-decimal format (i.e., two digits can be packed within a single byte) the byte requirement per record is eleven."

Relation 1: Product price relation

Field	Field Type	Number of Characters/Bytes
1. Product-number	numeric	5/3
2. Effective-price-date	numeric	6/4
3. Product-price	numeric	7/4
Total		18/11

SUMMARY

Once the difference between logical records and files and physical records and files is understood, the design of computer files is considerably simplified. Central to this understanding is the realization that data can be organized and stored on files in one way and organized and used in program processing in another way.

Record and file structure help clarify some of the options available to the systems design team in creating a record to be placed on a file and in organizing the data structure of the file itself. Of the three types of record structure, the fixed-length record is often required. It is the only type that can be handled without difficulty by either direct or indexed-sequential methods of file organization; it is required by the relational model.

Most designers realize that the details of relationships between different objects must be stored on computer files. To this end, entity-relationship (E-R) diagrams have proved most useful in modeling the relationship between different entity sets and their occurrences.

Relational file design has evolved as an alternative to other common methods of storing records on files. With relational design, the normalization of relations is required. Normalization eliminates all repeating nonkey groups from a record and restricts storage to items that cannot be derived.

Following the modeling of file relationships, the design team can determine file and database processing requirements. This step is usually straightforward, provided E-R diagrams and file and database considerations have been carefully thought through. The only remaining steps are to specify specific file relations and records, to make file projections, and to determine the impact of the design on file storage devices.

REVIEW QUESTIONS

11-1. What is meant by blocking records? What is a blocking factor?

11-2. What is the difference between a physical record and a logical record? Between a physical file and a logical file?

11-3. What is the difference between a fixed-length and a variable-length record structure?

11-4. What is an interrecord gap?

11-5. How does the search for a record stored on a sequentially organized file differ from the search for a record stored on a directly organized file?

11-6. What are the main advantages of indexed-sequential organization? What are the main disadvantages?

11-7. Why do systems designers construct entity-relationship diagrams?

11-8. What process is involved in the normalization of relations?

11-9. State the rules followed in relational file design.

11-10. From the design team's standpoint, what are the main objectives of system file design? What two steps are required in order to realize these objectives?

11-11. What is the difference between a master file and a transaction file?

11-12. How does a database differ from a database management system (DBMS)?

11-13. State the main difference between a hierarchical or tree DBMS and a plex or network DBMS.

11-14. How does the relational model allow data to be linked together?

11-15. What steps are taken by the design team in preparing the physical file design?

EXERCISES

11-1. Suppose a designer estimates the following:
- 25,000 35-byte logical records must be stored on an external file.
- Records of up to 1,200 bytes in size can be passed between the internal memory of the computer and the file storage device designated to store the 25,000 records.

What is the maximum blocking factor, given these estimates? What is the maximum size, in bytes, of each physical record?

11-2. A business has 100,000 business accounts numbered from 101,000 to 200,000; however, only 60 percent, or 60,000 accounts, are active at any one time. A single magnetic disk file storage device is available to store these business accounts. This disk will hold a maximum of 75,000 accounts.

Devise an indirect addressing hashing scheme to show where account 100,000, account 150,000, and account 200,000 would be stored on this disk.

11-3. Draw E-R diagrams to show the following:
- Several systems analysts are hired to work on Project Newsprint.
- John Turner goes to the library to check out several books.
- The musicians play several musical selections.
- The company is made up of several departments, with each department working on several projects. (Use a single diagram to show this joint relationship between company, departments, and projects. Do not show the relationship between company and projects).

11-4. What is wrong with the definitions written for the E-R diagram shown below?

Car (<u>Car-number</u>, car-type, model-number, year)
Is-driven-by (<u>Car-number</u>, <u>driver-name</u>)
Driver-name (Driver, <u>date-driven</u>, agency)

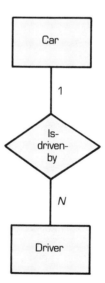

11-5. Suppose we are given a set of customer invoicing records (see page 393). Suppose further that we wish to transform this set of records into a normalized set of relations. Draw an E-R diagram describing the normalized structure. Create the normalized relations using the information contained in the customer records.

Customer Number	Customer Name	Street Address	City	State	Zip Code
6200	E. I. Jones, Inc.	1234 S. W. City Drive	Portland	OR	97400
6200	E. I. Jones, Inc.	1234 S. W. City Drive	Portland	OR	97400
6200	E. I. Jones, Inc.	1234 S. W. City Drive	Portland	OR	97400

Invoice Number	Date of Invoice	Extended Price
11246	9-26-XX	$34.95
11297	9-28-XX	$92.20
11424	9-28-XX	$22.10

11–6. The Martin-Montgomery company provides us with the following information about their active projects. They instruct us to transform this information into a normalized set of relations. Draw an E-R diagram describing the normalized structure. Design the normalized relations using the project information.

Project Number	Project Name	Start Time	Employee Number	Employee Last Name	Job Title	Hours Worked on Project
ABX	FANCORE	July	10	Johnson	Senior analyst	40
ABX	FANCORE	July	20	Martins	Junior analyst	10
JBZ	FLIGHT	August	10	Johnson	Senior analyst	30
JBZ	FLIGHT	August	20	Martins	Junior analyst	35

11–7. Of the records written to the product change file shown on Figure 11–18, suppose we know the following:
1. Each record requires 125 bytes of storage.
2. Three thousand records are to be stored.
3. The physical record size is 1,000 bytes.
4. Data are to be stored on magnetic disk.
5. Each track of the disk can store 7,264 bytes.
6. Each cylinder of the disk contains nine tracks for storing records.

What is the blocking factor? How many logical records can be stored on each track? What is the file size (in tracks)? What is the file size (in cylinders)?

WORLD INTERIORS, INC.—CASE STUDY 11
DESIGNING COMPUTER FILES

Introduction

After completing the preliminary designs of system inputs and outputs, Jane Strothers began to work on the design of computer files. Her conversations with John Welby informed her that WI would be acquiring a new computer that featured a powerful relational DBMS. With this in mind, Jane thought it would be best to model the data structure she would be considering. "Entity-relationship diagrams would help," she remarked. "After I have a firm understanding of how everything ties together, I can design relations and determine byte requirements for the system files."

The Open-Order File

Jane started her analysis by examining the logical contents of the open-order file. Drawing on the DFDs John had prepared, she reflected, "It is the one file that links order taking with our accounts payable processing. I'd better remember this when I begin working on computer programs."

In review of her data input layouts, Jane realized that the open-order file initially contained three types of information: customer information, customer order information, and line-item information (see case assignment 1, chapter 10). She also realized that the contents of each stored open order would be modified at least once and perhaps more than once. For example, when the factory invoice was processed, the open order would be modified. Besides the addition of the factory invoice number and the date of the invoice, payables information, such as the remittance number sent by the factory, would need to be added if these data were to be used later in processing. As another example, the open order would be modified if it became a problem order. A problem-order flag and the date of the problem order would be added in this instance.

Jane's review of the open-order file led to several other conclusions. First, each customer order resulted in a single customer invoice. To retrieve an invoice, three methods of retrieval would be possible: by customer invoice (order) number, by customer key, and by date of customer invoice. Second, each customer invoice typically contained several line items. These line items would need to be saved for possible display. Third, each customer invoice led to one or more factory invoices. If one factory could not fill an entire order, another factory might be asked to make a partial shipment. This led to two factory invoices for a single order. Fourth, each customer order could lead to one or more customer service sheets. The customer might receive the wrong count on the first shipment and the wrong finish on the second. Fifth, the customer key would need to be retained. When customers contacted WI, they generally could not remember their invoice number or when they had placed an order. Finally, a record code was required. This code would separate customer debit amounts from customer credit amounts.

Jane decided to simplify the design somewhat. She would match factory invoice line items against open-order line items; however, she concluded that it would be sufficient to store only the factory invoice summary data. These data consisted of the factory invoice number, date of the factory invoice, supplier number, remittance number, record code, invoice dollar total, due date, and factory terms code. Line-item details, such as item number, quantity shipped, finish number, and price each, were not to be stored for each factory invoice.

Once this decision had been made, Jane felt she understood how all of the relations would be defined except the ties between the customer invoice and customer service sheets. John helped her on this one. "Here, too, we want to keep the design simple," he stated. "Let's examine the type of output we desire. Since we want to list only those customer invoices with problems, let's set the problem-order flag and record the date of the problem. Let's not store the details of customer service sheets at this time."

The Supplier File

Jane's review of the logical contents of the old supplier product file led her to conclude that this file needed substantial change. "What we could not do with the old system was determine the quantity sold of each item by finish number and name," she commented. "We would like to do so in the future, especially since the suppliers have a tendency to add a new finish and drop another off the item list."

Besides wanting a breakdown by quantity sold and the finish number, Jane wanted to keep track of unit price and cost by item number, along with the number of orders taken by finish number. For reporting purposes, Jane wanted to know month-end dollar sales, contribution to profit totals, and average order size, by three categories: supplier, item number, and finish number. After these were reported, the fields in relations would be reset to zero to allow for new sales and order figures to be collected.

CASE ASSIGNMENT 1

Prepare an entity-relationship diagram or diagrams showing the relationships between the following entities: customer invoice, invoice line items, and the factory invoice. Show the keys important to all entity and relationship sets.

CASE ASSIGNMENT 2

Prepare a record format analysis of the open-orders file, showing the relations to be created for the DBMS (and diagrammed for case assignment 1). For each relation, provide a name. For each field within a relation

1. supply a name;
2. indicate whether the field is alphabetic, alphanumeric, or numeric;
3. indicate the number of characters required; and
4. indicate the number of bytes required, where numeric bytes are packed (i.e., two digits per byte). Estimate field lengths as necessary.

CASE ASSIGNMENT 3

Prepare an entity-relationship diagram or diagrams showing the relationships between the following entities: suppliers, items carried by suppliers, and available finishes. Show the keys important to all entity and relationship sets.

CASE ASSIGNMENT 4

Prepare a record format analysis of the supplier file, showing the relations to be created for the DBMS. For each relation, provide a name. For each field within a relation

1. supply a name;
2. indicate whether the field is alphabetic, alphanumeric, or numeric;
3. indicate the number of characters required; and
4. indicate the number of bytes required, where numeric bytes are packed.

Estimate field lengths as necessary.

REFERENCES

1. R. Ellzey, *Data Structures for Computer Information Systems* (Chicago: Science Research Associates, 1982).
2. P. Chen, "The Entity-Relationship Model—Towards a Unified View of Data," *ACM Transactions on Database Systems* 1 (March 1976): 9–36.
3. P. Chen, "The Entity-Relationship Model—A Basis for the Enterprise View of Data," *AFIPS Conference Proceedings* 46 (1977): 77–84.
4. D. Howe, *Data Analysis for Data Base Design* (London: Edward Arnold Publishers, 1983).
5. A. Eliason, *Online Business Computer Applications* (Chicago: Science Research Associates, 1983).

12
Computer Program Design

INTRODUCTION

ONCE the preliminary design of processing, the design of inputs and outputs, and the design of computer files and databases are complete, the systems design team can turn its attention to the *detailed design*—the design of computer programs to be coded and tested. Detailed design differs from the preliminary design in the following ways: data couples and flags are added to each module within the hierarchical structure of functions; program module specifications are prepared to show the *intent* of each module; and pseudocode is written to describe the *implementation* of each module.

Where to Begin?

The question of where to begin computer program design is answered in large part by preceding systems design activities. Consider the design of Mansfield's payroll system. Initially, the design team members divided the new system into three top-level modules: update hours worked, print payroll register, and print employee paychecks. Once they realized that their design would be a transaction-based design, they began to develop top-level structure charts for each of the three primary modules—a level-1 structure chart was designed for updating hours worked, a level-2 structure chart was designed for printing the payroll register (see Figure 3–13), and a level-3 structure chart was designed for printing employee paychecks.

Beginning at the uppermost level of a design and working toward ever-increasing levels of detail is called *stepwise refinement*—making a steady progression from general functions to detailed subfunctions. Principles important to this refinement process will be discussed first. Following this discussion, the factors important to describing the intent and the implementation of a programmed module will be considered.

When you complete this chapter, you should be able to

- describe principles that govern the design of program structure charts;
- describe rules important to the use of flags;
- construct a properly partitioned set of structure charts;
- write a nonprocedural and a procedural module specification;
- evolve a top-down program design.

PROGRAM STRUCTURE CHARTS

Structure charts, as defined earlier, provide a detailed graphic picture of the internal organization of a computer program. They are similar in design to VTOCs, as discussed in chapter 9, in that they show a hierarchical ordering of functions. Structure charts differ from VTOCs in their presentation of program modules and the linkages between these modules, however. Remember the four functions of a module (input, output, processing function, and internal data)? Structure charts feature data couples and flags to graphically illustrate which inputs are received by a module and which outputs are produced. Module specifications serve to provide a written description of the last two functions: processing function and internal data.

Just as there are guidelines for the construction of HIPO diagrams, so too are there guidelines for the design of structure charts:[1]

1. Principle 1—A large system should be partitioned into a hierarchy of smaller modules. This is called the principle of partitioning.
2. Principle 2—Each module should have little dependence on any other module. This is called the principle of coupling.
3. Principle 3—Each module should carry out a single processing function (also called the principle of cohesion).
4. Principle 4—Each module should indicate by its title what function is to be performed (also called the principle of clear labeling).
5. Principle 5—Each module should coordinate the functions of a reasonable number of subfunctions. This is called the principle of span of control.
6. Principle 6—Each module should be of reasonable size (also called the principle of reasonable size).
7. Principle 7—Two or more superior modules should call on the same subordinate module whenever the function to invoke is the same (also called the principle of shared use).

Principle of Partitioning

Of the various methodologies developed to assist in the process of systems analysis and design, those permitting large, complex problems to be partitioned into smaller, less complex problems are perhaps the most valuable. As we have seen, the construction of DFDs, a data dictionary, Warnier-Orr diagrams, HIPO diagrams, and now structure charts requires partitioning. With structure charts, the meaning of partitioning is especially clear: a large system is to be subdivided into a hierarchical collection of modules. Each module describes the make-up of part of the larger, more complex system.

Figure 12–1 provides another example of partitioning. The more complex function of making coffee can be partitioned into four more detailed modules: set up coffee maker, add water, add coffee, and run hot water through coffee.

Consider the third function, add coffee. The processing function of this module is clear; however, the input, output, and internal data required in processing require clarification. Thus, the designer could state: the input to this module is the coffee, output is a signal that coffee has been added, and internal data indicate whether there is sufficient coffee to add at this time. The designer might also provide an instruction set to show how the processing function, add coffee, is to be performed. Such an instruction set (in outline form) might be

1. Get coffee.

2. Get coffee scoop.

3. Add one scoop of coffee for each two cups of water.

4. Put away coffee.

5. Put away scoop.

As with HIPO diagrams, structure charts deal with get, make, and put instructions.

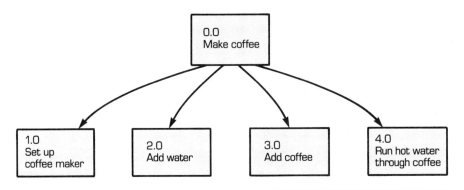

FIGURE 12–1 Partitioning a function into subfunctions

Principle of Coupling

A major difficulty with partitioning a system into modules lies in describing the interfaces or links that tie modules together. As a general rule, the systems designer attempts to keep any interface as simple as possible, so that there is a low dependence between each superior and subordinate module (i.e., each boss and worker module). Stated another way, the designer attempts to achieve "loose coupling" between all modules—between all bosses and workers.

Figure 12–2 illustrates the difference between loose and tight coupling. With *loose coupling*, the interdependence between modules is kept to a minimum. Why is less interdependence important? Because many errors in processing result from defective data being passed from one module to another. Study Figure 12–2a and 12–2b again. With loose coupling, the probability of passing defective data between modules is greatly reduced.

Four rules help clarify how coupling can be improved in the design of computer programs:

1. Pass neither too few nor too many parameters between modules.

2. Avoid passing superfluous data to lower-level modules.

3. Avoid passing data upward before they are needed by upper-level modules.

4. Check reporting relationships to ensure that subordinate modules report to the correct superior module.

Several examples will clarify the significance of these four rules. **Figure 12–3a,** for instance, illustrates a situation in which too few data are re-

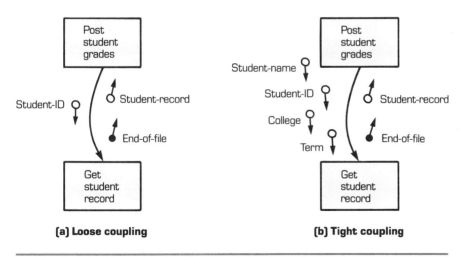

FIGURE 12–2 Loose (a) versus tight (b) coupling

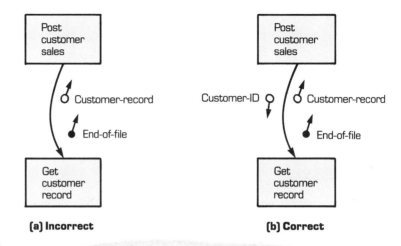

FIGURE 12–3 Passing too few parameters (a) versus passing the correct number (b)

ceived by lower-level modules. In reviewing this section of a structure chart, a designer might comment: "How does the module called get customer record know which customer to get?" **Figure 12–3b** shows the correct passing of parameters.

Figure 12–4a illustrates a situation in which superfluous data are passed downward to a lower-level module. The correct solution (see **Figure 12–4b**) is to send only the data needed by the module called key student grade. These data are limited to the student ID and the student name.

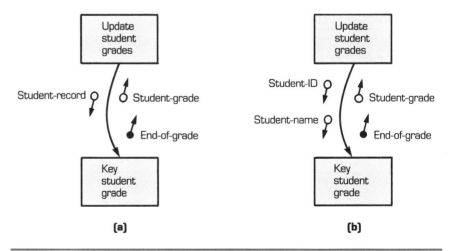

FIGURE 12–4 Passing superfluous data (a) versus passing correct identifiers (b)

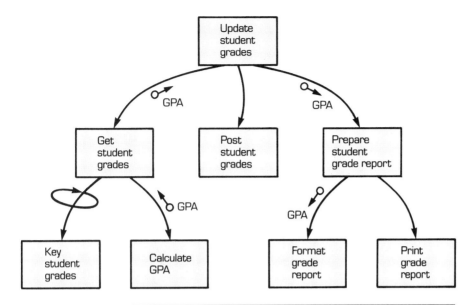

FIGURE 12–5 Allowing data to float around a design

Figure 12–5 shows a design that violates two rules: data are passed upward before they are needed, and a subordinate module reports to an incorrect superior module. Let's begin by examining the way in which this design works (note that not all data couples or flags are shown):

1. A student's grades are keyed to processing. Once they are keyed, the student's grade point average (GPA) is calculated.

2. The student's new grades are added (posted) to the student's grades from previous terms.

3. The student's grade report is formatted. Once formatted, the report is printed.

Figure 12–6 illustrates an improved design of processing. In this instance, the data couple GPA is not allowed to float or "tramp" around the design. The so-called tramp data can be avoided by calculating GPA just before it is needed in processing.

By changing the subordinate-to-superior reporting relationship of the module, calculate GPA—from get student grades to prepare student grade report—functional linkages are also improved. Calculating GPA has much more to do with preparing a grade report than with getting student grades.

Finally, we need to understand why tramp data are such a problem. Examine Figure 12–5 once again. Is it wrong to allow GPA to float

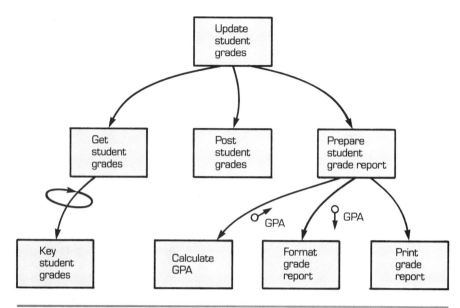

FIGURE 12–6 An improved solution

around the design if, in fact, the GPA was calculated correctly? Obviously not, provided the GPA was calculated correctly. However, if the GPA is erroneous, an effect better known as the *ripple effect* might be observed. The ripple effect occurs when an error that develops in one part of a system is permitted to spread to other parts of the system.

Principle of Cohesion

With coupling, the objective of systems design is to achieve a low degree of independence between modules. At the same time, the designer attempts to design each module so that its contents are highly cohesive, with *cohesion* being a measure of whether all instructions within a module contribute to the execution of a single, specified function. The module calculate GPA, for example, suggests that the instructions within it are limited to performing a single function—namely, calculating the GPA for a student. If the module calculated other ratios besides GPA, such as those shown by **Figure 12–7a,** the module would indicate rather poor cohesion.

Figure 12–7b illustrates how a multifunctional module can be transformed into a set of highly cohesive modules. The question of how many student ratios need to be calculated is now clear. Besides calculating the student GPA, a ratio of graded to pass/no pass credit hours and a ratio of upper-division to lower-division credit hours must be calculated.

Program Structure Charts **403**

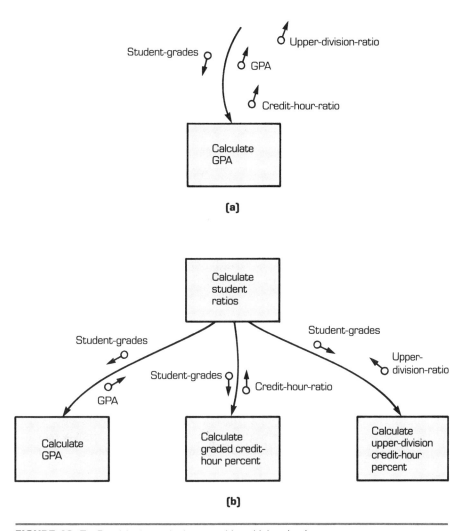

FIGURE 12-7 Partitioning a design to achieve high cohesion

Principle of Clear Labeling

The name of a module describes its function and specifies what the module must accomplish. With *clear labeling*, the systems designer can communicate the specific function of a module and its purpose relative to other modules. Consider Figure 12–7 once again. The label "Calculate graded credit-hour percent" communicates that this module performs one function, which is to compute the percentage of graded credit hours. It is much more informative than other labels might be, such as calculating credit ratio or performing credit ratio. An example of a poor label is

"Prepare display," since it begs the question: What type of display? "Prepare customer invoice remittance display" is much clearer. It communicates to others the module's specific function.

Principle of Span of Control

Span of control involves the internal organization of a structure chart and refers to the number of subordinate modules (workers) that report to a superior module (boss). With span of control, a boss should coordinate the activities of a reasonable number of workers. What is reasonable? Generally, seven is the maximum, although some designers prefer no more than a five-to-one ratio. Beyond five, design problems may arise, such as improper partitioning or incorrect coupling.

Let's return to the making coffee example to demonstrate a span-of-control problem. As shown in **Figure 12–8,** the process of making coffee can be defined in much greater detail than before. However, very fine details often mask the main functions of processing—which is what the designer attempts to define. Moreover, what happens if the number of modules is carried to the extreme? In this instance, there would be no need for instruction sets, since each module would consist of a single instruction. With this is mind, let's consider what is reasonable when it comes to module size.

Principle of Reasonable Size

While there are no hard and fast rules governing the number of instructions to be placed within a module, most designers agree that a module should be limited to fifty instructions (lines of code) or fewer; other de-

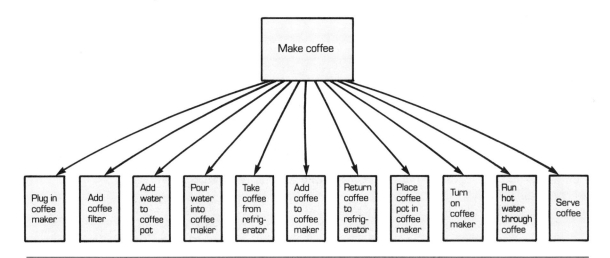

FIGURE 12–8 A span-of-control problem

signers are even more restrictive—limiting the number of lines to twenty or fewer so that each module can be displayed on a computer display screen. Besides an upper limit on size, most designers also favor a lower limit. They suggest that when a module becomes too small, say ten lines or fewer, then it should be combined with its immediate boss.

Figure 12–9 shows one way of combining modules and preserving the details of the structure chart. The small black triangle is called the *hat symbol*. It signifies that a separate instruction set will not be found for a module. Rather, the instructions for hatted modules will be combined with the instructions written for the superior module. Thus, the instructions for the modules calculate graded credit-hour percent and calculate upper-division credit-hour percent will be contained within the superior module calculate student ratios.

Principle of Shared Use

This principle is the inverse of the span-of-control principle. *Shared use* implies that it is perfectly acceptable to have several bosses call on a single worker. In practice, the principle of shared use can be stated most clearly by saying that whenever possible, modules should be developed for repeated use in a design.

Figure 12–10 illustrates a lower-level module that is invoked by two upper-level calling modules. The extra vertical lines for the module for calculating GPA show that this module is a *library routine*—a predefined procedure that is placed in a program library together with other library routines. (In some computer languages, a library routine would be represented by a subroutine or as a function). Observe how processing

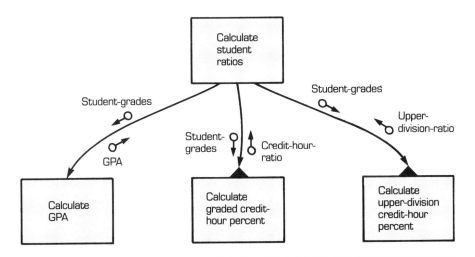

FIGURE 12–9 Combining modules to achieve modules of reasonable size

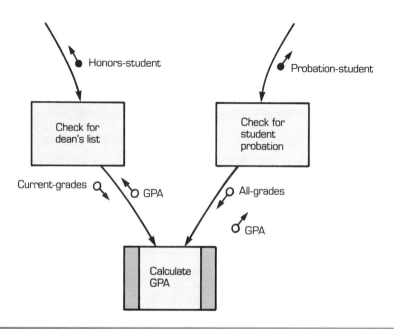

FIGURE 12–10 Shared-use library modules

takes place. In one part of processing, the determination must be made whether a student is to be placed on the dean's list. The module check for dean's list invokes the module calculate GPA and in so doing sends the student's current grades. The calculate GPA module determines the student's GPA and returns the computed value to the module check for dean's list. Once returned, the student's GPA is compared with a standard GPA, such as 3.50. If the computed GPA is equal to or greater than 3.5, a flag is set to indicate that the student is an honors student.

Observe next that in another part of processing, the determination must be made as to whether or not a student should be placed on probation. On receiving all grades for the student, the student's GPA is once again calculated, though for a quite different purpose. If the GPA is lower than a second standard, such as 2.0, a flag is set to indicate that the student should be placed on probation.

RULES GOVERNING THE USE OF FLAGS

A brief discussion of flags and their proper and improper use is in order. *Flags*, you will remember, are used to describe a condition (a *descriptive flag*) or to tell the boss to take some action (a *control flag*). Three rules important to the use of flags are:

1. Avoid the use of flags if possible.

2. Use descriptive flags rather than control flags.

3. Define all flags and place these definitions in the data dictionary.

Why should flags be avoided if possible? Flags might be viewed as on and off switches in a computer program. If a processing condition calls for the use of a flag, there is a tendency to turn on the switch and later fail to turn it off. Moreover, flags tend to muddy a design, as the following example illustrates.

Figure 12–11 shows before-and-after designs to compare the improper and proper use of flags. Before examining either design, consider the kind of processing that is taking place. The purpose of processing is

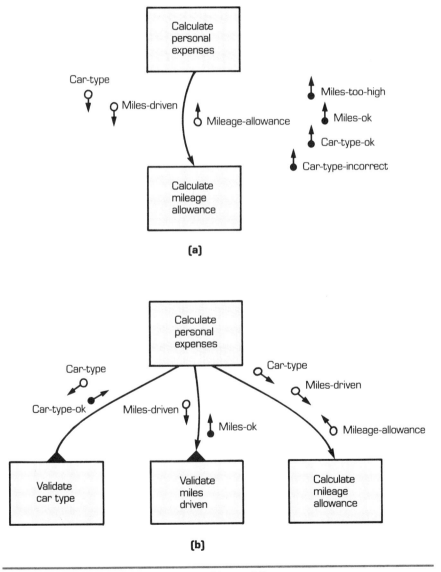

FIGURE 12–11 Improper (a) and proper (b) use of flags

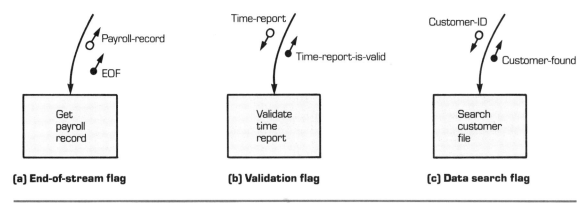

FIGURE 12–12 Three basic types of descriptive flags

to calculate a mileage allowance personal expense. The module calculate mileage allowance receives the type of car driven and the miles driven and is expected to return the mileage allowance. Next consider the before version of the design (Figure 12–11a). Not only is the mileage allowance calculated, but the module also reports whether the car type is OK or incorrect and whether the mileage figures are OK or too high. What is wrong? First, car type and miles driven should be validated *before* they are sent to the module calculate mileage allowance (see Figure 12–11b). Second, miles-too-high and car-type-incorrect are control flags rather than descriptive flags. They nag the boss to take some action. For example, the car-type-incorrect flag tells the boss module, "Do something. The car type is wrong."

Of the descriptive flags common to most designs, three types, *end-of-stream, validation,* and *data search* are the most common. **Figure 12–12** illustrates each type. In **Figure 12–12a,** an end-of-stream flag tells the boss that there are no more payroll records. In **Figure 12–12b,** a validation flag indicates that the time report is valid. In **Figure 12–12c,** a data search flag indicates that a customer is found. In practice, search flags can be set either on or off: with the customer found or not found. In either case, the message is clear: a search for an item was undertaken and the item could or could not be found.

MODULE SPECIFICATION Besides designing the structure charts for each program required by a system, the systems design team must prepare written specifications to describe the intent of each module and the way in which each module is to be implemented. What do we mean by describing the intent and implementation of each module? John Seevers, the lead systems designer at Mansfield, explained the difference between intent and implementation.

"By describing the intent of a module," he said, "a design team can clarify its primary function. The end result of this effort is preparing a *nonprocedural specification*—a clear statement of what each module is supposed to do. Describing the *implementation* of each module leads to a different type of specification—namely, a step-by-step *procedural specification*. This second type of specification indicates exactly how a module is to be coded and implemented."

Describing the Intent of Each Module

"There are several ways of describing a module's intent," John added. "A designer might provide a mockup of a report or display, stating that this is what the module is supposed to lead to. Or a designer might provide input and output parameters for a module, along with a brief description stating what the module is supposed to do."

Let's consider an example. Suppose that we wish to describe the intent of a module that involves building a conditional profits table. **Figure 12–13** illustrates that in processing, profit per sales unit, loss per unsold unit, salvage value per unsold unit, and sales range are required inputs to this module; a conditional profits table is the single expected output. In explaining what this output is all about, we might begin by preparing a picture of such a table. A picture greatly helps the programmer responsible for coding the module to visualize what the module is supposed to do. Consider the conditional profits table shown on **Figure 12–14.** Besides indicating the conditional profit figures for the entire sales and inventory range, the $10.55 expected profit total that results from selling twenty-two dozen when twenty-five dozen are stocked shows how a specific entry in the table is derived.

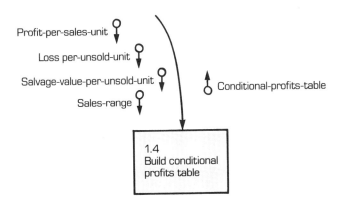

FIGURE 12–13 Building a conditional profits table module

Chapter 12: Computer Program Design

Sales Range (dozens)	\multicolumn{7}{c}{Inventory Range (Dozens)}
	20	21	22	23	24	25	26
20	10.00	9.85	9.70	9.55	9.40	9.25	9.10
21	10.00	10.50	10.35	10.20	10.05	9.90	9.75
22	10.00	10.50	11.00	10.85	10.70	10.55	10.40
23	10.00	10.50	11.00	11.50	11.35	11.20	11.05
24	10.00	10.50	11.00	11.50	12.00	11.85	11.70
25	10.00	10.50	11.00	11.50	12.00	12.50	12.35
26	10.00	10.50	11.00	11.50	12.00	12.50	13.00

Profit/dozen sold = .50
Loss/unsold dozen = .30
Salvage value/dozen = .15

Example: Profit on 22 dozen sold—25 dozen stocked

```
                 22 × $.50 profit/dozen     = $11.00
less:       (25-22 × $.30 loss/dozen)   =   (.90)
plus:  (25-22 × $.15 salvage value/dozen) =    .45
                                            -------
Expected profit:                            $10.55
```

FIGURE 12–14 Example showing how to build a conditional profits table

Besides a picture, we might prepare a brief abstract of an *input-output function description* of module 1.4, such as the following:

```
Module number:  1.4
Module name:    Conditional profits table
Receives:       Profit-per-sales-unit
                Loss-per-unsold-unit
                Salvage-value-per-unsold-unit
                Sales-range
Returns:        Conditional-profits-table
Function:       Build a conditional profits table for a
                given sales range (from 20 to 30 units a
                day) and for estimated profit and loss
                parameters.
```

Similar to the picture of the table, an input-output function description shows the intent of the module and not its implementation.

Finally, we will let John summarize the value of providing clear examples and abstracts of processing. "There are several good reasons for providing examples of specific outputs to be produced as a result of processing and brief abstracts to describe input, output, and function specifications," he stated. "First, examples clarify possible ways of translating formulas and expressions into computer code, without telling the programmer how to make the translation. This is why many designers prefer to write nonprocedural specifications rather than procedural specifications. They would rather clarify what is to be done, rather than how to do it. The how, they contend, is the responsibility of the programmer. Second, an example can be used later in testing the module. The test case might be to replicate the sample data, such as the data shown on Figure 12–14. Third, by writing an abstract to describe the input, output, and function of each module, the designer is able to clarify processing requirements. Up to this time, a module might be little more than a 'black box'."

Describing the Implementation of Each Module

Not all system designers feel as strongly about describing the intent of each module as John. Some believe that procedural specifications are preferable, and thus the techniques of writing *pseudocode* and preparing *program flowcharts* or *Nassi-Shneiderman charts* become important. Let's consider each of these techniques in more detail.

PSEUDOCODE

As stated in chapter 3, *pseudocode* is a high-level language used to specify the processing instructions to be completed by a module. What is the major advantage of pseudocode? Following its receipt, a programmer should find it relatively easy to transform the logic specified into computer code. Consider module 3.1 shown on **Figure 12–15** and the corresponding specification shown on **Figure 12–16.** As indicated by the structure chart, a new species type is to be keyed to processing and verified. If the name is correct and is not a synonym, a new species name is returned. How is this implementation to take place? The pseudocode shown in Figure 12–16 clarifies each major step. Next, observe how this more complete specification clarifies how module 3.1 is to be implemented:

1. The name of the module and its number are identified.

2. The parameters stating what the module receives and returns are identified.

3. The pseudocode shows that:
   —The user will have five attempts to enter a valid name. If attempts are exceeded, an error message and an abort will take place.
   —New species name consists of two parts: genus name and specific epithet.
   —New species name is verified by module 3.1.2.
   —The REPEAT-UNTIL loop continues until the attempt tally equals five, or until the is-synonym and species-name-ok flags are returned.

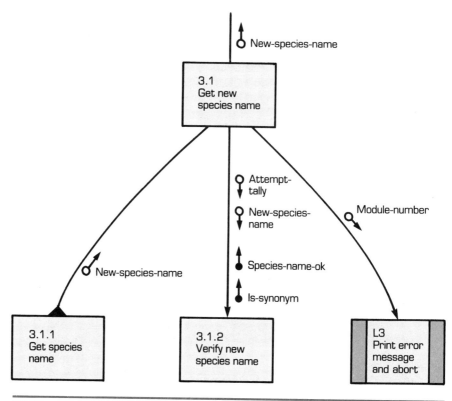

FIGURE 12–15 Getting a new species name

Finally, in addition to pseudocode, this specification sets process constraints, unit controls, and offers a unit test plan to be used later (see chapter 8 for a review of these three items). The process constraint sets an upper limit on the number of times the user can enter incorrect input, while unit controls limit the size of the genus name and the specific epithet. The unit test plan suggests several tests for this module later on. These tests include: using numeric characters, control characters, genus names larger than sixteen or specific-epithet names larger than two; making more than five input errors; adding a record that already exists; and adding a synonym for a record that already exists.

Program Flowcharts

Program flowcharts provide a graphic portrayal of the implementation of a module. As described in chapter 3, program flowcharts graph the flow of data in a module, much like circuit diagrams graph the flow of electricity in an electrical piece of equipment. Consider the partial program

Module Number: 3.1
         Name: GET NEW SPECIES NAME
_____

Receives: a prompt to begin
_____

Returns : new-species-name
_____

attempt-tally = 0
REPEAT
    IF attempt-tally < 5
        THEN Dispaly add-1 screen
        attempt-tally = attempt-tally + 1
        Retrieve genus-name
        Retrieve specific-epithet
        new-species-name = genus-name + specific-epithet
        Call VERIFY NEW SPECIES NAME [See 3.1.2]
    END-IF
UNTIL (attempt-tally = 5) or (not is-synonyn and
species-name-ok)
    IF (attempt-tally = 5) and (not species-name-ok)
        THEN module-number = "3.1"
        Call PRINT ERROR MESSAGE AND ABORT [SEE L3]
        ELSE
        Return new-species-name
    END-IF

_____

Process constraints: If input not valid after 5 tries, call PRINT ERROR MESSAGE AND ABORT to abort module. 0 <= attempt-tally <= 5.

Unit controls: For each new-species-name, check that only alphabetic characters are input. The genus-name must be no longer than 16 characters; the specific-epithet no longer than 2 characters.

Unit test plan: Try using numeric characters, control characters, or strings longer than maximum and/or more than 5 mistakes. Try to add a record that already exists or a synonym for an existing record.

FIGURE 12–16 Specifications for module 3.1, get a new species name

flowchart prepared for Module 3.1, get new species name. As **Figure 12–17** shows, sequence, decision, and repetition instructions are further clarified by the program flowchart, as are the different types of processing steps: input and output, data preparation, process, decision, and calls to external modules. (Compare this flowchart to the pseudocode shown in Figure 12–16 in tracing this flow of processing.)

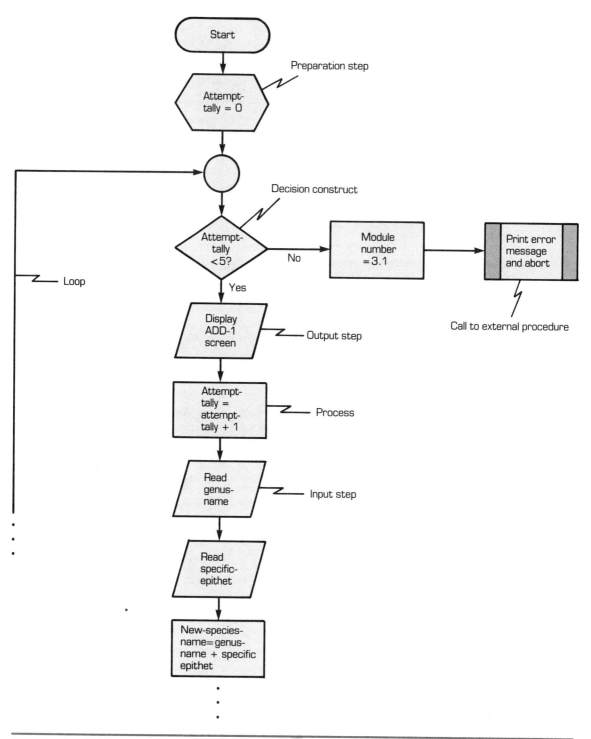

FIGURE 12-17 Partial program flowchart for module 3.1

A designer will often use program flowcharts to isolate a set of instructions called for by a design or to clarify the forks and paths of various decision steps. The disadvantages of program flowcharts, however, preclude their heavy use. Because program flowcharts are so time-consuming to prepare, they are rarely modified to reflect design changes. Moreover, for complicated designs, such as those embodying numerous nested decisions and loops, program flowcharts are simply too difficult to prepare.

Nassi-Shneiderman Charts

Nassi-Shneiderman (N-S) charts offer an alternative to either pseudocode or program flowcharts. Named after their authors,[2] N-S charts are much more compact than program flowcharts, include pseudocode-like statements, and feature sequence, decision, and repetition constructs. **Figure 12–18** illustrates an N-S chart designed for processing payroll checks. We will let John explain this design.

"First, we set the check dollar value (the checksum) to zero and enter the check dollar amount to be printed (the checktotal)," he explained. "By the way, we know the check dollar amount in advance. This amount

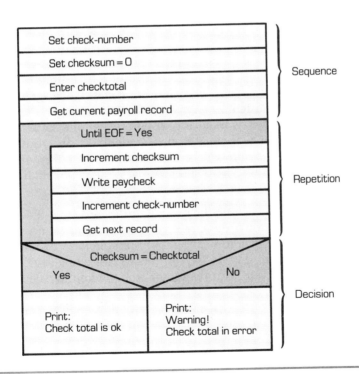

FIGURE 12–18 Nassi-Shneiderman chart

is calculated when we determine employee pay. It is shown on the payroll register."

He continued: "After we initialize processing, we get the first payroll record. This leads to a loop in which we increment the check-sum total, write the paycheck, increment the paycheck number, and get another record. The loop stops when we receive an end-of-file flag. At this point, we compare the checksum with the checktotal. If the two figures match, we can remove the checks from the printer and tell the treasurer that checks are ready for the company signature. If the two figures differ, we have a problem. Checksum figures must then be compared against the payroll register to determine why there is a difference."

EVOLVING THE TOP-DOWN PROGRAM DESIGN

Once the design team understands the principles governing the design of structure charts and the use of flags, along with how modules are to be specified, it can begin to evolve the top-down design of a system. Why evolve? Because team members may work through several drafts of a design before they agree that the features of one are best. To illustrate, let's return to the payroll system developed by the Mansfield staff. To begin, the design team determined that the design was a transaction-centered design and would consist of three subsystems: update hours worked, print payroll register, and print employee paychecks. The designers developed a top-level structure chart (see Figure 3–8) to show these three subsystems. Following this, they began to design structure charts for each subsystem. **Figure 12–19** shows a final draft of the chart prepared for module 2.0, print the payroll register. Once again, we will let John explain this design.

"First, we sum hours worked for an employee," he said. "Once we know the total hours worked (which incidentally are regular hours worked plus overtime hours worked), we calculate employee pay. This step is the most difficult part of processing, so let me explain module 2.2 point by point.

1. "We begin by retrieving the old payroll record from the employee master file. We send down the employee number and return either the old payroll record or a flag indicating that there is no record on file.

2. "Once we obtain the old payroll record, we extract the employee pay rate and send the rate with the total hours worked to calculate gross pay.

3. "Gross pay together with payroll deduction codes (including tax codes) are used next. We send them down to compute net pay and payroll deductions.

4. "Gross pay, net pay, and payroll deductions are then combined with the old payroll record to form the new payroll record. This record is written to the employee master file, to complete the update of this file. We also return the new payroll record to module 2.2, where it is

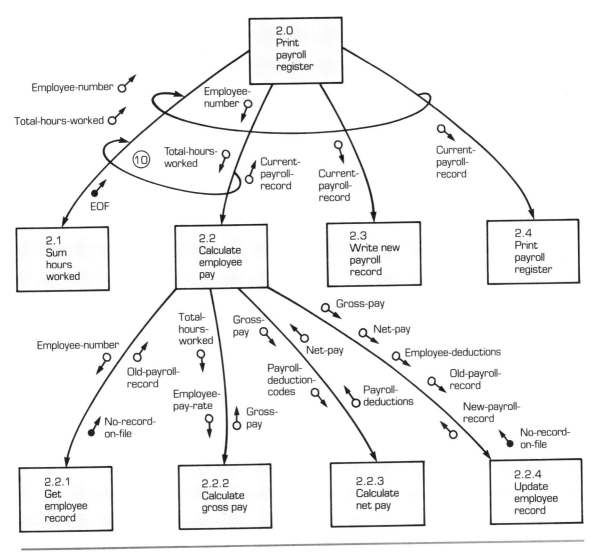

FIGURE 12–19 Level-2 structure chart

combined with the payroll period to become the current payroll record.

"The rest is easy," concluded John. "We send and write the current payroll record to the sequentially ordered current payroll file; we also print the current payroll record on the payroll register. Note that we added an extra loop to this design. The smaller loop was added when we decided to block records by a factor of ten. Thus, as indicated, we process module 2.1 and 2.2 ten times before we process 2.3 or 2.4. Once a set of records is printed, processing loops back to sum hours worked for the next employee."

The Mansfield, Inc., Case Study

The systems design group joined forces in working on the design of computer programs. John commented: "This is where all of our previous work begins to come together. It's one thing to design inputs, outputs, and files in the absence of processing routines; it's quite another thing to determine how the various pieces of the design are to be integrated."

John continued by instructing his group: "I want you to design structure charts for each and every program required by our new billing and receivables system. Besides showing the data couples and flags for each module, I want you to fully define each data couple and flag. You may find these definitions as difficult as drawing the structure charts themselves."

Defining the Data Flow for Top-level Modules

Walter began work on the design of program 2.0, enter customer payments. Using the VTOC prepared earlier, Walter reasoned that it would be best to sketch in the basic data flow for the top-level modules and fully define all required data couples and flags. "By following these steps," Walter stated, "I should have a clear picture of what information must be sent to lower-level modules."

Figure 12–20 shows Walter's top-level structure chart for program 2.0. Walter explained his design as follows: "First, I must receive remittance data from either an invoice remittance or a monthly statement remittance. In either case, remittance data must be keyed to processing and edited before being

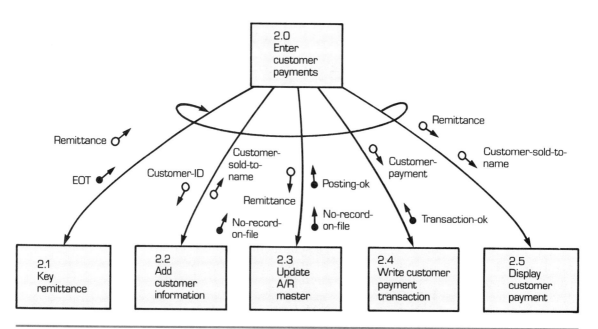

FIGURE 12–20 Top-level structure chart for program 2.0, enter customer payments

sent. Second, I plan to send only the customer ID to module 2.2 and, provided the customer record is on file, receive the customer sold-to name. Third, I will send the entire remittance to be posted to the accounts receivable (A/R) file. The only information to be returned is a message telling me that posting was successful (e.g., posting-ok) or that something went wrong, such as that the A/R record could not be found. Fourth, once posting is completed, I plan to write the customer payment to the customer payment file. Another flag is used to tell me that processing went as planned. Fifth, I need to display the details of the remittance keyed to processing and the sold-to name of the customer. Finally, after processing a remittance, I will loop back to enter another transaction. This looping back continues until an EOT (end-of-transmission flag) is set, indicating that all remittances have been keyed to processing."

Defining Data Couples and Flags

Before Walter continued his analysis of entering customer payments, he remembered that John wanted all data couples and flags defined. Walter said to himself, "I'd better get on with it by defining what I meant by remittance. Thank goodness Carolyn worked on the input design for entering a remittance [see chapter 10]. Her design will be of great help in explaining the difference between an invoice remittance and a monthly statement remittance."

Walter went on to define a remittance as follows:

```
Remittance = [Invoice-remittance/monthly-statement-
              remittance]
Invoice-remittance = Invoice-remittance-number + cus-
                     tomer-number + customer-code +
                     date-of-check + check-number +
                     invoice-number + cash-payment +
                     general-ledger (G/L)-code
Monthly-statement-remittance = Statement-remittance-
                               number + customer-num-
                               ber + customer-code +
                               date-of-check + check-
                               number + 1{invoice-num-
                               ber + cash-payment +
                               G/L-code}N
```

Walter next defined customer payment and customer ID:

```
Customer-payment = Remittance + customer-sold-to-name
Customer-ID = Customer-number + customer-code
```

Finally, Walter provided comments to describe the flags important to his top-level structure chart.

```
EOT = *end of transmission*
No-record-on-file = *cannot find record even though it
                     should exist*
```

```
Posting-OK    = *posting successfully completed*
Transaction-OK = *transaction successfully written to
                  the file*
```

Adding Data Couples and Flags to Lower-Level Modules

Figure 12–21 shows the completed structure chart for program 2.0. Walter explained how processing would be designed. "Module 2.1 coordinates the keying of remittances. If an invoice remittance is to be keyed and transmitted,

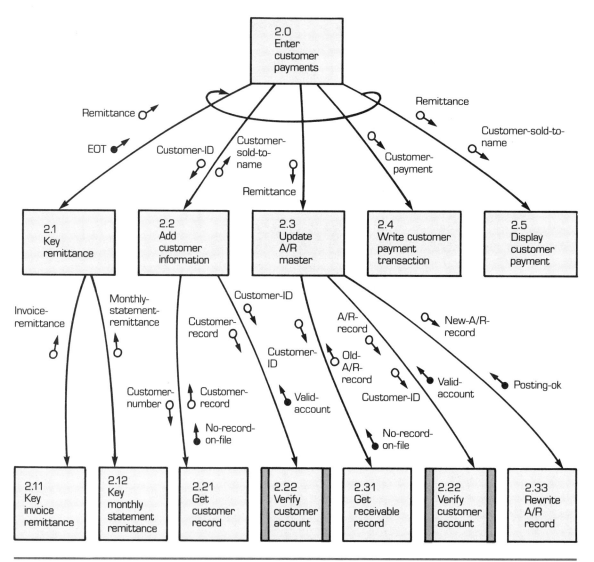

FIGURE 12–21 Structure chart for program 2.0, enter customer payment

module 2.11 is invoked. If a monthly statement remittance is to be keyed and transmitted, module 2.12 is invoked.

"Module 2.2 supplies the customer sold-to name and checks whether the customer is valid. How is this check made? We match the customer code (the first three letters of the customer's name) to the first three letters stored on file. If we sold goods to Murphy Products, Inc., the customer code would be keyed as MUR.

"Module 2.3 coordinates the retrieving of the old receivables record, determines whether the A/R record is valid by matching customer codes, and adds the remittance to the old A/R record to form the new A/R record. This is the most complicated part of processing.

"Module 2.4 simply writes the customer payment to the customer payment transaction file. Module 2.5 coordinates the displays required in processing."

SUMMARY

The purpose of the detailed design is to provide a clear work statement for the programmers assigned to the task of translating a system into computer code. To this end, structure charts provide a hierarchical structure of functions, expressed as modules, and clarify how these functions are to be linked together. Program module specifications provide additional clarification. They describe the intent of each module or show how each module is to be implemented.

Seven principles guide the designer in the construction of structure charts. First, the principle of partitioning urges that a large system be subdivided into a hierarchy of smaller modules. Second, the principle of coupling deals with how modules are linked together. In design, it is best if one module has little dependence on another module. Third, the principle of cohesion considers what goes on within a module. The design guideline is to limit the activities of a module to a single function. Fourth, the principle of clear labeling indicates that designers will be better able to communicate the details of their designs if the title of each module is clear. Fifth, the principle of span of control is analogous to how many workers a boss should supervise. In design, no boss module should coordinate the functions of too many worker modules. Sixth, the principle of reasonable size cautions the designer to limit the number of instructions per module to no more than fifty. Seventh, the principle of shared use encourages the construction of library routines to allow instruction sets to be shared by as many calling modules as possible.

Besides principles pertaining to the design of structure charts, three important rules govern the use of flags. These are: avoid the use of flags if possible, use descriptive flags rather than control flags, and define all flags and place these definitions in the data dictionary.

Module specifications are prepared in addition to structure charts to clarify what each module is supposed to do (the intent of the module) and how each module is to be coded and implemented (the implementation of the module). Describing the intent of the module leads to a written nonprocedural description of processing. A specification limited to input, output, and function, for example, indicates what the module is supposed to do, but not how implementation is to take place. Describing the implementation of a module leads to a procedural specification. Pseudocode, program flowcharts, and Nassi-Shneiderman charts are three different techniques for showing, step by step, how an instruction set is to be coded and implemented.

A top-down program design implies beginning at the uppermost level of a design and working toward ever-increasing levels of detail.[3] Structure charts simplify the steps that are important to this process. Even so, the design of structure charts is an evolutionary process. Several trials may be needed to evolve an acceptable design of processing.

REVIEW QUESTIONS

12–1. How does detailed systems design differ from preliminary systems design?

12–2. What is meant by stepwise refinement?

12–3. Name the basic attributes (properties) of each program module.

12–4. What is the difference between tight and loose coupling? Which type is preferred?

12–5. Briefly explain the ripple effect.

12–6. What is the difference between strong and weak cohesion? Which is preferred?

12–7. Why is "prepared output" an example of a poor label?

12–8. With span of control, a boss module should coordinate the activities of a reasonable number of worker modules. What is a reasonable number?

12–9. Program modules should contain no more than a reasonable number of instructions. What is a reasonable number?

12–10. How does a library routine differ from other program modules?

12–11. What three types of descriptive flags are common to most structure charts?

12–12. How does a nonprocedural specification differ from a procedural specification?

12–13. What is a major advantage of pseudocode?

12–14. Why are program flowcharts rarely used to describe the design of an entire system?

12–15. How do Nassi-Shneiderman charts offer an alternative to either pseudocode or program flowcharts?

12–16. What is meant by the following statement: A systems design team must often evolve an acceptable design.

12–1. What is wrong with the structure charts shown below? How should each of these charts be changed? (If necessary, draw a new chart to document your findings.)

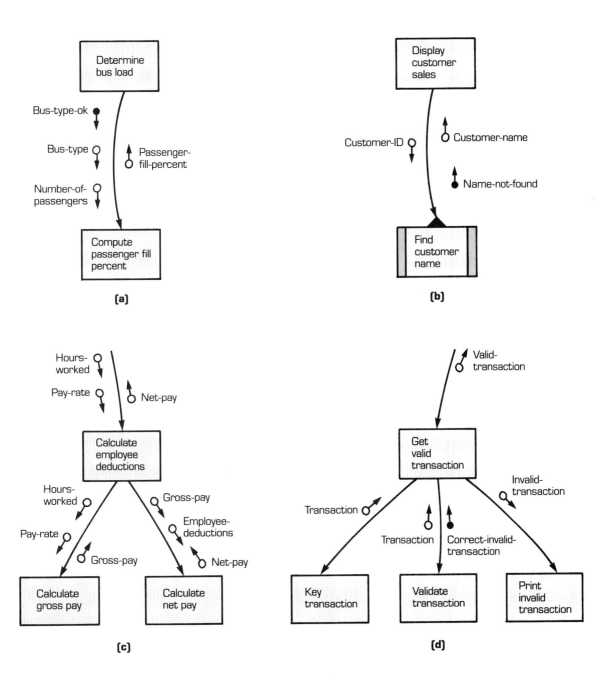

12–2. Improve the structure chart shown below. As indicated, the purpose of this design is to obtain a product record. In order to do this, it is necessary to key into processing the product number, product description, unit of measure (e.g., dozen, pounds), color, bin number (showing where the product can be found), warehouse number (showing the warehouse in which the product is located), product (selling) price, product cost, and quantity discount (2 percent, 100–199; 4 percent, 200–399; and so forth).

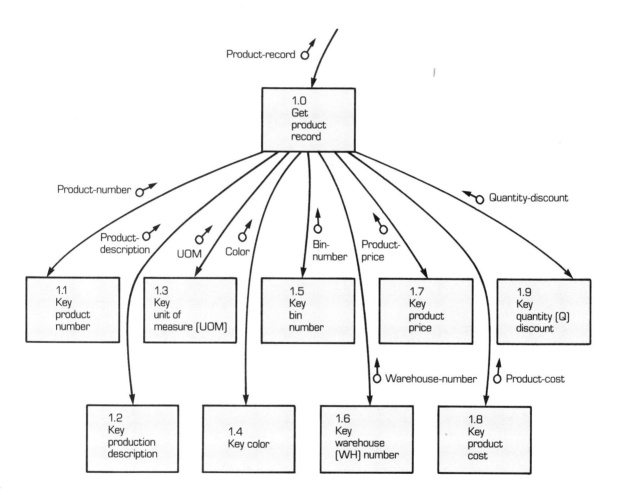

12–3. Once a field is keyed to processing, it must be verified. If the results of this verification indicate that the field is in error (e.g., it contains nonnumeric data, or it does not contain a decimal point), this error condition must be reported. Using the figure prepared for exercise 2, draw lower-level structure charts (e.g., 1.1.1, 1.1.2) to show how product number and product cost would be entered into processing

and verified. In drawing these two charts, make the following assumptions:
1. Library routines are labeled L1, L2, and so on.
2. Error messages for both product number and product cost errors would be the same.
3. The data verification test for product cost will also be used to test for product price.

WORLD INTERIORS, INC.—CASE STUDY 12
DESIGNING PROGRAM STRUCTURE CHARTS

Introduction

John Welby took on the task of transforming the HIPO charts he had prepared earlier (see chapter 9) into program structure charts. Before writing computer code for the various computer programs, John wanted to show how the modules for each program were to be tied together. He especially wanted to explain the data couples and flags important to the passage of data between computer programs. He also thought that this would be the best time to attempt to combine the designs for input, output, and computer files with program designs. "This is one of the most difficult aspects of design," he remarked.

Processing Customer Orders

John began his program design assignment by drawing the top-level structure chart for program 1.0, entering customer orders. As shown in **Figure 12–22,** processing first requires getting and

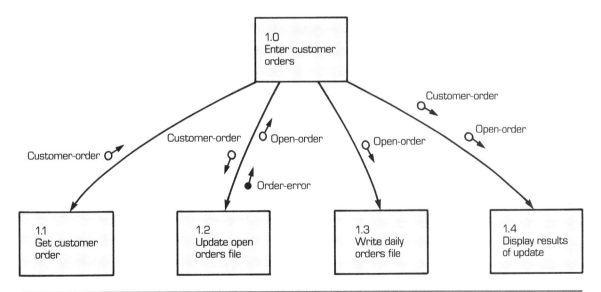

FIGURE 12–22 Level-1.0 structure chart for program 1.0

sending customer order input. These key-entered data were needed to enter new order information to the open-orders file. The other main part of the program consisted of forming the open-order record and of writing it to two files. The record was created by module 1.2, updating the open-orders file. Following the update, the open-order record was to be sent to two other modules: 1.3, writing the daily orders file, and 1.4, displaying the results of the update.

One difficulty in processing that John found was in defining the meaning of the various data couples. Using a copy of Jane's input analysis and her input display layout, John defined an open order as:

```
Open-order = Customer-information +
             customer-order-infor-
             mation + line-item-
             information
```

He knew that one of his next steps would be to define what each of these other terms meant.

The most difficult part of entering customer orders lay in the update of the open-orders file. For each valid update, several actions were required. These included:

1. The key-entered supplier number had to lead to the return of a supplier's table storing product information (e.g., item numbers, item number descriptions, finishes, prices, and so forth).
2. Each key-entered item number had to match an item number stored on the table.
3. Each key-entered finish number had to match a finish number stored on the table for the item ordered.
4. Each key-entered price had to match a price stored on the table for the item ordered.
5. The key-entered sales tax had to match the calculated sales tax.
6. The key-entered customer payment had to match the calculated sales total.

Processing Factory Invoices

John began his design of the top-level structure chart for program 6.0, entering factory invoices (see **Figure 12–23**), by passing the data couple called factory invoice. He commented, "Once I can define the content of the factory invoice, I can determine how to update the open-order and sales history files; I can also determine what information is to be written to the filled-orders file."

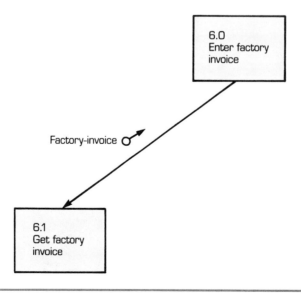

FIGURE 12–23 Factory invoice data couple

"Probably the most difficult parts of the design will be the passing of data to the open-order file and to the supplier," John continued. "I really do not want to pass data that are not important to processing. Jane's decision to simplify the design [see chapter 11], while helping file storage, does not affect processing. I will still have to match line-item charges from the factory against customer line-item amounts."

CASE ASSIGNMENT 1

Using John's definition of an open order, decompose line-item information to show the entries to be made in the data dictionary. Suggested new terms for this assignment are:

    price-extended
    sales-subtotal
    customer-invoice-line
    item-number
    item-description
    customer-invoice-totals
    quantity-ordered
    sales-tax
    price-each
    sales-total
    no-tax
    finished-ordered
    customer-payment

CASE ASSIGNMENT 2

Prepare a program structure chart for module 1.2 of program 1.0, entering customer orders. The names of the data couples and the flags suggested for this assignment are:

    customer-order
    open-order
    no-supplier
    item-number
    sales-tax-error
    payment-error
    supplier-number
    order-error
    sales-total
    price-extended
    price-each
    customer-payment
    quantity-ordered
    item-no-error

    sales-tax
    finish-ordered
    price-each-error
    product-table
    sales-subtotal
    finish-no-error

These names are not in correct order and will need to be rearranged. You may change the data couples and flags shown above. If changes are made, please explain your reasons.

CASE ASSIGNMENT 3

Decompose the factory invoice to show the entries to be made in the data dictionary. Suggested terms for this assignment are:

    terms-code
    factory-cost-extended
    cost-each
    factory-invoice-ID
    factory-invoice-line
    item-number
    supplier-number
    customer-key
    quantity-ordered
    factory-invoice-charges
    month
    day
    year
    finish-ordered
    date-of-factory-invoice
    remittance-number
    invoice-number
    invoice-due-date
    payables-information

CASE ASSIGNMENT 4

Prepare a top-level 6.0 structure chart for program 6.0, entering factory invoices. Unless directed, do not attempt to show how the open-order or sales history files are updated. Your completed chart should be similar to Figure 12–22.

The names of data couples and flags suggested for this assignment are:

    factory-invoice
    supplier-number
    quantity-ordered
    filled-order

factory-invoice-charges
sales-tax-table
finish-number
sales-update-problem
factory-invoice-ID
problem-order
item-number
payables-information
sales-update-total
sales-and-tax-totals

CASE ASSIGNMENT 5

Prepare a level-6.2 structure chart limited to modules 6.2.1, 6.2.2, and 6.2.3 of program 6.0, entering factory invoices. The names of data couples and flags suggested for this design are:

line-item-error
supplier-number
filled-order
no-open-order
supplier-number-error
sales-tax-table
factory-invoice-charges
open-order
customer-key-error
low-profit-margin
customer-invoice-number
order-OK

Complete case assignment 4 before starting this assignment.

REFERENCES

1. Dolan, K., *Business Computer System Design* (Santa Cruz, Calif.: Mitchell Publishing, 1984).

2. I. Nassi and B. Shneiderman, "Flowchart Techniques for Structured Programming," *ACM SIGPLAN Notices* (August 1973): 12–26; see also Schneyer, R., *Modern Structured Programming, Program Logic, Style, and Testing* (Santa Cruz, Calif.: Mitchell Publishing, 1985).

3. Koffman, E., *Problem Solving and Structured Programming in Pascal* (Reading, Mass.: Addison-Wesley Publishing, 1985).

13
Processing Control Design and the Technical Design Specification

INTRODUCTION

THE design of processing controls begins with the development of the logical design and is closely tied to the design of computer inputs, outputs, data files and databases, and computer programs. As stated in chapter 8, *processing controls* verify the correctness of processing. Where possible, they should be specified for every program module; they must exist for the system considered as a whole. John Seevers of Mansfield, Inc., explained his view of processing controls. "Processing controls help us verify that each step in processing leads to correct results," he said. "We typically design both computer- and human-directed types of controls. For example, we design processing controls to

- verify that all data have been correctly processed;
- block data in error from entry into processing;
- make it possible to reconstruct data if data transmission is faulty;
- prohibit tampering with the system by unauthorized personnel; and
- prevent employee fraud, embezzlement, and theft.

"Whenever we are working on a system involving dollars, the design of processing controls becomes critical," John continued. "Why? We need to make sure that outsiders, in general, and employees, in particular, cannot profit from processing. Consider the payroll system. Printing too high a paycheck or two paychecks, while storing only the details of one on file, is a situation we must avoid."

This chapter continues the description of the design process by considering various types of processing controls and data validation tests.

The objective underlying the use of controls and tests is, as John stated, much more than merely testing input data before they enter processing. In the design of processing controls, the systems designer must be concerned with overall system security. Systems should be designed so that it is extremely difficult to tamper with or somehow violate them.

When you complete this chapter, you should be able to

- describe the various types of processing controls;
- design a variety of edit tests;
- design a check-digit method of testing;
- explain different types of audits; and
- identify the main components of the technical design specification.

TYPES OF PROCESSING CONTROLS

In developing a new system's design, the systems designer usually develops four types of processing controls: source-document controls; input (transmission) controls, including input from computer files; output controls, including output written to computer files; and computer program controls. Both transaction and batch control procedures are important for all four types. *Transaction control procedures* are developed to verify that information specific to a single transaction is correct. For example, a procedure is needed to determine whether data from a single timecard have been correctly processed. *Batch control procedures*, in contrast, are designed to verify that information specific to an entire set of transactions is correct. A procedure could be followed to determine whether all timecards have been correctly processed. With this difference in mind, let's consider the four types of processing controls in more detail.

Source-Document Controls

Source-document control procedures are designed to verify that all data have been entered into processing and that source documents can be recovered should the content of data transmission be questioned or need to be repeated. **Figure 13-1** illustrates how a designer can account for all payroll transactions entered into processing. John explained this source-document control procedure.

"Before entering timecard information into the computer," he explained, "we count the number of timecards and sum the total number of hours worked for all employees. We enter and date these batch totals on the payroll processing log. The next step is to key-enter the individual timecard totals into the computer. Once entered, the computer sums the number of timecards processed and the number of hours worked. These sums are displayed. The third step is to compare the totals in the

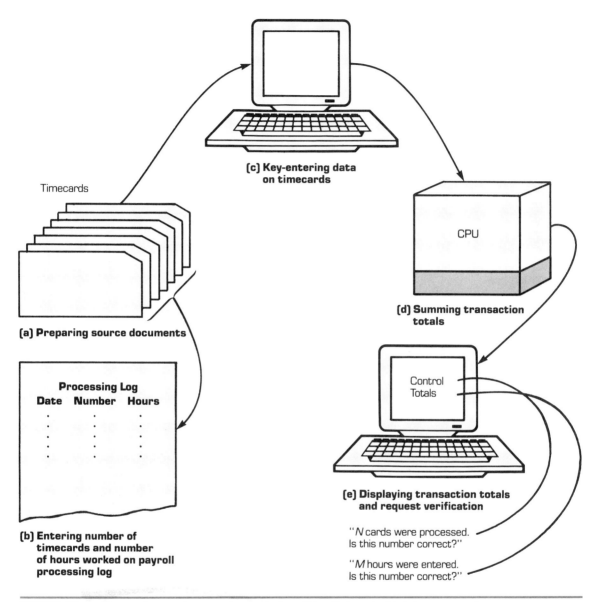

FIGURE 13–1 Source-document processing control: verifying that control totals are correct

log with the totals displayed on the screen. If the figures are the same, we can be confident that all data have been entered into processing."

Carolyn questioned John's explanation: "John, is it necessary to manually total the number of hours worked? Shouldn't the computer do this simple task?"

John responded, "The procedure I explained is probably more elaborate than what we'll need in the long run. Once the payroll system is stable, we will continue to count the number of timecards; however, we will most likely sample the correctness of hours-worked totals. If done correctly, sampling will permit us to spot most errors."

WITHIN-DOCUMENT CONTROLS

Within-document controls are designed to verify that data specific to a single transaction have been correctly entered into processing. They provide one way of avoiding the need for more elaborate source-document controls. **Figure 13–2** provides an example of a within-document control. Initially, line-item totals are entered into processing (step 1), as well as the grand total for all line items (step 2). To check the accuracy of data entry and of the data contained on the source document, the computer sums the line-item totals (step 3) and compares this computed total with the grand total. If a difference is discovered, the user is informed (step 4).

Carolyn comments once again. "This within-document control procedure is more to my liking. Some work is redundant (namely, the keying of the grand total), but this extra work is minimal."

FORMS CONTROL

Developing effective office procedures for keeping track of source documents is an essential part of source-document control. The systems designer should be aware of the following forms control considerations:

1. Forms identification—all forms should be clearly identified to avoid misunderstanding or mix-ups. A time-recording form, for example, should be clearly marked as such. Likewise, the number of the form and

	Project	Hours Worked
Step 1: Enter line items	XYZ	10
	XXY	6
	XXC	12
	XXD	6
Step 2: Enter total	Total	34

Step 3: Compare computer-summed hours-worked total to key-entered total

Step 4: Display error message if totals do not agree.

FIGURE 13–2 Within-document control procedure

the date it was first placed in service should be indicated. These markings generally appear on the lower left-hand corner of a form.

2. Date of completion—all forms should specify when they were completed and by whom. Employee timecards, for instance, generally provide space for the employee signature and date. Customer order forms specify when the order was placed and by whom.

3. Authorization—all forms should contain a unique authorization number or a signature designating responsibility. Employee timecards provide space for a supervisor's signature, for example. Purchase requisitions must provide an authorized budget number, such as a number to charge the purchase to.

4. Filing of processed forms—procedures for filing source documents after they have been used is a step many designers miss, often to their later embarrassment. Consider timecards once again. After data have been key-entered, the cards should be filed, together with the register indicating that processing was approved. Generally, short-term storage of source documents is required. If long-term storage is needed, a picture of the document is often recommended to save space.

5. Disposal of processed forms—besides filing procedures, disposal procedures are also required, especially if forms contain sensitive information. Shredding equipment is often an excellent way to prepare sensitive material for disposal.

Input (Transmission) Controls

Input controls are designed to verify that data keyed or read to processing from files are received by the computer. Most input controls are built into computer hardware devices, such as computer terminals or transmission and receiving data sets. Pressing the return key on a terminal, for instance, tells the computer that a message has ended.

The systems designer should consider several types of input controls, including flashback checks, journaling, and file-balance controls.

Flashback (echo) checks tell the user that the computer did receive the transmitted data. The user may enter a customer number, and the computer responds by returning the customer's name. In some designs, a light-emitting diode is turned on to indicate that a message has been received. In still other designs the computer duplicates the keyed information. The user might type PAUL R. ROY as the customer name. The computer might respond with the message: THE NAME OF THE CUSTOMER IS PAUL R. ROY.

Journaling is a procedure that permits recovery of data, should data transmission be faulty. Online cash registers, for example, incorporate small cassette tapes to record all data transmission. If data transmission proves faulty, the tape is read to repeat the transmission.

File-balance controls verify that data read from files are accurate and complete. These controls require the designer to add a control table or record to each file. A *control table* is a small file that contains control totals (see the Mansfield case at the end of the chapter for an example). A *control record* is typically a small header record appended to the beginning of a file. It is used to store the number of records contained on the file and other important control totals, such as the total dollar amount stored. With a control record in place, the control procedure is straightforward (see **Figure 13–3**):

1. The control record is read.

2. Individual records are read and totals are summed.

3. Summed totals are compared with control record totals.

4. If differences are found, they are displayed.

Walter explained how the use of file-balance control was useful in the payroll processing design: "We placed a control record in the employee master file to store the number of employees and the total wages paid, quarter to date and year to date. Before calculating employee pay and printing the employee register, we read the control record into memory. After each update of an employee's record, we also updated the

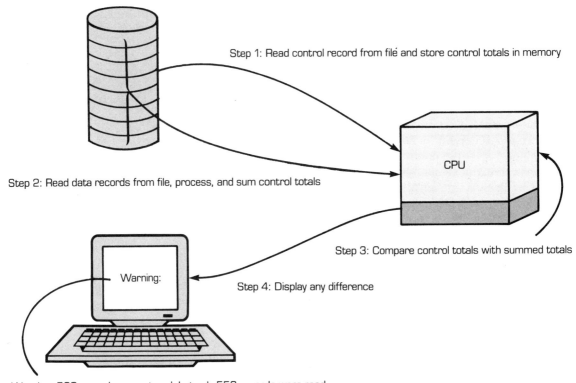

FIGURE 13–3 File-balance control procedure to verify the record count

control record. This process continued until the last employee record was updated, at which time we once again stored the control record on file.

"The control record was helpful in several ways," he continued. "First, the count of employees was useful in verifying whether all employees on file had been accounted for whenever we printed a register of our employees. Second, the quarter-to-date wages-paid total was invaluable in producing the quarterly payroll tax program for the government. Third, the year-to-date wages-paid total was essential in producing year-end payroll reports for the government and in printing out W-2 forms for all employees."

Output Controls

Similar to source-document controls, output controls are designed to verify that all data have been sent from processing (that is, printed, displayed, or written to output files) and output materials cannot be obtained by unauthorized personnel. Typical output control measures include assigning authorized personnel account codes and passwords. Unless users remember their codes and passwords, they are not allowed access to processing.

Two traditional types of output controls are printed registers and output-control totals. *Printed registers* show the contents of processed output and files. Consider the registers featured in the payroll design for Mansfield. In the system design, a payroll register was printed. Why? To allow users to visually review computer output before it would be printed on custom forms (e.g., employee paychecks).

Output-control totals are totals added to registers to show such items as the number of checks to be printed and their dollar amount or the number of invoices to be printed and their dollar amount. These stored control balances are compared against actual counts and amounts resulting from a custom-form print run. Consider Mansfield's printing of payroll checks.

John explained: "Before printing payroll checks, we review the payroll register to detect anything suspicious. Finding the register to be in order, we begin the payroll check print run. As part of the run, we void the first and last check. On the first check we print control totals—the expected number of checks and their total dollar amount. On the last check we print the actual number of checks printed and their amount. Any difference between the first and last set of control totals must be resolved before paychecks can be released to employees."

Computer Program Controls

Computer program controls, as discussed in the previous chapter, validate the accuracy of programmed procedures. Most program controls deal with verifying the accuracy of input data. Because of the importance of data validation procedures, the next section of this chapter will consider this topic in detail. A second type of program control involves the setting of flags to indicate errors or unusual conditions. End-of-data, end-of-file, out-of-range, totals-do-not-agree, and so forth are examples of program controls specified by flags. Still another type of program control consists of messages that warn users of some impending danger. For example, the message DISK FILE 90 PERCENT FULL warns the user to either remove data from the disk or to stop adding data to the disk. The message MEMORY 95 PERCENT FILLED, PLEASE FILE TO FREE SPACE illustrates a similar condition. The user is given advance notice of a potential problem.

Adding warnings to program modules is one of the more difficult design aspects. Typically, the need for warning messages begins to surface during program and system testing (see chapters 14 and 15).

With interactive systems in particular, it is important to incorporate tests that prohibit unauthorized users from gaining access to programs. How are these tests built in? The logic specified by the dialogue tree shown earlier (see Figure 10–18) illustrates one way of adding access-control instructions. In this test, the user was given three chances to enter a correct employee number. After three tries, further attempts were blocked.

Tests of user identification are common with many types of computer systems. Most automatic bank teller machines work this way. Once again, a user is given three tries to enter a correct personal identification number (PIN). After three incorrect tries, further attempts are blocked.

DATA VALIDATION

Data validation represents a special type of program control. In this instance, programmed routines are written to verify that all data presented to a system (e.g., keyed or fed) are correct and complete.

Another name for data validation is *data checking*. What the designer seeks to determine is whether the data to be used in processing are feasible—that is, whether they are likely to satisfy the conditions specified by the computer program.

There are two main types of built-in data validation procedures: edit tests and check-digit tests.

Edit Tests

As the name suggests, *edit tests* review data to determine whether a definite error or suspicious condition exists. While these types of tests are used primarily to screen data keyed to processing, in some cases output data may also be edited. Let's consider several different types of edit tests.

FIELD (INPUT) TESTS

Numerous tests can be performed to determine whether data fields have been keyed correctly. These include tests to determine whether a field contains numeric data only, alphabetic data only, or blank data only, or if a field is positive or negative. Suppose that a numeric field is required. If in testing an alphabetic character is discovered, a valid flag could not be set. Instead, the field would be rejected.

PICTURE CHECKS

Picture checks are a special form of field (input) tests. They are designed to match each character in a field to its prescribed field layout pattern. Suppose the field layout pattern is coded as follows: Picture = AA9A999, where A represents a character and 9 represents a digit. Suppose next that the number YB66C32 is keyed to processing. Because this number does not agree with the prescribed pattern (AA99A99 is not equal to AA9A999), the keyed data would be rejected.

RANGE TESTS

This type of test determines whether a value assigned to a variable falls within an acceptable range—that is, within an acceptable upper or lower limit. Overtime hours greater than thirty, for example, might lead to the message: OVERTIME HOURS APPEAR TO BE HIGH. ARE YOU SURE (Y/N)?

Messages such as these do not lead to the rejection of entered values, but query the user as to whether the values are acceptable. Moreover, these tests can be used for both input and output data. Suppose the computer calculates an employee's weekly wage as $9,145.00 (instead of $914.50). Following the calculation, the computer compares the wage with an upper limit, such as $2,000. Finding the employee's pay to be too high, the account would be flagged. A flag leads to a please verify or warning message, which is printed on the payroll register. The message might read: EMPLOYEE PAY EXCEEDS MAXIMUM RANGE. PLEASE VERIFY.

COMBINATION TESTS

At times it is important to combine the values of several fields to determine whether their combined value is valid. An employee who works four regular hours, but reports eight overtime hours, provides an illustration of a combined value to be flagged. While the data might be valid, they nonetheless look problematic. As another illustration, if gross margin (product price less product cost) is too high in relation to sales volume, this condition could be flagged and later printed. The message might read: VOLUME TO GROSS MARGIN IS 10 TO 1, PLEASE VERIFY.

Combination tests are sometimes referred to as *tests for reasonable-*

ness of data.[1] As with range tests, they are important because they help determine whether data fall within "normal" boundaries.

CONSISTENCY TESTS

A special form of combination test is a consistency test—a test that compares values in two or more fields within the same record to determine whether their values are consistent. Suppose a student is classified as a senior, but his record shows only ten earned credit hours. This inconsistency would be flagged for review.

Consistency tests can be built into a design to ensure proper matching as well. Suppose that a computer program is to match pairs for a golf tournament. In matched play, a person with a handicap of twenty-five, for example, should not be paired with a person whose handicap is only five.

PROBABILITY TESTS

Checks that examine the likelihood of an occurrence are called probability tests. They are conducted by comparing an input or output value against a table of probabilities. In this way, the probability of each value could be reported; however, usually only values that are some distance from the mean are significant. For example, the tolerance of accepted parts might be keyed to processing and compared against a parts tolerance probability table. Any reported value above or below three standard deviations from the mean would be flagged and reported.

Check-Digit Tests

Check-digit tests, the second main type of data validation test, use self-checking numbers. For such a test to be conducted, a check digit must be appended to a numeric number before the test. The test itself creates a hash total, whose remainder must always be zero.

The most widely used check-digit method is called *modulus-11*. With this method, several steps are required to create the check digit:

- Step 1—Assign weights to a numeric account, beginning with a weight of 2 for the rightmost digit.

Account number	3	6	1	2
Weight	5	4	3	2

- Step 2—Multiply the numeric account by its weight and add the sum.

Account number	3	6	1	2
Weight	5	4	3	2
Product and sum	+15	+24	+3	+4 = 46

- Step 3—Divide the sum by the modulus number.

$$46/11 = 4 \text{ remainder } 2,$$

- Step 4—Subtract the remainder from the modulus number to determine the check digit.

$$\text{Check digit} = 11 - 2 = 9$$

- Step 5—Add the check digit to the account number.

$$\text{Account number (with check digit added)} = 36129$$

- Step 6—Test the correctness of the check digit by repeating steps 1 through 3. Add a weight of 1 to the check digit. The remainder must be 0.

Account number	3	6	1	2	9
Weight	5	4	3	2	1
Product and sum	+15	+24	+3	+4	+9 = 55
Remainder			55/11 = 5, remainder 0		

How does a check-digit test work in practice? Suppose a user keys the account number shown above as follows: 36219. Before this number is accepted, the check-digit test is run. This particular run would return a value of 5, remainder 1. Since the remainder is greater than 0, this number would not be accepted. Instead, the following error message would appear: ACCOUNT NUMBER IS IN ERROR. PLEASE REENTER.

Problems with modulus-11 occur when the remainder determined by step 3 equals 0 or 1. When the remainder is 0, the check digit is set to 11. Fortunately, this remainder can be handled by substituting a check digit of zero. Append a check digit to the number 983 to test this out. The summed value is 66, which leads to a remainder of 0 and a check digit of 11. Next, try adding a check digit of 0 to account 983 and testing this number. The remainder, as expected, is 0.

While a number with a check digit of 0 can be handled, a number with a remainder of 1 (see step 3 once again) leads to quite a different problem. In this instance, a code, such as the letter X, is added to the account number. The better alternative is to discard any number that results in a remainder of 1. This works well when numbers carry no special meaning other than to designate the identification of an account.

How important are check-digit tests? Clifton[2] reports that of all the errors made in copying numbers, *transcription errors* (writing down a number incorrectly) account for 86 percent of all errors, *transposition errors* (interchanging two digits of a number) account for 8 percent of all errors, and all other errors (including the omission of digits or double transposition) account for the remaining 6 percent of all errors. Check-digit tests are especially good at identifying transcription and most (but not all) single transposition errors. They are especially effective when coded numbers contain six digits or more and are transcribed frequently.

Check-digit tests are also important when used with different types of data input equipment, such as hand-held light pens or other types of optical scanners. As part of the input process, a check-digit test is conducted to determine whether coded input is valid. If the test reveals a problem, the input system generally emits a sound, such as a beep, instructing the data entry operator to rescan (reenter) the number.

Some designers question the use of check digits especially if files have been carefully formatted in advance, with no provision to store an extra digit. Fortunately, this is not a major problem. Consider how Mansfield dealt with the question of check digits.

"We attach a check digit to all employee numbers," Walter stated. "However, we do not store the extra digit in the employee master file. Instead, we only use the check digit to verify the correctness of our six-digit employee numbers. **Figure 13–4** illustrates how our system works in practice. First, we key in the employee number, which includes a check digit. Second, we verify the check digit. Third, if the verification is positive, we proceed to retrieve the employee record, using the employee number minus the check digit. Since the check digit has served its purpose—to validate the correctness of input to processing—there is no need to store and retain it."

AUDIT CONSIDERATIONS

As part of the process of designing tests to ensure that all data presented to a system are correct and complete, systems designers must work with internal and external auditors to determine whether built-in verification and processing control tests are sufficient to satisfy company audit requirements. What is an audit? An *audit* is an in-depth review of the processing procedures and financial records of an organization to verify their correctness. *Auditors* (personnel responsible for conducting audits) have a clear stake in the design and development of new systems. Unless they understand how these systems work, they have little chance of determining whether what the system produces is accurate and complete.

Types of Audits

There are four main types of audits: audits around the computer, audits through the computer, audits using the computer, and physical audits. Of these, only the fourth, physical audits, resembles traditional audit practices. With *physical audits,* a manual accounting is made of such items as counts of stock carried in inventory or fixed assets carried on a company's books. These manual counts are compared with the counts stored on computer files. Once the correct count can be determined, it can be shown on the official records of the company. The other three types of audits are quite different from physical audits: they verify that the design of computer software is accurate and able to handle routine

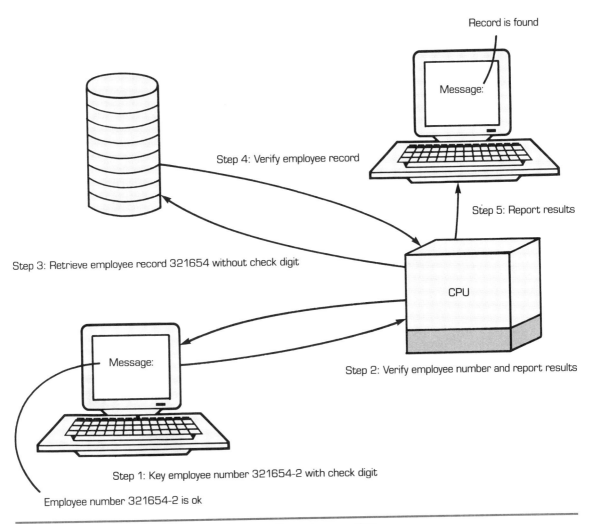

FIGURE 13-4 Check-digit control procedure to verify employee number

and exceptional processing runs. Let's examine each of these four types of audits in detail.

AUDITS AROUND THE COMPUTER

Audits around the computer determine whether outputs produced by processing match inputs read to processing, without regard to processing procedures that take place in between. The objective of these audits is to determine whether the *audit trail* is proper—whether computer outputs can be traced back to their source. Consider the audit trail used by individuals in reconciling their checking accounts. First, the bank sends the bank statement for review. If a charge is suspicious, the charge on the bank statement can be traced back to the canceled check. If the check looks suspicious, the date the check was written and its description can be traced back to the check register. If the check register provides insufficient information, it should be possible to conduct a search of manual files to find the billing statement that led to the writing of the check.

When the computer is involved in processing, an audit can be conducted much like the audit in the checkbook example. Figure 8–10 provided one illustration of how an audit trail is constructed. If $500 of new bills enter the system and $200 of bills are paid, then the file that contains bills to be paid should increase by $300.

Figure 13–5 provides a similar example. In payables processing (see 13–5a), the decision must be made regarding which bills to pay. Once that is done the file of unpaid bills (outstanding payables) is split into two subfiles: a file of bills to be held for payment at a later time (payables held over) and a file of bills to be paid (bills to be paid this period).

The auditor's role becomes important in determining whether the split was made correctly (see 13–5b). In making this determination, the auditor must have access to two different registers: a register showing the contents of the entire outstanding payables file before processing (input to processing) and the register of the contents of the payables-held-over and the payables-to-be-paid-this-period files (outputs from processing). If the totals on these registers agree (i.e., outputs equal inputs), the auditor can warrant that processing is correct. This warranty is made without regard to the processing steps in between—namely, those that transform inputs into outputs.

AUDITS THROUGH THE COMPUTER

Audits through the computer involve being able to trace processing paths from input to output through the computer-based parts of a system. With this approach, *test data* (sets of input data that present a variety of transactions) are presented to processing to determine whether invalid results can be obtained and if processing control totals are accurate.

The use of test data often leads to unusual results. One firm discovered that if an employee indicated that he or she had worked on 500 projects during a day, that this bogus number would not only be accepted, but appear later in several types of summary data. Another firm discovered that if more than 200 timecards were entered into processing, its payroll system would stop. It was designed for only 199 employees.

Like audits around the computer, audits through the computer are economical to perform. In addition, the tests can be conducted by individuals with little technical training. Even so, there are drawbacks to this approach. Mair et al.[3] state that test data are limited to preconceived situations. They fail to anticipate unusual circumstances and, at best, are overly simplistic.

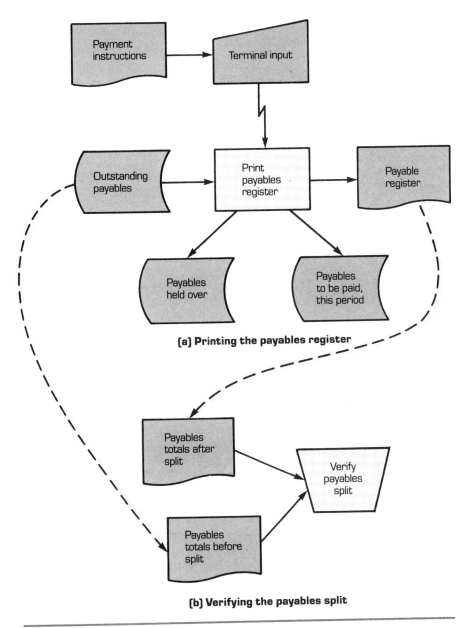

FIGURE 13-5 An audit around the computer

AUDITS USING THE COMPUTER

Audits using the computer feature auditing software tailored to the needs of the audit team. This software is perhaps typified by database management systems (DBMSs) software (software that enables auditors to make inquiries of data stored on file).

The options included in audits using the computer are several and include:

1. Data extraction—records can be extracted from files according to specified criteria. The query SELECT ALL EMPLOYEES WHO HAVE RECORDED MORE THAN 25 HOURS OF OVERTIME illustrates how an auditor might extract data from an employee master file.

2. Totaling—records can be extracted from files and values can be summed. This procedure permits file subtotals to be compared with file totals. Consider the following queries: SELECT ALL EMPLOYEES AND TOTAL HOURS WORKED BY DEPARTMENT followed by SELECT ALL EMPLOYEES AND TOTAL HOURS WORKED. An auditor would sample these two sets of totals to determine whether they agreed.

3. Aging—records can be extracted from files and aged. This procedure permits transactions patterns to be analyzed according to specified time periods. The query SUM CASH RECEIVED BY DAY OF THE WEEK provides a list of cash totals added to processing for each day of the week. Why is this list important? It can be compared with the list of bank deposits to determine whether the two sets of totals agree.

4. Sampling—records can be extracted at random from files based on a specified sample size. This practice permits the auditor to check the correctness of all transactions based on the correctness of the sample.

5. Frequency distribution—records can be extracted and arranged according to specified distribution categories. For example, invoices might be extracted for purposes of determining the number of invoices whose value is less than $25.00, between $25.01 and $50.00, between $50.01 and $75.00, and so forth. The product of processing—a frequency distribution—permits the auditor to note any disturbing changes in pattern and most certainly any extremes, such as invoices larger than $100,000.

PHYSICAL AUDITS

Even though audits around, through, and using the computer have significantly altered the concepts and practices of auditing, they have not done away with the need for physical audits. A computer file might show, for instance, that 50,000 tons of grain are stored in several silos; moreover, all computer audits show is that this total is accurate. Imagine discovering that not only is the grain nonexistent but so are the silos. Why would fictitious numbers be found in a computer system? Several reasons contribute to artificial listings, including financial benefits from false listings of a company's assets.

The taking of physical audits has improved with computer technology. Consider an inventory or stockkeeping example. As shown in **Figure 13-6,** a physical audit procedure begins by receiving inventory count cards. These cards are prepared following a check of each item stored in an inventory file to determine whether the date set for a physical audit is past (see step 1). The inventory count card, showing the name and location of the item to be physically audited, contains space

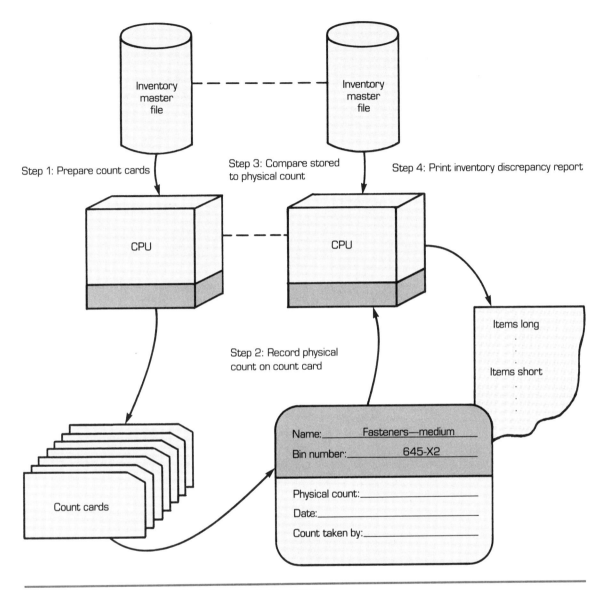

FIGURE 13-6 Procedure used to take a physical inventory

for the auditor to record the physical count (see step 2). Following the count, the number of items found is keyed to processing and compared with the number stored on file (see step 3). If a difference is found, it is printed on an inventory discrepancy report (see step 4). All differences shown on this report must be investigated; the contents of the computer file must be adjusted to show the true count of goods carried in inventory.

THE TECHNICAL DESIGN SPECIFICATION

Once processing controls are finalized, designers must address the technical design specification. This specification, as defined in chapter 1, is a document that shows how computer programs are organized and how they are written; it specifies all input, output, database and data file, and processing control requirements. For the most part, all of this is done in the design of the new system. What remains is packaging the materials by placing structure charts, pseudocode, input/output (I/O) layouts and requirements, database and data file I/O layouts and requirements, and processing control requirements into a single document.

Figure 13–7 outlines the material contained in the technical design specification. Much like the logical design specification (see chapter 8), the technical specification begins with an executive summary and ends with a description of hardware and software and user support requirements. The body of the specification now deals with system requirements—a topic only briefly considered in the logical specification. Likewise, the materials important to the logical specification—DFDs, data store descriptions, transform descriptions, and the data dictionary—are no longer included in the body of the report, but instead are placed in appendixes. With this organization in mind, each section of the technical specification is summarized briefly.

I. Executive summary

II. System organization

III. System requirements
    A. Processing requirements
        1. Structure chart
        2. Pseudocode
        3. Process constraints
        4. Unit controls
        5. Unit test plans
    B. Output requirements
    C. Input requirements
    D. Database and data file requirements
    E. Processing control requirements

IV. Coding and test plan

V. Test schedule and budget

VI. Hardware and software specification

VII. User support requirements

Appendixes
    A. Data flow diagrams
    B. Data store descriptions
    C. Transform descriptions
    D. Data dictionary

FIGURE 13–7 Suggested table of contents for the technical design specifications

Executive Summary

As in the logical specification, the executive summary here describes the problem investigated, reviews the purpose of the study, considers processing alternatives, and reviews the proposed design. However, this new summary also recaps the results of the design process: the type of design proposed, the number of programs required, important features of inputs and outputs, the type of file storage needed, and important program and system controls. The purpose of the summary, as before, is to introduce managers and users to significant design findings; it is written to encourage readers to review and to accept all design recommendations.

System Organization

This section provides an overview of the more detailed design specification. Top-level structure charts and processing menus (as appropriate) are described, together with statements describing why the design was organized in its particular way. By reviewing the essence of the design—at this particular point in the specification—the reader can better understand the details of the technical design that follow.

System Requirements

This section provides the final set of design specifications proposed by the design team. It includes structure charts for all computer programs, input and output layouts and processing requirements, database and file layouts and processing requirements, and processing control requirements. The central part of this section (and of the entire document) is the material detailing processing requirements. Besides structure charts, these materials include pseudocode, process constraints, unit (module) controls, and unit (module) test plans. In this way the design team can document for each module how computer code is to be written, what design assumptions and restrictions are important, where controls are built in, and what tests are needed. Following this discussion of processing requirements, other design specifications follow. In each instance, references to structure charts can be made, since these were discussed earlier.

Coding and Test Plan

The test plan describes how the design will be coded and tested following its approval. *Coding* or *programming* is the actual writing of computer programs and involves translating the design specifications into computer code. *Testing consists of running the various coded pieces of a design to verify that all coded instructions are correct.* Rather than testing on a hit-or-miss basis, the testing process works best if it is carefully organized. To this end, the *test plan* generally provides some provision for three types of testing: unit (module), program, and system testing. In addition to showing how coded modules are to be tested as individual units, the test plan reveals how larger coded sections, such as computer programs, are to be tested; it must also describe ways of testing the entire system.

Test Schedule and Budget

This section is similar to the schedule and budget section prepared for the project and the logical design specifications. It compares actual and estimated times and costs for the project thus far and projects future time and cost requirements for completing the project.

What the design team must emphasize in this section is that considerable work remains in the next phase of development—namely, the implementation of the new design. Fifty percent of total development time is generally required for system implementation. Of this time, program and system testing requires the greatest share—ranging from 25 to 40 percent of the total.

Hardware and Software Specification

The hardware and software specification extends the preliminary specification provided earlier (see chapter 8) by documenting

1. the types of hardware and software needed to transmit, transform, and store the data specified by the design;
2. the degree to which current hardware and software can be expected to handle new processing requirements;
3. the types of hardware and software to be acquired, from which vendors it can be purchased (and at what cost), and when delivery is expected; and
4. the types of hardware and software interfacing problems anticipated, especially those resulting from new hardware and software.

In many instances, the design team may prepare a separate, more detailed hardware and software specification to record the steps leading to final hardware and software requirements, followed by the vendor selection process, the vendor contract, and equipment installation schedules. This kind of specification (which falls outside the scope of this book) cannot be undertaken lightly. Designers fully realize that proper equipment must accompany a design if the design is to be successful.

User Support Requirements

This section revises and expands the user and operations documentation outlines provided by the systems analysis team (see Figure 8–12). By this point, much is known about how users are expected to use the system, as well as how the system is to be installed and maintained. Consequently, the design team is expected to provide a complete outline of the user's guide, describe the types of user training required, and indicate what other types of user support are needed. **Figure 13–8** illustrates an expanded outline of the user's guide for a computer terminal scheduler. Compared with Figure 8–12's outline, this new outline indicates that certain tutorials are advised, beginning with one on confirming the schedule.

Revised operations documentation is equally important. Since the designers by this time have spent considerable time thinking about the value of system parameters, the use of error messages, and so forth, they can begin work on operations support materials, explaining such items as how to modify system parameters, add or modify individual privileges, add or modify group privileges, change user passwords, modify transaction logging procedures, and so forth.

Volume 1: User's guide
   I. Introduction to the computer terminal scheduler

   II. How to use the manual
      A. About this manual
      B. What you should know
      C. Tutorial: Confirming your schedule
      D. Visualizing the example
      E. Summary

   III. How to schedule times
      A. Reviewing the current schedule
         1. Defining the schedule
         2. Paging from one period to another
         3. Tutorial: Reviewing the schedule
         4. Printing the schedule
      B. Reserving terminal time
         1. Modifying the schedule
         2. Tutorial: Checking the revised schedule
         3. Printing the revised schedule
      C. Canceling a reserved time
         1. Modifying the schedule once again
         2. Tutorial: Double-checking the schedule
         3. Printing the revised schedule
      D. Summary

   IV. How to interpret error messages
      A. Program versus system errors
      B. Tutorial: Tests to run for yourself
      C. Asking for help
      D. Summary

FIGURE 13–8 Expanded user's guide for the computer terminal scheduler

The Mansfield, Inc., Case Study

John believed strongly in the need for controls, especially when dollars were involved. "The one thing I do not want questioned," he stated, "is where the dollars went. With customer payments we have to be especially careful. We don't want a system that encourages our employees to pocket and cash customer checks. Likewise, we have to make sure that records stored on computer files are not tampered with—either by accident or on purpose. We need to forestall the possibility of someone adjusting a customer's receivable account without proper authorization or without leaving a clear audit trail. Can you imagine what could happen if our system permitted unauthorized personnel to adjust a customer's bill, leaving no trace?"

Within-Document Control

Pamela was asked to check I/O routines to determine whether processing controls were adequate. She commented, "We designed the monthly statement remittance display to require within-document cash payment matching [see Figure 10–21]. We require each invoice payment to be keyed as well as the total customer payment received. We then use the computer to determine whether the sum of the keyed invoices equals the payment received. If it doesn't, the keyed totals are blocked from entry into processing."

Batch and Output Control

When John questioned Pamela about the controls designed for processing invoice remittances, she added: "At present, we only count the number of invoice remittances processed and compute total invoice payments [see Figure 10–20]. We print these totals when we prepare the payments register. One way to improve this procedure is to count the number of invoices and to add their dollar amounts before processing. Later we can compare the summed amount computed manually with the summed amount computed during processing."

Although John felt that this procedure was weaker than he would like, he decided to go along with it—at least on a trial basis.

Passwords

Unauthorized entry to the system will be blocked with user account numbers and passwords," John continued. "We will need to change passwords occasionally to help keep users aware of the need for tight system security."

File-Balance Controls

John asked his staff whether they had added file-balance controls to the design. When they looked puzzled, shaking their heads to suggest that they had not, John explained: "We need to create a control table showing the summary totals for the receivables account, the invoice summary, and the payment relations of the accounts receivable database [see Figure 11–20]. The layout of the receivables account table entry should read

```
Receivables-account-control (Control number, control-
   name, balance-forward-total, YTD-finance-charge-
   total, number-of-accounts-total, date-of-last-change)
```

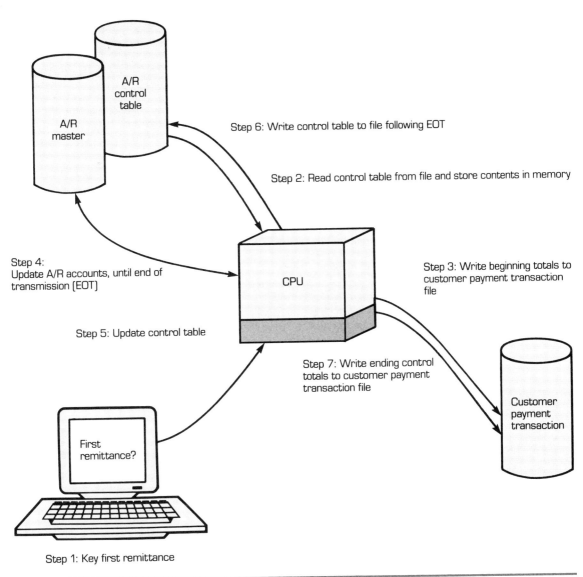

FIGURE 13–9 File-balance control procedure in which a control table is used to verify counts and amounts

"Likewise, the layouts of the invoice summary and the payment table entries should read

```
Invoice-summary-control (Control number, control-name,
   total-product-charge, total-sales-tax, total-cash-
   discount, total-number-of-invoices, date-of-last-
   change)
Payment-control (Control-number, control-name, total-
   invoice-payment, total-statement-payment, number-of-
```

```
invoice-remittances, number-of-statement-remittances,
date-of-last-change)
```

"These table entries contain all the important counts and amounts. If we were to print the contents of the A/R relations, the summed dollar values should equal the values recorded on the control table."

John continued by tracing the file-balance control procedure design for the processing of customer payments (see **Figure 13-9**). "The procedure is activated once we key the first remittance to processing (see step 1). This first remittance tells us to read the control table from the file and store the contents in computer memory (see step 2). Next, we write the beginning control total to the customer payment transaction file (see step 3). This establishes the 'before condition'—what we started with before transactions were processed."

John paused for a moment, asking for questions. Finding none, he continued: "The update of the control table and the A/R file occurs with each remittance keyed to processing (see steps 4 and 5). If the transaction is an invoice remittance, for example, we change the following control totals: the total invoice payment amount, number of invoice remittances processed, and date of the last change."

The last two parts of the procedure are quite straightforward, John said: "On receipt of an end-of-transmission (EOT) signal, we store the updated control table (see step 6); we also write the ending control totals to the customer payment transaction file (see step 7). This establishes the 'after condition'—what we wind up with once all the transactions are processed. All that remains is to print the contents of the transaction file. The payment register provides us with a hard-copy record of the before condition, a listing of all processed transactions, and the after condition."

"Any questions?" John asked again.

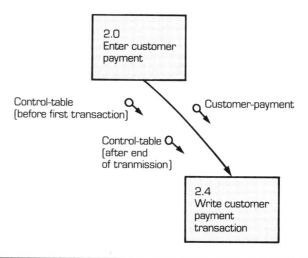

FIGURE 13-10 Revised structure chart showing passing of file control totals

"Only one," stated Walter. "How difficult would it be to change our structure charts to show a control table?"

"Not difficult at all," replied John. "It's a simple matter of adding control table data couples to structure charts [see **Figure 13–10**]. Remember to add one data couple for the before and another data couple for the after condition, however."

SUMMARY

The design of processing controls runs parallel to the design of program module specifications. Besides ensuring that each module is designed correctly, the designer must add controls to provide overall system security.

Four types of processing controls must be designed: source-document controls, input controls, output controls, and computer program controls. Source-document controls verify that all data have been entered into processing and, as important, that source documents can be recovered should the content of data transmission be questioned or need to be repeated. Within-document controls and forms control are important aspects of this kind of control. Input controls verify that data keyed or read from files are actually received by the computer. Flashback checks, journaling, and file-balance controls help determine whether all input was received. Output controls verify that all data have been sent from processing and cannot be obtained by unauthorized personnel. Account codes, passwords, printed registers, and output-control totals help determine whether all output was sent. Computer program controls validate the accuracy of programmed procedures. Data validation procedures, the use of flags to indicate errors or unusual conditions, warning messages, and tests of user identification all represent ways of checking the correctness of processing.

Data validation is an important aspect of computer program control. Edit tests (including field tests), picture checks, range tests, combination tests, consistency tests, and probability tests are programmed procedures that check input data for error. In contrast, check-digit tests use self-checking numbers to determine whether data keyed to processing are keyed correctly.

Auditing goes hand in hand with the design of processing controls. Audits around the computer, which determine whether system outputs are consistent with system inputs, also verify the completeness of source-document, input, and output controls. Audits through the computer primarily determine the completeness of computer program controls. These help determine whether all forms of test data can be processed successfully. Audits using the computer help verify all types of processing controls, though they tend to favor the checking of input and output controls, including file-balance controls.

Physical audit procedures must be designed along with computer-assisted audits. With physical audits, counts and amounts determined manually are compared with counts and amounts stored on computer files.

Once processing control requirements are finalized, the technical design specification can be completed. This specification documents all features of the detailed design. Besides providing clear design requirements, it includes a coding and test plan, a test schedule and budget, a hardware and software specification, and a section that outlines user support requirements.

REVIEW QUESTIONS

13–1. Name several reasons for designing processing controls.

13–2. What are source-document control procedures designed to verify?

13–3. What purpose is served by a within-document control procedure?

13–4. Name several considerations important to forms control.

13–5. What is the difference between a flashback check and journaling?

13–6. What is a control record?

13–7. What do output controls help verify?

13–8. What is another name for data validation? What purpose does it serve?

13–9. What is the difference between a combination edit test and a consistency edit test?

13–10. How do check-digit tests differ from other types of data validation tests?

13–11. What types of errors do most check-digit tests identify correctly?

13–12. What is an audit?

13–13. How do physical audits differ from audits through the computer?

13–14. What options are included in audits using the computer?

13–15. How does the technical design specification differ from the logical design specification?

EXERCISES

13–1. The number 694329 is keyed to processing and checked by the self-checking method called modulus-11. What would happen? Prove your answer.

13–2. Suppose you wish to add a check digit to the account number 96524, using the modulus-11 method. What digit would you add? Prove your answer.

13–3. Suppose you wish to append a check digit to the number 16875, using once again the modulus-11 method. What number would you append? Prove your answer.

WORLD INTERIORS, INC.—CASE STUDY 13
DESIGNING PROCESSING CONTROLS

Introduction

On review of the detailed system design, John Welby and Jane Strothers both realized that processing controls represented a potential design weakness. "We need to review all of our input, output, database, and computer program designs to determine whether we have incorporated adequate controls," commented John. "Jane, why don't you begin work in the I/O area, since you're so familiar with these processing requirements. I want to begin work on a control table to handle file-balance controls.

"Before I forget," he added, "let's document as we go along. I'm afraid that some of our Warnier-Orr diagrams and structure charts may need to be revised. We will also need to define the composition of each processing control and to add those definitions to the data dictionary."

I/O Controls

Jane started her analysis by reviewing the input displays prepared for processing customer orders and factory invoices (see chapter 10). She muttered to herself: "My designs look complete. I established a within-document control procedure for the customer order display and I did what I could with the factory order display. Perhaps John doesn't understand what I set out to accomplish. I'd better take the time to describe in writing how my design works."

Jane was also puzzled by John's questioning of her output controls. "The daily register, for example, provides a complete listing of the events of the day. John is certainly not interested in printing more detail. Perhaps not enough summary information is provided on this register."

File-Balance Controls

While John was confident in Jane's ability to handle I/O processing controls, he was not sure that she understood control table requirements fully. "We need to be able to control totals important to the relations which make up the open-orders database," he thought. "We cannot lose track of summary information such as the number of open orders and their dollar total, the number of problem orders and their dollar total, and sales tax total amounts. By keeping before-and-after information we can check the accuracy of the daily order register, the filled-order register, and the problem-order register, as well as the register printed to show the contents of the entire open-order database.

"Line-item control," John continued, "while not as important as customer invoice control, is also required. We need to store in the control table counts so that one relation can be compared with another. We can also print before-and-after counts on the daily order register."

"Finally, factory invoice control is important," John concluded. "Here, too, counts are required to permit us to show filled- to open-order relationships and the number of partially filled customer orders. Some dollar controls are also advised. These will allow us to better audit counts and amounts printed on the filled-order register."

CASE ASSIGNMENT 1

Using the displays prepared for the World Interiors case study assignment in chapter 10 and the structure charts prepared for the World Interiors case study in chapter 12, describe the within-document control procedure designed by Jane for the customer order input display. Which flags are returned when an error in processing is found?

CASE ASSIGNMENT 2

Using the displays prepared for the World Interiors case study in chapter 10 and the structure charts prepared for chapter 12, describe the within-document control procedure (if any) designed by Jane for the factory invoice input display. If a control procedure does not exist, what design change would you recommend? How would you modify the level-6.0 structure chart (see Figure 12–23) to document this design change?

CASE ASSIGNMENT 3

Review the layout of the daily order register (see Figure 10-24) and the Warnier-Orr diagram prepared for chapter 10. What, if any, output controls are missing? Revise the Warnier-Orr diagram to show your recommended modification.

CASE ASSIGNMENT 4

Design control table entries for the open-orders database by defining the contents of the customer invoice control, line-item control, and factory invoice control. Use the following notation in completing this assignment:

```
customer-invoice-control (. . .
       insert contents here . . .)
```

CASE ASSIGNMENT 5

Once again review the design of the daily order register (see Figure 10-24) and the Warnier-Orr diagram prepared for chapter 10. Suppose you now wish to print control table information (see the last assignment) on this register. Modify the Warnier-Orr diagram to add control-table information. (You do not need to draw the entire diagram. Show only where changes are needed.)

REFERENCES

1. G. Shelly and T. Cashman, *Business Systems Analysis and Design* (Fullerton, Calif.: Anaheim Publishing, 1975).

2. H. D. Clifton, *Business Data Systems, A Practical Guide to Systems Analysis and Data Processing* (Englewood Cliffs, N.J.: Prentice-Hall International, 1983), p. 244.

3. W. Mair, D. Wood, and K. Davis, *Computer Control and Audit* (Altamonte Springs, Fla.: The Institute of Internal Auditors, 1976).

14
Programming and Program Testing

INTRODUCTION

THE development of a preliminary system design followed by a detailed system design leads to a fully documented *structured design specification*. This specification not only describes how each computer program is to be designed; it also specifies how all controls for input, output, files, and processing are to be built into the new system. The work that remains is *systems implementation*—the translation of the technical design into usable computer software. The life cycle activities important to this final step in systems development include programming, testing, and conversion. This chapter examines the activity known as programming and one part of testing: program testing. *Programming* involves the writing of computer programs, with a program consisting of a series of coded instructions. *Program testing* consists of putting together the various coded pieces of a program, testing those pieces, and correcting parts that are incorrect or inappropriate. Chapter 15 will describe concepts important to system testing and conversion. These two activities go hand in hand in making new software operational.

When you complete this chapter, you should be able to

- describe the concept of team programming and explain why it is useful;
- understand rules and principles important to program design;
- understand constructs important to structured programming;
- design Nassi-Shneiderman diagrams to show how structured programs are organized;
- conduct a unit test for a coded module;

- explain the advantages and disadvantages of an incremental approach to testing; and
- design stubs for a computer module.

Moving from Systems Design to Systems Implementation

Systems implementation begins once the design specification has been approved. Who approves the specification and what does that approval entail? Much like the process of obtaining approval for the logical design, management, user groups, and members of the systems staff must approve the design specification. In their review, the following factors become important:

1. Are the technical requirements clear and complete? Can they be understood by the personnel responsible for implementing the design?
2. What ideas are offered to increase the probability of achieving a successful design? Can the design be implemented in stages? Will users and management participate in the implementation of the design?
3. Have the operational aspects of the design been fully communicated? Do users and management know what they can expect from the new system?
4. Are resources sufficient to fully implement the design? Does top management understand how these resources will be spent?

In brief, systems implementation should begin once the implications of the software design are fully understood.

A major difficulty in making the transition from systems design to systems implementation is that different technical personnel are generally added to the systems development team. For implementation purposes, several programmers are often combined with one or more chief programmers and one or more systems designers to form a systems implementation team. Before beginning work, these new team members must be briefed on the section of the design they will be asked to complete. This briefing generally takes the form of a structured walkthrough.

Why are systems designers usually members of the implementation team? In part, they are there to explain design features and characteristics. Likewise, some designers become the chief programmers of a project. They assume the new role of coordinating the work of programmers in preparing the computer code for the new system.

Team Programming

Team programming involves a special type of programming team. With team programming, more than one person is responsible for writing and testing a single computer program. Why should many people have that responsibility instead of one? To encourage what Weinberg calls *egoless programming*.¹ Instead of a program being "my program" or "your program," programs become known as "our programs."

Several advantages are associated with team programming, including the following:

1. Program bugs do not reflect on a single individual, since no one owns the program.
2. Typically, fewer program bugs arise. This is often a direct result of extensive internal review of code by programming team members.
3. Software design is typically more efficient. Higher cohesion and looser coupling (see chapter 12, pages 399–402) are characteristics of programs written by groups.
4. Programmers tend to be more productive. In a team effort, programmers usually help one another, especially during the debugging of a program.
5. Programmers enjoy flexibility. With team programming, programmers can shift from work on one program to another.

Even though team programming offers clear advantages, many project managers continue to assign complete program responsibility to a single individual. Why? Primarily because teams are more difficult to manage and evaluate. They also pose more of a threat to the manager. Consider the following problem of evaluation: Suppose that a programming team were highly successful, completing a project three months earlier than expected. Suppose next that management decided to promote the individual responsible for this fine achievement. Management would soon discover that the success of the project was the result of effort by all members of the team—not the talents of a single individual. Feeling frustrated, management finally would decide against promotion. Instead, they would decide to break up the team.

Examples such as this one point to the problems inherent in managing a group process; however, they also help shed light on the kinds of procedures that can improve programmer productivity. As the concept of team programming suggests, shared understanding of how computer programs are written often leads to dramatic increases in programmer productivity. Next we will consider the subject of programming. This section is written to describe good programming techniques that, in turn, help others understand how a computer program is written.

PROGRAMMING

The work of translating design specifications into computer code is better known as *programming*. While some continue to hold that programming is an art more than a science, others consider programming as a process that can be highly structured.[2,3,4,6] For our purposes, computer programs, the result of programming, should hold to the following conditions:

1. They should be easy to read.
2. They should be written using structured constructs.
3. They should embody the same rules as those used in designing structure charts.
4. They should be well documented.
5. They should be tested thoroughly.

Making Programs Easy to Read

Readability is a major factor in the design of computer programs. Generally, if a program is easy to read, it will be easier to follow logically. As a test of this statement, consider the code shown in **Figure 14–1.** This program consists of perfectly good Pascal code; however, it is an illustration of very poor code in terms of readability. Suppose this code contained an error. Finding the source of the error would be one problem. Fixing the error would be another.

Readable code is organized very differently from that example. Consider the code (a single procedure from a COBOL program) shown in

```
PROGRAM AA(INPUT,OUTPUT); CONST BB=10; VAR CC:ARRAY[1..BB]OF
STRING;DD:ARRAY [1..BB]OF INTEGER;EE,FF,GG:TEXT;FUNCTION HH:REAL;VAR
II:INTEGER;JJ:REAL; BEGIN JJ:=0;FOR II:=1 TO BB DO
JJ:=JJ+DD[II];HH:=JJ/BB;END;PROCEDURE KK; VAR II:INTEGER;BEGIN
RESET(FF,LL);RESET(GG,MM);FOR II:=1 TO BB DO BEGIN
READLN(FF,CC[II]);READLN(GG,DD[II]);WRITELN(EE,CC[II]:15,DD[II]:3);END;
CLOSE(FF);CLOSE(GG);END;PROCEDURE NN;VAR II,OO,PP:INTEGER;QQ:BOOLEAN;RR:
ARRAY[1..BB]OF STRING;BEGIN  ;PP:=BB;REPEAT QQ:=TRUE;PP:=PP-1;FOR
II:=1 TO PP DO IF DD[II]<DD[II+1] THEN BEGIN OO:=DD[II];DD[II]:=DD[II+1];
DD[II+1]:=OO;RR[II]:=CC[II];CC[II]:=CC[II+1];CC[II+1]:=RR[II];
QQ:=FALSE;END; UNTIL QQ;END;PROCEDURE SS;VAR TT:REAL;II:INTEGER;BEGIN
TT:=HH;WRITELN(EE);WRITE(EE,'    UU  ');WRITE(EE,'VV');
WRITELN(EE,'    WW');FOR II:=1 TO BB DO BEGIN IF DD[II]<(TT-10) THEN
BEGIN WRITELN(EE);WRITE(EE,CC[II]:15,DD [II]:9);WRITELN(EE,'    XX');
END;IF DD[II]>(TT+10) THEN BEGIN WRITELN(EE);WRITE(EE,CC[II]:15,
DD[II]:9);WRITELN(EE,'    YY');END;IF ((DD[II])>=(TT-10))
AND((DD[II])<=(TT+10))THEN BEGIN WRITELN(EE);WRITE
(EE,CC[II]:15,DD[II]:9);WRITELN(EE,'    ZZ);END;
END;WRITELN(EE);WRITELN(EE,'THE HH VV IS:',TT:4:1);END;BEGIN
REWRITE(EE,'AAA:');KK;NN;SS;END.
```

FIGURE 14–1 Difficult-to-read code: an extreme example

```
S U B    A D M I N Q  COBOL 12B(1131) BIS   13-Mar 00:40
T25.      12-Mar 18:40

1095   /
1096   *  ** THE NEXT SECTION VALIDATES USER ACTIVITY CATEGORIES ***
1097   *  ** TO ANSWER PATTERNS QUESTIONS FOR.       ***
1098
1099
1100          2-5-1-3-1-GET-CHOICE.
1101              ACCEPT PAT-CHOICE.
1102              MOVE "Y" TO VALID-CHOICE.
1103              IF ((PAT-CHOICE = "A") OR (PAT-CHOICE > "A"))
1104                  AND ((PAT-CHOICE = "Q") OR PAT-CHOICE < "Q"))
1105                  MOVE "Y" TO VALID-CHOICE
1106              ELSE
1107                  IF PAT-CHOICE = "R" MOVE "Y" TO QUIT-PATTERNS
1108              ELSE
1109                  IF PAT-CHOICE = "S" MOVE "Y" TO QUIT-PATTERNS
1110                                      MOVE "Y" TO DONE
1111              ELSE
1112                  IF PAT-CHOICE = "T" MOVE "Y" TO QUIT-PATTERNS
1113                                      MOVE "Y" TO QUIT-PATIENT
1114                                      MOVE "Y" TO DONE
1115              ELSE
1116                  IF PAT-CHOICE = "U" MOVE "Y" TO QUIT-PATTERNS
1117                                      MOVE "Y" TO QUIT-PATIENT
1118                                      MOVE "Y" TO QUIT-TRENDS
1119                                      MOVE "Y" TO DONE
1120              ELSE
1121                  MOVE "N" TO VALID-CHOICE
1122                  DISPLAY CX-COORD
1123                      "NOT A VALID SELECTION. PLEASE TRY AGAIN:"
1124                      WITH NO ADVANCING.
```

FIGURE 14–2 A COBOL procedure: an example of readable code

Figure 14–2. This more readable code contains considerable white space, or space between lines of code. The programmer decided to break up the code into sections and indent to improve overall readability. Such a practice is of great assistance to the reviewers of the code. It allows them to focus quickly on the organization and structure of the coded instructions.

Besides using white space and indentation, programmers use a wide variety of common-sense rules to improve the readability of their code. Consider the COBOL program shown on Figure 14–2 once again. In writing this code, the programmer decided to:

1. Allocate one line for each statement.
2. Select meaningful variable names (e.g., pat-choice, quit-pattern, and valid-choice).
3. Label the procedure (or module). This procedure is numbered 2-5-1-3-1 and is entitled GET-CHOICE.
4. Describe the purpose of the module.
5. Keep the procedure to less than fifty lines.
6. Set a flag to a valid choice to begin with. The flag is reset to an invalid choice only if an incorrect pattern choice is entered.

How important is improved readability? Probably not very to the person writing a computer program, since he or she understands the procedure well. However, readability is very important to the person who must review the program and perhaps correct its mistakes. When program review is considered, the following rules become basic tenets of program design: making one assignment per line, using meaningful variable names, indenting to show where logical contructs begin and end, clearly identifying important parts of the program, and setting flags in a realistic way. By following simple rules, programmers can communicate to others the systematic way in which they developed the computer code.

Designing Programs Using Standard Constructs

Central to the concept of improved readability is the concept of *structured programming*. With structured programming, individuals can write efficient, relatively easy-to-read computer programs by dividing the tasks to be performed into well-defined units or modules. Programs developed in this way tend to minimize the complexity of program flow, especially the transfer of control. In addition, each section of a program is kept small through the use of modules.

Dijkstra[4] is often credited as the founder of structured programming; however, its prominent use today is the result of the efforts of many. Böhm and Jacopini,[5] for example, proved that any algorithmic process could be represented as a combination of exactly three basic constructs: a sequential construct, an IF-THEN-ELSE (decision) construct, and a DO-WHILE (repetition or iteration) construct. Why should these constructs be familiar? Because they are identical to the constructs used in writing structured English and pseudocode.

Structured programming generally includes two other structures besides the three structures advanced by Böhm and Jacopini. These are a DO-UNTIL and a CASE construct. As discussed next, these additions are more special cases than new logical structures.

Let's examine each of these five constructs using Nassi-Shneiderman charts to explain their differences. In this way we can understand how modular programs are written. As important, we can demonstrate how

| Pour cake mix into bowl. |
| Combine mix with eggs, oil, and water. |
| Beat mixture. |
| Place in cake pan. |
| Bake in oven. |

FIGURE 14–3 Sequence required to bake a cake

programmers can define any algorithmic process using a combination of these five constructs.

Figure 14–3 illustrates a set of sequential constructs to describe the process of baking a cake. A vital part of the programming lies in defining what action to take first, second, and so forth.

Next, we will change the example to include an IF-THEN-ELSE construct. We will question whether the bowl for the cake batter is clean (see **Figure 14–4**). If it is, the bowl can be used immediately. If it is not clean, the bowl must be washed.

We can also add a DO-WHILE loop to this process. As shown in **Figure 14–5,** several cakes will be baked, depending on the number of cake mixes in stock. This new algorithm might be called "Make all cakes."

A DO-UNTIL loop can also be added. As shown in **Figure 14–6,** the general form of the DO-UNTIL is

```
         loop action
UNTIL test for action.
```

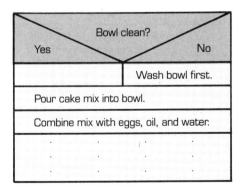

FIGURE 14–4 Adding an IF-THEN-ELSE construct to a sequential construct

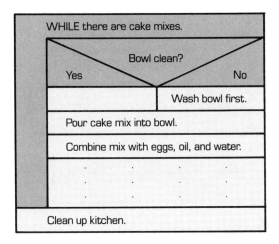

FIGURE 14-5 Adding a DO-WHILE loop to the process of baking a cake

As indicated, the cook would continue to beat the cake mixture until the cake batter was smooth.

Finally, a CASE structure could be added to this example, in which one of several *mutually exclusive* actions is possible. **Figure 14–7a** shows the general form of the CASE structure, while **Figure 14–7b** ap-

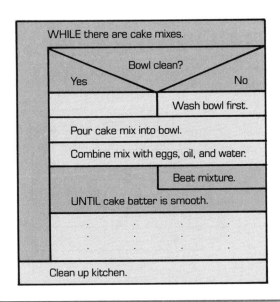

FIGURE 14-6 Adding a DO-UNTIL loop to the process of baking a cake

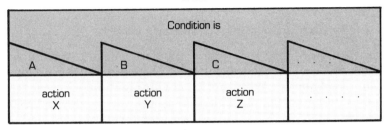

(a) General form of CASE structure

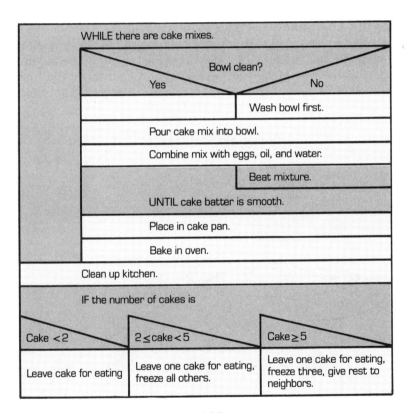

(b) Using a CASE structure

FIGURE 14-7 General form of the CASE structure (a) and adding the CASE structure to the process of baking a cake (b)

plies the case structure to the cake-baking example. As shown, the number of cakes must satisfy one and only one condition in order for a particular action to be taken.

Designing Programs Following a Top-Down Design

Besides restricting program code to standard program constructs, programmers are well advised to follow a procedure known as *top-down program development*. With this procedure, the uppermost levels of a structured design are implemented first, followed by lower-level modules. For example, modules 0.0, 1.0, 2.0, and 3.0 of the Mansfield payroll system were programmed before any other modules (see Figure 3–8). Following their implementation, the top-level modules shown on the level-1, level-2, and level-3 structure charts were programmed. Let's consider the procedure followed by the programmer in preparing the code for module 2.0 (printing the payroll register) for the moment. First, the programmer reviewed the structure chart to isolate the processing requirements for this module. In review (see **Figure 14–8**), the programmer determined that this high-level module would coordinate the functions of four lower-level modules. Second, the programmer determined the

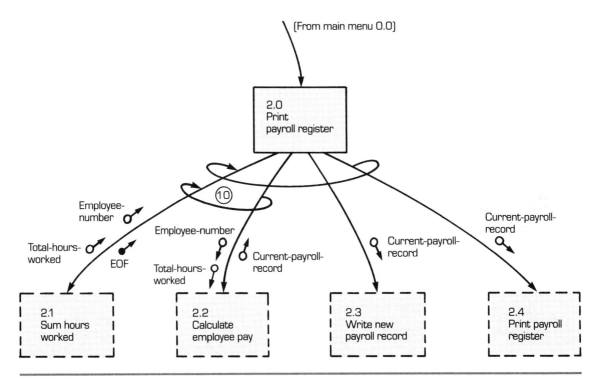

FIGURE 14–8 Isolating the processing requirements of module 2.0

coordinating functions to be performed by the module. The structure chart indicated that it was important to

1. receive employee numbers and total hours worked; and

2. keep a count of the number of times hours-worked data were received. If the count exceeded ten, write and print requirements were to be initiated.

Even though they were not shown on the structure chart, the programmer also determined what, if any, processing controls were to be maintained. The required controls in this instance were a count of the number of employees and the sum of employee hours worked.

Figure 14–9 illustrates the program developed for module 2.0, once again using a Nassi-Shneiderman chart to clarify the logic of the code. As indicated, the first part of the code is designed to initialize all counts and amounts. Following this, the program begins a loop that continues until all records are read. Within the loop are calls to modules 2.1, 2.2, 2.3, and 2.4. Modules 2.1 and 2.2 are called several times (up to ten) before modules 2.3 and 2.4 are invoked. The value returned by module 2.1, the current payable record, is placed in an array sized to equal the count. Once an end-of-flag record is received (that is, all records have been read), a final check of the count is made. If the count is greater than zero, the contents of the current payroll record array are printed. Finally, processing controls are printed. These consist of the number of employee records processed and the number of hours worked by all employees.

This example should clarify two things: how the transition is made from a structure chart to a computer program and what is meant by top-down program development. However, let's suppose next that a module consists of a single line of code and is hatted on a structure chart. How would this be incorporated into computer code?

Figure 14–10 shows a structure chart for a module that calculates student average deviation. As indicated, the module must coordinate several functions: getting the grades, calculating the grade point average (GPA), calculating deviations, and calculating the average deviation. Moreover, each lower-level module is hatted, indicating that one procedure is to be written for all five modules.

Figure 14–11 shows the Nassi-Shneiderman chart written to describe the top-level design of this module. Through the tracing of the logic of this procedure, the role of the boss module and the worker modules should become clear. Observe, for example, that the boss module (1.0) does little more than initialize variables. Observe next that each worker module performs a single processing function. These functions can be summarized as follows:

1. Module 1.1 gets a student grade and places the grade in an array called GRADE(N). In this instance a grade is a numerical score.

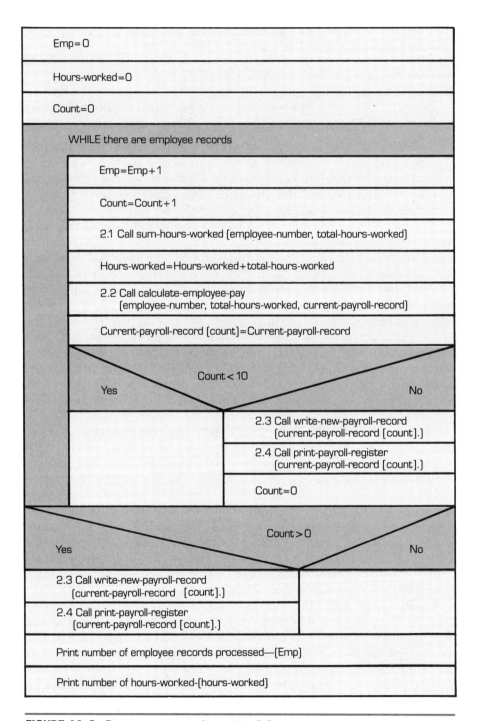

FIGURE 14–9 Program structure for module 2.0

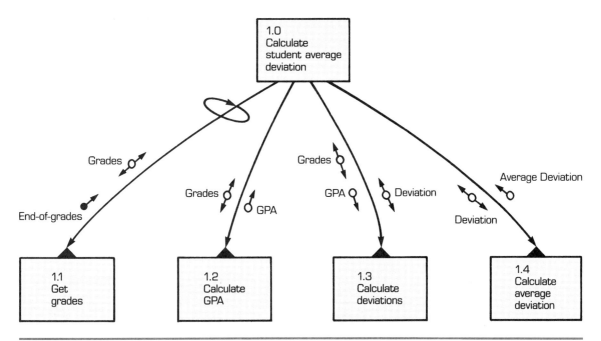

FIGURE 14–10 Structure chart showing steps needed to calculate students' average grade deviation

2. Module 1.2, limited to a single instruction, calculates the GPA.

3. Module 1.3 computes the deviation from the GPA for each student grade.

4. Module 1.4 computes the average deviation.

These last two examples clarify some interesting parallels between structure charts and top-down program logic. We can summarize them as follows:

1. Structure charts guide the programmer in the development of programs by identifying the processing functions to be encoded and indicating how these functions are interconnected.

2. Structure charts encourage the top-down development of computer programs.

3. Structure chart logic and computer program logic, though similar, are not always identical. The programmer typically modifies the logic shown on structure charts as it becomes clear that there are better ways to code a procedure.

4. Structure charts by themselves do not show how data are received, validated, transformed, or sent by a module. These decisions continue to be made by the programmer.

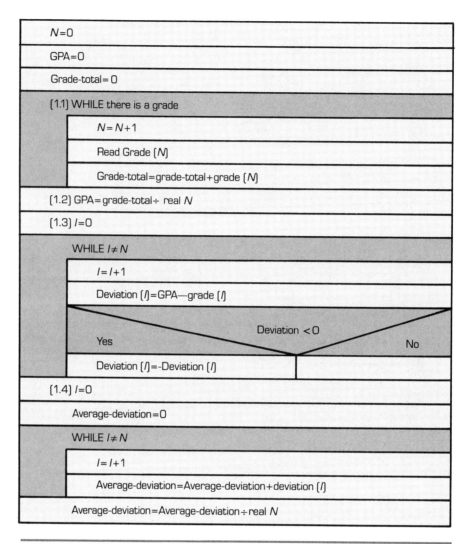

FIGURE 14-11 Program structure chart for module 1.0, calculate student average deviation

Documenting Computer Code

Besides writing code to be processed by the computer, programmers must carefully document computer programs—to identify and explain to others the steps important to processing. Careful documentation of programs also improves their readability. How is this done? Consider three rules of documenting computer code.

First, each procedure should begin with a program identification section. This section should

1. show the number of the module and its name;
2. indicate the name and number of the boss module;
3. list all modules called by the procedure; and
4. identify the names and meanings of pertinent variables required in processing.

Figure 14–12 illustrates such an identification section. The number of the module (5.2) and its name (compute FIFO) are documented, followed by the number and name of the boss module (5.0—inventory valuation) and the numbers of all modules called by the procedure. The next documented section explains the purpose of the module. As illustrated, "This module determines the accounting figures for each item in the bakery's inventory at the end of the latest period." The third documented

```
**************************************************************
* Module (5.2) — Compute FIFO                                 *
* Called From: (5.0) Inventory valuation                      *
* Calls: 5.2.1, 5.2.2, 5.2.3, 5.2.4, 5.2.5, 5.2.5, Library 5.A *
**************************************************************
* This module determines the accounting figures for each item in the *
* bakery's inventory at the end of the latest period. This is the    *
* main module of the inventory valuation procedure and it calls all  *
* subordinate modules.                                               *
**************************************************************
* Pertinent variables to the main module:                     *
*     access:    number of order records processed            *
*                                                             *
*     found:     indicates whether a block of purchase records exist *
*                for the inventory item being evaluated       *
*                                                             *
*     num_mast:  number of types of raw materials and bags in inventory *
*                                                             *
*     num_order: number of records in the orders file         *
*                                                             *
*     proc_pur:  dollar value of purchases make of a inventory item *
*                in the period being evaluated                *
*                                                             *
*     proc_qty:  number of units of the inventory item at period end *
*                                                             *
*     proc_val:  dollar value of the inventory item at period end *
*                                                             *
*     trip:      number of inventory records that have been processed *
*                                                             *
**************************************************************
```

FIGURE 14–12 Identification section of a coded procedure

section defines important variables. Consider the variable named "num_mast," for example. Unless defined, the meaning of this variable would be difficult to determine. Even with a clear definition, someone might question why the programmer selected "num_mast" when "num_type" or "inv_type" might be better. Naming such variables is not an easy task. What may be clear to one programmer is not always clear to others. Thus, programmers must document the meaning of important variables within a procedure.

Second, documentation should be added to indicate when a procedure begins and ends. **Figure 14–13** shows how comment statements can be used to document the beginning and end of a procedure, which in our example is a hatted procedure numbered 5.2.3. As with larger procedures, the hatted procedure is numbered and named, shows call-from and call-to relationships, and describes the purpose of the procedure. (In this example, no calls are made to other procedures.) Finally, the end of the module is documented by number and name.

Finally, documentation should be added within the procedure to clarify important flags, decisions, special assignments, and loops. **Figure 14–14** provides an example in Pascal to illustrate the importance of inserting comments within a program. As illustrated, one variable named SWITCH_SCORES and another named SWITCH_NAME are placed in the program to temporarily store the score and the name of a student.

Testing Computer Code

Thoroughly testing computer code is also essential in the design of computer programs. As an initial test, programmers should determine whether each coded procedure is easy to read, uses structured constructs, follows a top-down organization, and is well documented. If this

```
*******************************************************************
* Module (5.2.3) — Get inventory record            hatted *
* Called from: embedded in (5.2) Compute FIFO             *
*******************************************************************
* reset values for next inventory item                   *
*******************************************************************
trip     = trip + 1
found    = .T.
store 0 to proc_pur, pur_qty, proc_val
proc_qty = end_qty/UNIT_DES->unit_cnver
match    = inv_number
* End: Module (5.2.3), get inventory record
```

FIGURE 14–13 Using comments to indicate the beginning and the ending of a procedure

```
PROCEDURE SWITCH_TWO_ENTRIES (VAR SORTED : BOOLEAN);

(*----------------------------------------------------------------
THIS IS CALLED FROM THE PROCEDURE SORT_TWO_LISTS, IT DOES THE SWITCH-
ING INSIDE THE SORT ROUTINE.
----------------------------------------------------------------*)

VAR
    SWITCH_SCORES   : INTEGER;
    SWITCH_NAME     : ARRAY [1..MAXIMUM_PEOPLE] OF STRING;

BEGIN (*---------------SWITCH_TWO_ENTRIES---------------*)

    SWITCH_SCORES                := SCORES_OF_STUDENTS[I];
    SCORES_OF_STUDENTS[I]        := SCORES_OF_STUDENTS[I + 1];
    SCORES_OF_STUDENTS[I + 1]    := SWITCH_SCORES;

(*THE ABOVE USES A TEMP. VARIABLE SWITCH_SCORES TO HOLD THE VALUE*)
(*TO BE SWITCHED*)

    SWITCH_NAME[I]               := NAMES_OF_STUDENTS[I];
    NAMES_OF_STUDENTS[I]         := NAMES_OF_STUDENTS[I + 1];
    NAMES_OF_STUDENTS[I + 1]     := SWITCH_NAME[I];

(*THE ABOVE USES A TEMP. VARIABLE SWITCH_NAME TO HOLD THE VALUE*)
(*TO BE SWITCHED*)
    SORTED                       := FALSE;

END; (*---------------SWITCH_TWO_ENTRIES---------------*)

(*----------------------------------------------------------------
PROGRAM LISTING CONTINUED ON THE NEXT PAGE.
----------------------------------------------------------------*)
```

FIGURE 14-14 Adding comments within a procedure to clarify important steps

is the case, unit testing and program testing can begin, followed by system testing. (Because of the importance of testing, the last section of this chapter is devoted to unit and program testing. Chapter 15 examines the subject of system testing.)

This description of programming and program development, though brief, should lead you to conclude that without some type of formal specification, programming styles will vary considerably from one programmer to another. To this end, *program specifications*, the rules or

PROGRAM SPECIFICATIONS

standards to be followed in the design of a computer program, are generally prepared by most systems and programming organizations. What rules are included in such a specification? While the rules vary from one organization to another, most sets of rules include the following categories: program design and development, program instruction, and program documentation.

Program design and development specifications provide standards for organizing the way in which software is coded. Consider the following rules:

1. Develop one program procedure for each module (not hatted) shown on the structure charts. To do otherwise would defeat the purpose of structured analysis and design.

2. Number and name each procedure to correspond with the number and name shown on the structure charts.

3. Revise a structure chart if a better way of coding a module is discovered.

4. Minimize the use of global variables (i.e., variables that can be used by any module in a program).

5. Factor the entire programming project from top to bottom, coding and testing uppermost modules first.

6. Limit each procedure to a maximum of fifty lines of code.

7. Where possible, provide a single entrance to and exit from each module.

In brief, program design and development rules help ensure that the organization of computer code is consistent with the organization of the system as defined by the structured design.

Program instruction specifications deal with the way in which instructions are written. Consistency of programming style is as important as is efficiency of processing. In other words, computer code must be clean and efficient.

The lack of consistency in style from one program to another is often not realized until it is too late. However, the results of this practice are all too clear: programmers have great difficulty in understanding other programmers' code. Why? Largely because most programs are written in a disorganized manner.

In an attempt to make programs look alike, many firms have decided to require a program structure similar in form to the structure provided by the COBOL language. As shown in **Figure 14-15,** the structure of a COBOL program consists of four divisions: an identification division, an environment division, a data division, and a procedure division.

Another advantage of COBOL is that is is more or less self-documenting. For example, consider the following code:

```
Identification division
    Name of program
    Written by whom
    Written, when, where

Environment division
    Written for what specific equipment

Data division
    Requires what types of data
    With what format

Procedure Division
    Name of procedure
    Purpose of procedure
    Listing and definition of important variables
    Processing instructions to execute
```

FIGURE 14–15 Structure of a COBOL program

```
SET-WORK-AREA.
    IF ACCOUNT-NUMBER IN MASTER-INPUT-AREA
        IS EQUAL TO CURRENT-KEY
        MOVE MASTER-INPUT-AREA TO WORK-AREA
        PERFORM READ-A-MASTER-RECORD
        MOVE "Y" TO IS-MASTER-RECORD-IN-WORK-AREA
    ELSE
        MOVE "N" TO IS-MASTER-RECORD-IN-WORK-AREA.
```

Because this code is English-like and follows a structured notation, it becomes relatively easy to determine what the code was written to accomplish. Even with COBOL, however, style-related rules are important. Consider the following guidelines:

1. Indent to separate and define different logical parts of the program.

2. Space between important parts of the program.

3. Avoid GO-TO statements by limiting coded instructions to structured programming constructs.

4. Select meaningful program, module, constant, and variable names.

The rules governing the writing of efficient code are more difficult to comprehend than are the rules pertaining to stylistic consistency. Consider the following rules:

1. Block records when possible.

2. Use codes to avoid processing large strings of data.

3. Use table look-up procedures to avoid storing computed values.
4. Incorporate fast sort procedures, especially when the list of items to sort exceeds twenty.
5. Limit the testing of innermost loops.

Rules to improve processing efficiency are more important than they may initially appear. Yet most experienced programmers realize that inefficient programs may make even a well-defined system impossible to use. (In chapter 16, we consider the writing of efficient programs in more detail.)

Program documentation specifications explain how computer instructions, once written, are to be commented on by the programmer. As such, these rules deal with the way in which computer code is explained, not with the way it is written. Several rules are required to specify how programs are to be documented. Examples of these rules (including some that were discussed earlier) include:

1. Label each procedure to provide a direct reference to a structure chart.
2. Describe the purpose of each module.
3. Show where internal and external calls are made within a module.
4. Show where exits are made from a program.
5. Define the meaning of confusing variable names.

How many comments should be placed within a computer program? A general rule is one comment for every three program instructions.[6] While this number may appear high, in practice it is not. Count the number of comments shown on Figure 14–14. This module contains as many comments as program instructions.

Specifications such as these—even though they are subject to some interpretation—are quite essential to organizations that must maintain several hundred computer programs. When carefully written, these specifications do not tell programmers how to program. Rather, they instruct programmers on how to present their coded results.

PROGRAM TESTING

Once computer programs have been written, they must be tested. *Testing* is a multistep process with a designated purpose: to uncover errors and flaws in a coded design. Testing typically begins with unit testing, followed by program testing, which in turn is followed by system testing and acceptance testing. Let's clarify what we mean by these terms. *Unit tests* are conducted to remove syntax and logic errors from a single module or unit of a computer program, while *program tests* are conducted to remove syntax and logic errors from an entire computer program. Both unit and program tests are conducted by the programmer or programmers who wrote the code. *System tests* are tests conducted to determine whether any logic errors exist when the system is tested as a whole. These tests are conducted by someone other than the person who wrote the program—typically members of a test team. Finally, *acceptance tests* are tests conducted to determine whether the final version of the design is acceptable. These tests are conducted by users.

This broad range of testing—from unit testing to acceptance testing—can take considerable time to complete. It is often reasonable to expend up to 30 to 40 percent of all systems development time on this single stage in development. With so much at stake, most organizations encourage their programmers to undertake as much unit and program testing as possible as they code their programs.

Where does unit and program testing end? While there is no specific end point, programmers must be satisfied that their computer code is relatively free of bugs before they agree to turn it over for system testing. What system testing personnel wish to avoid is receiving code for testing that has never worked well, if at all.

Several strategies are open to the programmer in conducting unit and program tests. The following strategies are of particular importance:

1. Test each coded module. Combine modules to test each computer program. Combine computer programs to test the entire system. (This approach is called the *traditional approach*.)

2. Beginning at the uppermost module in the overall system, code and test a single module before coding and testing the next lower-level module. (This approach is called the *incremental approach*.)

3. Code and test all upper-level modules in the overall system first; however, once system interfaces are operational, code all critical lower-level modules. (This approach is called the *critical path approach*.)

The Traditional Approach

The traditional approach to testing works from the bottom up and is sequential in nature. The steps important to this process are (see **Figure 14–16**):

1. Code, test, and debug each module. This test is called the *unit test*.

2. Group units into programs and test and debug each program. This is called the *program integration test*.

3. Group all programs and test and debug the entire system. This is called the *system integration test*.

Even though this approach is traditional, it has many critics. The major disadvantages of this approach are twofold: *system interfaces* (i.e., the bridges between programs) are tested too infrequently, and bugs discovered during system testing often lead to debugging the entire system, which is both time-consuming and extremely frustrating for the programming team.

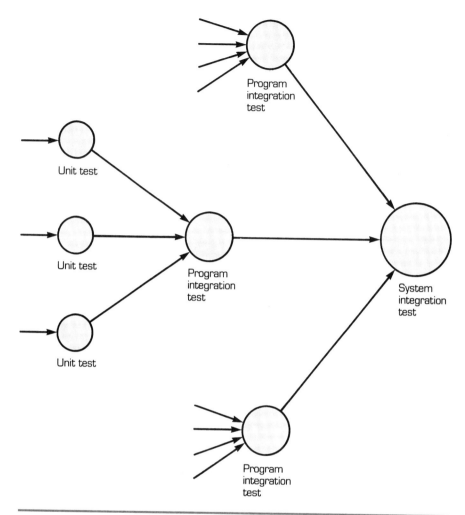

FIGURE 14–16 Traditional approach to testing

The Incremental Approach

The incremental approach to testing is iterative in nature. Page-Jones[7] describes this approach as follows:

1. Code, unit test, and debug the uppermost module shown on a structure chart.
2. Code, unit test, and debug the next highest module.
3. Add this next highest module to the previously tested and debugged module(s).
4. Test and debug the system.

5. Repeat steps 2 through 5 working with one module at a time until all modules have been tested and debugged and the entire system has been tested.

Unlike the traditional approach—which waits until all programs are tested and debugged before testing the entire system—the incremental approach stresses that testing should be top-down: upper-level parts of a system should be tested before lower levels. As shown in **Figure 14–17,** system interfaces are tested frequently: outputs from programs 1 and 2 (A and B) must be received as inputs by programs 3 and 4.

Besides testing system interfaces frequently—a main advantage of the incremental approach—this approach has several other advantages:

1. Members of the programming team must agree on data types in advance, so that data passed from one program to another are consistent. For example, if a record is to be passed from program 1 to program 3, the content and layout of the record must be the same for both programs.

2. The programming team gains a better appreciation of how the system will work once it becomes fully operational. System inputs and outputs in particular must be defined early on.

3. Programmer productivity and morale tend to improve. Overall test time, for instance, is often reduced. Moreover, programmers can see the results of their efforts almost immediately and, as a consequence, are less likely to become frustrated when bugs appear, especially since problems can usually be traced to a single module.

4. With improved productivity, the entire project is more likely to be completed on schedule.

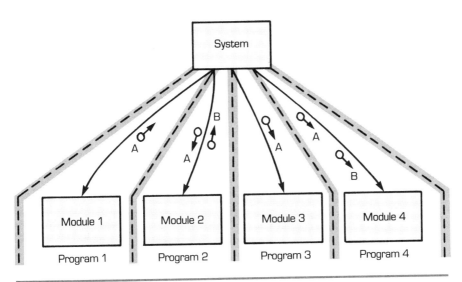

FIGURE 14–17 Incremental approach to testing

5. Attention can be focused on especially sensitive or error-prone modules. "Hot spots" in a design can be identified and closely monitored.

6. Users can be asked to review a design before it is completed. If they find fault with features of the design, such as the way in which information is presented, only the upper levels of the design will need modification.

Nevertheless, the incremental approach also has its disadvantages. A main disadvantage is that the results produced by a lower-level module must be simulated by a *stub* or dummy module. Deutsch defines a stub as "a dummy component that simulates the functioning of the next component subordinate to the component that is the present testing target."[8] How do stubs work in practice? Consider the following example.

Suppose that we wish to test module 3.14 shown on **Figure 14–18.** As illustrated, the test module receives as input, data couple A and flag B; it sends as output, data couples A and C. To test this module, it would be necessary to simulate the values of data couple A and flag B. A value can be assigned to data couple A, for instance. This value would be sent to simulate the behavior of module 3.141. Once received by the test module, the value assigned to A would be sent to another simulated module, module 3.142. Module 3.141, for example, might consist of a single instruction, such as

$$A = \$3.65$$

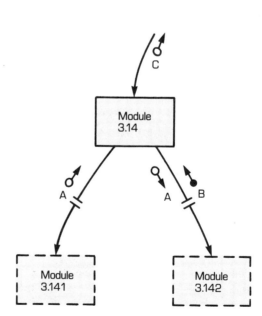

FIGURE 14–18 Adding stubs to a program

Since the value is simulated, the coded module would not indicate how A was derived. Likewise, Module 3.142 might be coded as follows:

> IF A IS POSITIVE
> THEN B IS VALID.

Why do stubs pose a disadvantage? At times it is difficult to estimate the value of a parameter such as A. Suppose A turns out to be thirty-six trillion rather than a value close to zero. In addition, stubs require programmer time to develop and take computer resources to store and execute. The only way to minimize these disadvantages is to keep the stubs themselves simple.

The Critical Path Approach

The critical path approach combines the incremental, top-down testing approach with bottom-up integration. With this third approach, only the top two or three levels of a design are implemented and tested using the incremental approach. Following this, the modules critical to the overall success of a project are identified by the programming team. Once identified, these critical modules are implemented and tested and are added to the top-level portion of a design. In brief, the critical path approach can be summarized as consisting of two stages. Stage 1 represents top-down implementation:

```
REPEAT UNTIL the top levels of the design are com-
  plete.
      Implement and unit test a module.
      Add the module to the system.
      Test and debug the system.
```

Stage 2 represents bottom-up implementation:

```
REPEAT UNTIL all critical modules have been imple-
  mented.
      Implement and unit test a critical module.
      Add the module to a critical cluster of modules.
      Test and debug the critical cluster.
      Add the critical cluster to the system.
      Test and debug the system.
```

Which module or cluster of modules represents a critical path in a design? In practice, the critical path includes those sections of code judged by the programming team to be the most difficult to code. It may represent a cluster (see **Figure 14–19**) in which the outputs of processing are difficult to estimate, or it may represent a function that is viewed as difficult to code—such as describing the properties of an irregular object

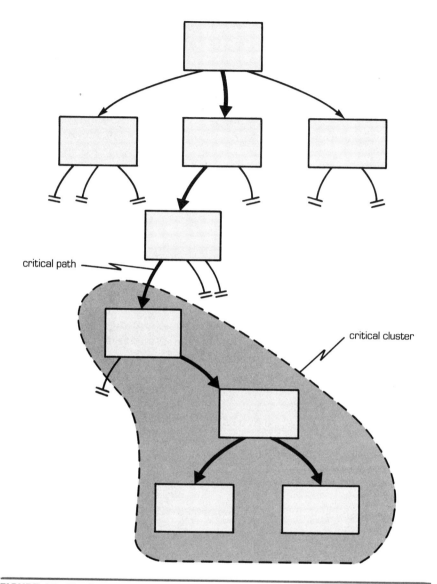

FIGURE 14-19 Critical path approach to testing

or coding a student schedule for a statewide round-robin speech tournament. In any event, critical path testing combines the advantages of both the incremental and the traditional approach. It requires first that the top levels of a design be implemented and tested; however, it requires next that the most difficult aspects of a design be implemented and tested.

The Mansfield, Inc., Case Study

The coding of Mansfield's new integrated billing and receivables system was simplified by the mere fact that it was a menu-driven design. John Seevers described the coding and testing process.

"Because we are dealing with a menu-driven design, the first thing to do is to build a *driver program*—the upper-most program in the system that calls all other programs," he explained. "Once the 'driver' is in place, the coding of the system programs can begin [see Figure 7–17 for a partial listing of these programs]. Let's start with an incremental approach to coding and testing. I want the three of you [Carolyn, Pamela, and Walter] to code and unit test individual modules. Once you are satisfied with your code, send the completed procedure to me. I will add the new module to the system and test each new combination."

Programming

"We used a modified form of team programming in writing the code for the new system," remarked Walter. "Before the first line of code was written, we had a long discussion on program specifications and procedures. We agreed that all of our code would follow a structured notation, include a large number of comments, be properly labeled, and avoid 'cute' coded instructions and comments. In addition, we decided to swap code before sending it to John for testing. This practice helped us determine whether our code was readable and efficient."

"We discovered that our code could be made more efficient if more than one person examined it," added Pamela. "Let me give you an example [see Figure 12–20 for additional clarification]. Instead of keying a remittance, adding customer information to it, updating the accounts receivable (A/R) master, and writing customer payment records to the customer payment transaction file, one transaction at a time, we decided to block and sort customer payment transactions. This simple decision permitted us to write a block of sorted records to the customer payment transaction file. How did this help? Later on, when we wanted to print the customer payments register, we no longer needed to sort the entire file. Instead of a sort, we were faced with a merging of previously sorted lists."

Incremental Development and Testing

"The incremental approach to coding and testing also helped us more than we expected," continued Pamela. "We were able to produce clearer and more efficient code once we fully understood system and program interface requirements. Consider the code (pseudocode) written for module 2.0, enter customer payments. As illustrated by **Figure 14–20,** we tended to isolate the flags and the exits important to top-level modules before starting to consider the processing requirements of lower-level modules. This particular section of code reveals several things. First, flags were initialized together with loops. Second, when an error flag was received, such as NO-RECORD-ON-FILE = TRUE, this told us that we needed to display the error and abort (discontinue processing) or display the error and continue. Module 2.0 presented us with exactly this situation. If there was no record on file or if posting was not correct, then we needed to abort the transaction. If, however, there was a record on file and posting was correct, but the writing of the transaction to the payment transaction file was not correct, we needed to display the error and con-

```
*MODULE 2.0-ENTER CUSTOMER PAYMENTS*
*CALLED BY DRIVER PROGRAM 0.0*

*INITIALIZE*
EOT = FALSE
LOOP = 0
BLOCKING-FACTOR = 20

*BEGIN ENTERING CUSTOMER PAYMENTS*
DO-UNTIL EOT = TRUE

    CALL 2.1 KEY-REMITTANCE
    CALL 2.5 DISPLAY-CUSTOMER-PAYMENT
    CALL 2.2 ADD-CUSTOMER-INFORMATION

        *CHECK SETTING OF FIRST FLAG*
        IF NO-RECORD-ON-FILE = TRUE
            THEN DISPLAY ERROR AND ABORT
            ELSE CALL 2.5 DISPLAY-CUSTOMER-PAYMENT
        END-IF

    CALL 2.3 UPDATE-A/R-MASTER

        *CHECK SETTING OF SECOND FLAG*
        IF NO-RECORD-ON-FILE = TRUE
            THEN DISPLAY ERROR AND ABORT
        END-IF

        *CHECK SETTING OF THIRD FLAG*
        IF POSTING-OK = FALSE
            THEN DISPLAY ERROR AND ABORT
            *INCREMENT LOOP IF PROCESSING IS SUCCESSFUL*
            ELSE LOOP = LOOP + 1
        END-IF

    *DETERMINE IF BLOCK IS FULL*
    IF LOOP = BLOCKING-FACTOR
        THEN CALL 2.4 WRITE-CUSTOMER-PAYMENT-TRANSACTION
            *RESET LOOP*
            LOOP = 0
            *CHECK SETTING OF FOURTH FLAG*
            IF TRANS-OK = FALSE
                THEN DISPLAY ERROR AND CONTINUE
            END-IF
    END-IF

END-DO-UNTIL
IF LOOP > 0
    THEN CALL 2.4 WRITE-CUSTOMER-PAYMENT-TRANSACTION
END-IF
*END MODULE 2.0-ENTER CUSTOMER PAYMENTS*
```

FIGURE 14–20 Pseudocode for module 2.0, enter customer payments

tinue. Why not abort here, too? The top-down loop told us that it was too late to abort once the update of the A/R file had taken place. The problem was not in the update process. Rather, it was limited to writing the result of the update to the transaction file."

SUMMARY

The final step in systems development is the implementation of the systems design. The activity known as systems implementation consists of three primary activities: programming, testing, and conversion.

In making the transition from the design of a system to its implementation, technical requirements must be clear. In addition, the operational implications of the completed design must be fully understood by users.

Team programming is often suggested as a means of organizing the programming of the new system. The main advantage of team programming is that it is ownerless and thus egoless, since no single person is responsible for writing the program.

While computer programming continues to be an art, the completed written code should nonetheless meet a set of prescribed conditions: it should be easy to read, be written using structured constructs, embody the same rules as those used in designing structure charts, and be thoroughly tested.

Five constructs are important in writing structured computer programs: the sequential construct, the IF-THEN-ELSE construct, the DO-WHILE construct, the DO-UNTIL construct, and the CASE construct. Used in combination, these five constructs permit any computer program to be written.

Top-down program development goes hand in hand with the design of structure charts. With a top-down approach, the uppermost modules of a system are coded and tested before lower-level modules are coded and tested.

Program specifications—the rules or standards to be followed in the design of computer programs—attempt to make the practice of computer programming less variable. Three program specification categories are important: program design and development, program instructions, and program documentation. Collectively, these specifications encourage consistency of programming style, greater efficiency of processing, and improved documentation of processing instructions.

Unit and program testing are often integrated with the coding of a computer program. The incremental approach to program development—in which the uppermost modules are coded, tested, and implemented before the lowermost modules—requires such integration. The advantages of top-down coding and testing are several: they test system interfaces frequently, require data types to be determined in advance, give the programming team an opportunity to see early results, improve programmer productivity and morale, keep projects on schedule, focus attention on "hot spots" in design, and allow users to review a design in stages. The main disadvantage of this approach is the need to stub lower-level modules during implementation. Stubbed modules may be difficult to code and may require extra code.

The incremental and critical path approaches to program coding and testing offer alternatives to the traditional approach. With the critical path approach, the modules considered to be most critical to the system are implemented early on. In this way, the overall success of a project can be determined before resources are expended on coding and testing easy-to-handle portions of a system.

REVIEW QUESTIONS

14–1. What is meant by systems implementation, and which life cycle activities are important to this stage?

14–2. Why is team programming referred to as egoless programming?

14–3. What rules are important to making computer code more readable?

14–4. Why should programs be designed using the five standard structured programming constructs?

14–5. What is meant by top-down program development?

14–6. How do structure charts guide the programmer in program development? How do programmers change the logic shown on structure charts?

14–7. What types of information are contained in the program identification section of a computer program?

14–8. What are program specifications? Which three types of specifications are generally developed by organizations?

14–9. What is the difference between a program test and a system test?

14–10. Explain the difference between the traditional and the incremental approach to program testing.

14–11. What is a program stub?

14–12. Why is the critical path approach sometimes the best test strategy?

EXERCISES

14–1. What is unusual about the following logic shown by the Nassi-Shneiderman diagram?

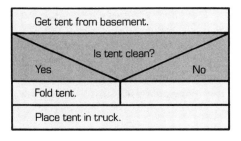

14—2. A friend of yours tells you the following: "What I want you to do is call all members of the team to find out who can make the game this Saturday. For those who can make the game, add their name and number to the visiting team list. Oh, one other small matter. After you find out how many people are going to the game, reserve school transportation. If fewer than five people plan to go, reserve a car. If more than ten plan to go, reserve a large van. For a count somewhere in between, we should be able to get by with a small van."

Draw a Nassi-Shneiderman chart to show the logic of this statement.

14—3. The module 6.0, print stock reorder report, as shown below, can be interpreted in the following manner. A stock record is obtained from module 6.1, get stock record. Following this "get," a short procedure (6.2, determine stock reorders) is required to determine whether the reorder field in the stock record is set to on or off. If on, the stock number and the stock to reorder (the quantity of stock to reorder) are sent to module 6.3, write stock reorders.

This procedure continues until all records have been read. At that time the reorder count (the total number of items to reorder) and the reorder amount (the total quantity of stock to reorder) are sent to module 6.3.

Draw a Nassi-Shneiderman diagram to show the logic of module 6.0.

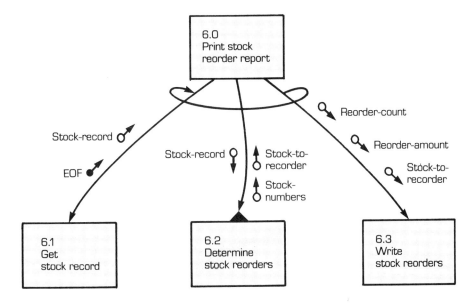

14—4. Study the structure chart shown on page 488. As indicated, the module receives N. It must then initialize all counters, get data N times, and write the accumulated sum. Design a Nassi-Shneiderman diagram to show the program logic for this structure chart. Assume that the value of N is ten.

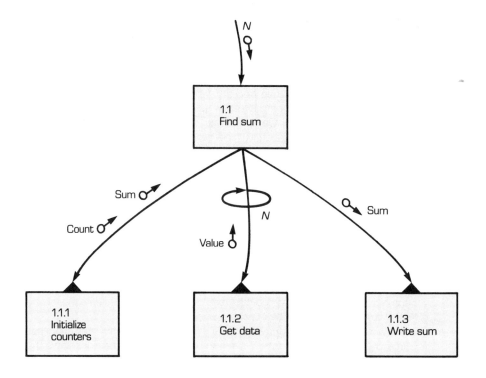

14–5. Suppose that in testing module 8.0, process student record, you decide to stub module 8.1, get student grades (see below). Suppose further that the condition you wish to simulate is as follows:
1. The four grades for a student are A, B, B+, and A.
2. These grades are placed in an array named GRADES(I).
3. The array is to be sent provided the student number is 6254. All other numbers are invalid.

Design the logic of the stubbed module 8.1, using pseudocode. Add comments to indicate the beginning and end of your procedure, much like the comments shown on Figure 14–13.

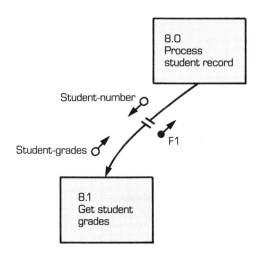

14–6. Suppose that you decide to adopt the incremental approach in testing module 2.0 (see below). In order to test this module in advance, you must stub modules 2.1, 2.2, 2.3, and 2.4. Suppose next that you know that for A to be valid, A must be greater than 5,000 and B must be greater than 2A.

Using pseudocode, write the procedures for these four stubbed modules.

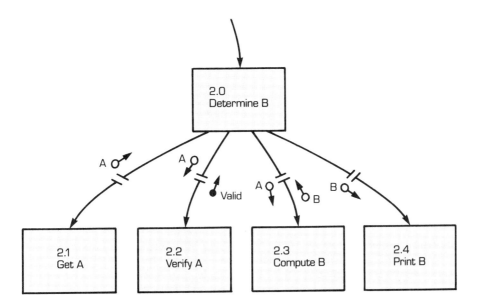

WORLD INTERIORS, INC.—CASE STUDY 14
PROGRAMMING AND TESTING THE NEW SYSTEM

Introduction

John Welby was eager to begin the coding of the new design. "The design looks sound," he remarked to Rena. "However, its true test will come when we begin coding and testing. In all likelihood the key to successful implementation will lie in our ability to write clean and efficient code."

Program Specifications

John discussed his concept of program specifications at length with Jane. "We need to prepare computer code that is easy to read, follows a structured notation, is top-down, and is well documented," he said. "Moreover, we need to work at writing efficient code. I've discovered that some of my code in the past has been quite readable but slow to execute. One of the things I've been working on is writing code that is both clean and efficient.

"There are several ways of improving code," continued John. "One way is to always initialize such things as counts and amounts at the beginning of a procedure. Oh, and flags, too! It often works best if we set flags off at the beginning of a procedure, turning them "on" whenever a problem in processing is discovered. Suppose we want to report that a payment is late. In coding this procedure we would write:

```
PAYMENT-IS-LATE = FALSE
IF DATE-OF-PAYMENT > DUE-DATE
   THEN PAYMENT-IS-LATE = TRUE
END-IF
```

"In this way, the flag is set to on only when a test condition tells us that something is wrong."

Program Coding and Testing

When Jane began to question John regarding how to organize the coding and testing process, he stated: "We should use an incremental approach to program coding and testing. By coding and testing top-level modules first, we will be able to test system interfaces frequently. The only difficulty with this approach is that immediate lower-level modules must be stubbed. As a rule, I add IF-THEN statements to stubbed modules to test the turning on and off of flags. A problem we have had in the past is not testing flags early enough. We discovered—much to our dismay—that we might turn on a flag but then fail to reset it to its off position."

Jane realized both the wisdom and the difficulty of John's concept of program and testing. "Flags are of great importance to our design," she thought to herself. "Consider the processing of customer orders [see Figure 12–22 to review the functions of module 1.0, enter customer orders]. Our design will work only if orders are entered properly. This means that the order-error flag must be false or we decide to abort the order. When we decide to abort, the processing of the transaction is stopped. We set it aside for correction and begin processing another order. We never store a transaction in error."

CASE ASSIGNMENT 1

Write pseudocode to show the logic important to module 1.0, enter customer orders (Figure 12–22 shows that this module performs four functions: get customer order, update open-order file, write daily order, and display results of updates). To improve processing efficiency, records written to the daily order file should be blocked (by a blocking factor of ten). Only valid updates are to be written to the daily order file. If an update is invalid—an order error is discovered—the error must be corrected or the transaction is to be aborted. Finally, a count of the number of transactions processed and the total dollar amount of the customer orders processed are to be written to the daily order file once the processing of orders is completed.

CASE ASSIGNMENT 2

Design a Nassi-Shneiderman diagram for module 1.2.1.2, calculate line-item charges. As discovered in chapter 12, this module is needed to verify that the item number, finish number, and price each are valid for each unit number keyed to processing. Validity is determined by searching the product table prepared for a supplier to match numbers. This module also calculates the price extended for each item ordered.

CASE ASSIGNMENT 3

Use pseudocode to show how Jane might stub module 1.2.1.3, calculate sales tax. As developed in chapter 12, this module must match keyed sales tax to the computed sales tax.

CASE ASSIGNMENT 4

Use pseudocode to describe how Jane might stub module 6.1, get factory invoice. This module would typically send key-verified information to module 6.0, enter factory invoice (see Figure 12–23).

REFERENCES

1. G. Weinberg, *The Psychology of Computer Programming* (Van Nostrand, 1971).
2. M. Bohl, *Tools for Structured Design* (Chicago: Science Research Associates, 1978).

3. D. Galletta, *COBOL with an Emphasis on Structured Program Design* (Englewood Cliffs, N.J.: Prentice-Hall, 1985).

4. E. W. Dijkstra, "Programming Considered as a Human Activity," *Proceedings of the 1965 IFIP Congress* (Amsterdam: North-Holland Publishing), 1965, 213–217.

5. C. Böhm and I. Jacopini, "Flow Diagrams, Turing Machines, and Languages with Only Two Formation Rules," *Communications of the ACM* 9 (May 1966): 366–371.

6. E. Yourdon, *Techniques of Program Structure and Design* (Englewood Cliffs, N.J.: Prentice-Hall, 1982).

7. M. Page-Jones, *The Practical Guide to Structured Systems Design* (New York: Yourdon Press, 1980).

8. M. Deutsch, *Software Verification and Validation: Realistic Project Approaches* (Englewood Cliffs, N.J.: Prentice-Hall, 1982).

15
System Testing and Conversion

INTRODUCTION

FOLLOWING the completion of program testing, system testing begins. With system testing, a test team determines whether the quality of the software developed and tested by computer programmers can be ensured. If it can—thus indicating that the software is acceptable from the systems staff point of view—the process of conversion begins. *Conversion* is the installation of the new, tested software, making it fully operational. System installation is part of this conversion process. It consists of two activities: database creation and system changeover. Besides system installation, conversion consists of writing new work procedures, completing all system documentation, training users, and conducting acceptance tests of the system by user groups. The end result of this conversion effort is the release of the new system for day-to-day use.

Both system testing and conversion are difficult activities, though for different reasons. With system testing, programmers must release the code they have designed and tested, allowing others to question, probe, and poke at their work in an active attempt to discover logical errors. Such a practice is disturbing, especially if programmers are belittled by the test team for every suspected error.

With conversion, users are asked to undergo an often frightful experience. They are asked to make the transition from an existing to a new system—which they have no experience with and barely understand. Project managers tend to be especially nervous during this phase. After all, what if users dislike the new system and begin to sabotage it without giving it a chance? How can users damage a new system? Damage can range from the destruction of equipment to improper software use. Passive resistance is often evident. Users typically continue to follow procedures designed for the old system, stating that they cannot be expected to fully use the new system when they do not understand how it works.

Besides completing the discussion of system testing, this chapter will consider topics essential to conversion. It ends with a brief description of acceptance testing. When you complete this chapter, you should be able to

- understand the difference between black-box and white-box testing;
- devise a test plan that provides for equivalence partitioning;
- describe the activities important to system conversion;
- describe the types of system documentation that need to be completed by the systems implementation team; and
- define acceptance testing criteria for a new system.

SYSTEM TESTING

One way to minimize user resistance to a new system is by providing a software product that is easy to use and free from errors. The objective of system testing is to ensure users that the software is of high quality. To this end, system testing, while an extension of program testing, is organized differently and is more formal in its design.

System testing is usually performed by a *test team*: two or more people who did not write the computer code and who are responsible for determining whether the software can do what is required of it. In larger organizations, the members of the test team are drawn from a special division of the systems and programming organization, which might be called the *software quality assurance division*. The role of this division is to test all software before its release to ensure its overall quality. In smaller organizations, system testing is not as formalized. Members of the test team are typically members of the programming staff who were responsible for coding the design. In system testing, programmers are asked to swap code and, in so doing, to test the quality of their fellow programmers' work.

Regardless of which method is utilized, the review and testing of code require special administrative rules. These include the following:

1. All system tests are to be carefully documented. With documentation, tests that fail (that is, they fail to identify software errors) are documented along with tests that succeed.

2. Program bugs discovered in testing are to be fixed by members of the test team. The team cannot require the original programmer to handle all errors, saying "Here's another error! Fix it!" However, the team can seek the programmer's help in making repairs.

3. Program bugs should not reflect negatively on the programmer who wrote the code, nor is finger pointing or buck passing allowed. If, for example, a test team program modification leads to a subsequent error, the team cannot point to the programmer and say, "What could be expected! The design was coded poorly to begin with."

4. All changes to the design must be clearly documented. These changes include modifying structure charts, adding new terms to the data dictionary, and revising record layouts, I/O mats and descriptions, and processing control descriptions.

Types of System Tests

System testing is generally organized to include two main types of testing: black-box testing and white-box testing. With *black-box testing*, the test team makes no attempt to determine what goes on inside a module. Much like audits around the computer, black-box testing concentrates on system inputs and outputs, holding to the view that one need only look at inputs to a module to determine which test to run (see **Figure 15–1a**). With *white-box testing*, the opposite view holds; the test team must look inside a module to determine which tests to run (see **Figure 15–1b**).

To fully understand the system and program specifications, the test team usually begins with white-box testing. Once the team is satisfied that the design handles usual processing conditions, testing is expanded to include black-box testing.

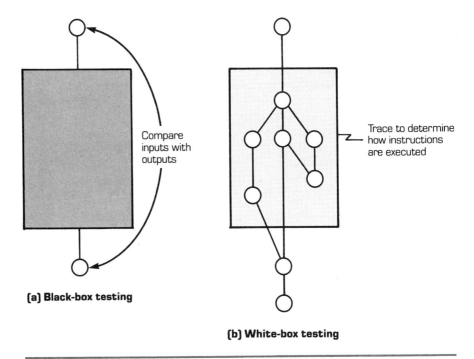

FIGURE 15–1 Black-box versus white-box testing

WHITE-BOX TESTING

The immediate concern with white-box testing is achieving adequate *coverage*: the degree to which all instructions in a system have been tested by a series of tests. There are three types of coverage—which are identical to the constructs important to systems programming. These are sequential coverage, decision coverage, and loop coverage.

Sequential coverage is a measure of the total number of sequential segments executed by a test team, with a sequential segment being a set of instructions that are always executed together without a decision of any kind. In practice, sequential coverage should approach 100 percent in white-box testing. Moreover, few errors should be found, provided each module was carefully tested by the programmer responsible for unit testing.

Decision coverage includes sequential coverage and adds the requirement that each decision step (including case statements) be tested in a system. In undertaking this type of testing, the test team must isolate each IF-THEN-ELSE construct and test to determine whether each possible outcome is in fact possible.

Loop coverage includes decision coverage and adds the requirements that all system loops be executed and tested. In this instance, the test team must isolate all DO-WHILE and DO-UNTIL statements, testing each type as follows:

1. For DO-WHILE statements, conduct three tests:
- skip the loop entirely,
- loop only once, and
- loop more than once before exiting.

2. For DO-UNTIL statements, conduct two tests:
- loop only once and
- loop more than once before exiting.

BLACK-BOX TESTING

With white-box testing, only 50 to 70 percent of all software errors are generally discovered. The remainder must be detected with one or more types of black-box tests. With black-box tests, "blind spots" in a design can be identified. What are blind spots? These are areas in a design that work well under normal circumstances yet result in errors whenever data happen to be somewhat unusual.

While there are several different types of black-box tests, three types in particular help programmers identify the blind spots in a software design. These are stress tests, normal path tests, and error path tests.

Stress tests involve pushing a software design to its limits to determine whether it will fail. Suppose the software to be tested is a student registration system. Suppose next that the system was designed to handle 2,000 students. One stress test would be to simulate the number of students to the system to determine whether it could handle the 2,000 students called for by the software specification. If it could, another stress test would be to increase the number of students to levels of 2,001, 2,500, and so forth, to pinpoint the number at which the software would fail. Questions regarding the point of failure could then be addressed. For instance, What happens when the software fails? Do error messages exist? Are they clear? Is the user warned before actual system failure?

Besides testing for capacity, stress tests are also useful in determining throughput and response time limitations of the software. Consider the student registration system once again. Suppose the system is allowed to grow from 500 to 1,500 students. By comparing these two processing levels, the test team can determine whether processing bottlenecks become evident and what kind of changes in response time result. A test might reveal, for example, that slow response time at high processing volumes renders the new system unusable.

Normal path tests determine whether a design can handle valid data, so that all valid input leads to valid results. Most normal path tests are limited to testing the input, output, and functional boundaries of a system. Why the boundaries? Because data on the boundaries are more likely to be in error than are data somewhere in the middle. Suppose that as part of testing input, a character string is defined as between one and forty characters in length. Normal path testing would be used to determine whether the software could handle a string size of one, thirty-nine, and forty. If it could, it would be assumed that data in the middle, say a name of twenty characters in length, would also be processed correctly. Suppose next that as part of testing output, a positive, six-digit numeric code is to be printed. Normal path testing would be used to determine whether the software would accept 000001 and 999999 as valid codes. Finally, testing of functional boundaries, such as arrays and files, would be done similarly. Suppose that a module named TABLE_SORT were written to sort a numeric table (array) of nonnegative items. Normal path testing would be used to determine several items: that the table values were all numeric and positive, and that the sort would be performed even when the table was empty, consisted of a single item, was half-filled, was completely filled, was in a sorted order to begin with, was filled with entries whose values were the same, and was filled with entries with paired values.

Error path tests determine whether a design can handle invalid data, so that invalid input leads to error messages or a graceful exit from processing. As with normal path testing, error path testing is often limited to the testing of system boundaries. Consider the same examples once again. Error path testing of a character string defined as between one and forty characters would call for two error path tests: one to determine what would happen if a string length of zero were entered and another to determine what would happen if a string length of forty-one or greater were entered. Likewise, with a positive, six-digit code, three error path tests are possible: one to determine the result of entering a negative number, another to determine the result of a number greater than six digits, and a third to determine the result of the code 000000. Finally, error path testing of a fixed-length table, such as the table used in procedure TABLE_SORT, might consist of a test to determine the result of sorting a table that contains nonnumeric entries and a test to determine the result of sorting a table that contains negative values.

In preparing error path tests, the test team is permitted to be devious. For example, tests might be conducted to determine whether it would be possible to print a payroll check for $1,000,000 (in a payroll system), to enter a stock with a dividend rate greater than the purchase price (in a stock analysis system), to enter identical file information into processing (in a file updating procedure), to supply ratings that collectively exceed 100 percent, where 100 percent is the maximum (in a group rating system), or to enter identical schedules in which alternative schedules are required (in a student registration system). Why test conditions such as these? To determine whether the software can handle exceptions to processing, especially those that might be attempted by people who are trying to beat the system.

Equivalence Partitioning

Another term for normal and error path testing is *equivalence partitioning*. What the test team attempts to do in system testing is to divide all input into valid and invalid classes and to proceed to test each class. **Figure 15–2** illustrates how the data dictionary term EMPLOYEE-LAST-

```
Data Dictionary Specification:

    EMPLOYEE-LAST-NAME = 2 {characters} 20

Notes:
    1. Employee-last-name is alphabetic.
    2. Employee-last-name is 2 to 20 characters in
       length.
```

```
        Valid
1. Employee-last-name is alphabetic
2. 1 character< employee-last-name <= 20 characters

        Invalid
3. Employee-last-name is not alphabetic
4. Employee-last-name < 2 characters
5. Employee-last-name > 20 characters
```

FIGURE 15–2 Equivalence classes for testing employee-last-name

NAME is partitioned into valid and invalid equivalence classes—the two parts that must be in compliance with the formal specification. As indicated, if the employee last name is specified as alphabetic, it cannot be alphanumeric or numeric. If the length is specified as between two and twenty characters, it cannot be less than two or greater than twenty characters.

Equivalence classes can be specified for all input conditions, including contiguous and noncontiguous values. **Figure 15–3** illustrates a noncontiguous condition. As shown, the valid case is defined as DATE-OF-MONTH = [JUNE/JULY]. This means that date-of-month is invalid for all other months.

Test Plan

One of the most important steps in system testing is the writing of the *test plan:* an overall plan and schedule indicating which specific tests are to be conducted and how test results are to be documented. **Figure 15–4** illustrates the outline of the test specification.

The *objectives of testing* section indicates what ends are to be achieved by system testing (e.g., error-free operation under stated conditions for a specified number of test runs).

```
Data Dictionary Specification:

    DATE-OF-MONTH = [JUNE/JULY]

Notes:
    1. Date-of-month is either June or July.
    2. Date-of-month is alphabetic.
    3. Date-of-month is 4 characters.
```

    Valid
1. Date-of-month is June
2. Date-of-month is July
3. Date-of-month is alphabetic
4. Date-of-month = 4 characters

    Invalid
5. Date-of-month is any month except June or July
6. Date-of-month is not alphabetic
7. Date-of-month< 4 characters
8. Date-of-month > 4 characters

FIGURE 15–3 Equivalence partition of a noncontiguous condition

```
                         Test Plan

              I. Objectives of testing

             II. Test completion criteria

            III. Test strategy

            IV. Test requirements
                A. Coverage tests
                B. Black-box tests
                   1. Stress tests
                   2. Normal-path tests
                   3. Error-path tests
                C. Other

             V. Test log
```

FIGURE 15-4 Sample outline of a test specification

The *test completion criteria* section describes the extent of the logical path coverage to be achieved and what types of stress, normal path, error path, and other performance tests are to be conducted.

The *test strategy* section examines how the design will be tested (e.g., top-down, program by program), the schedule to be followed, and team member divisions and responsibilities.

The *test requirements* section describes the actual tests to be conducted. As such, this section is the most important part of the test specification. Before testing, team members are asked to determine which tests (or experiments) are to be run. They are specifically required to provide a description of the test to be conducted, indicate the test data required to perform the test, and describe the expected results.

The *test log* section describes how each test is to be documented. **Figure 15-5** shows a page from an actual test log. As illustrated, the log initially serves to identify the test date and the team responsible for conducting the test. Next, the module to be tested is identified by number and name; the calling module is also referenced by number and name. A description of the test, the test data required, and the expected results of the test are then indicated. In this example, the test team decided to enter a character instead of a number in response to a menu prompt for a school number. The test data consisted of a character, from A to Z. The expected results were not clear to the team—they expected an error of some sort.

Next, the actual results are shown, following the running of the experiment. As shown in the figure, the test team entered a character and received a beep. No error message was indicated.

Then, the action recommended by the test team is recorded. In our example, no action was recommended. It was felt by the test team that

Test Documentation

Test date: 3/13	Tested by: J.R.H.	Author(s): S.H., T.O.
Module number: 2.1		Module name: Register school
Calling module #: 2.0		Calling name: Set up files

Parameters: Menu-choice

Test description: Enter character instead of number in response to prompt.

Test data: A....Z

Expected results: Error of some sort

Actual results: Got beep, no error message

Action recommended: No action recommended. User should understand that a beep represents an invalid entry.

Action taken:

Remarks:

FIGURE 15-5 Page from a test log

---

the user would be able to acknowledge that a beep meant that an invalid entry had been attempted.

Finally, the action taken by the test team and any remarks are entered on the test log. In our example, no action was taken. However, suppose that an error message had been necessary. The action taken would describe the error message placed in the computer code; the remarks section might go on to explain that structure chart and data dictionary modifications were made to document this design change.

CONVERSION

During the later stages of system testing, the process of system conversion begins. As defined earlier, *conversion* consists of installing the systems software and making it fully operational. The difficulty with this activity lies in making the transition from the old to the new. Since the new system replaces something that existed before—either an existing computer-based design or a set of manual procedures—the project manager and staff members responsible for implementation are faced with a new role, namely that of a *change agent*, an individual responsible for managing change in an organization. Acting as such an agent, the project manager seeks to minimize user resistance to the new system.

Two activities occur simultaneously during conversion: making new software operational (and, in so doing, replacing the existing system) and helping users use the new software. Which of these activities is the most difficult? John Seevers of Mansfield, Inc., explained:

"Generally, we have less difficulty making software operational," he said. "By the time we decide to install the new system, we are familiar with both the strong and the weak points of the software design. We have more difficulty predicting how users will react to the software and what special needs they may have. As a result, we tend to tread slowly. In the final analysis, we want the new system to be accepted and used effectively."

Several activities must be completed by the systems staff in making the transition from the old to the new. The staff must create the actual database, install the new system (by using one of several changeover methods), complete all work procedures, complete all system documentation, and train users in operating the new system. Only after all this is done can the final activity of acceptance testing begin.

Database Creation

New files must be built, tested, and made operational before the new system is installed. If the number of files is high, or the records to be stored are extensive, database creation can take considerable time. Consider the Mansfield case once again. Before the company could install its new payroll system, the employee master file had to be built, tested, and reviewed. This meant the following:

1. *Utility program software* was needed to allow records to be stored on the employee master file. The software was designed so that employee records could be added, modified once stored, and deleted from the file.

2. *Special file conversion software* was required to convert records from the old payroll system to the new payroll system. As shown in **Figure 15–6,** this software made it possible to merge old records with new employee information to create the new employee database.

3. *File conversion system tests* were needed to determine whether the conversion was successful. These tests included printing a complete register of both old and new versions of the system and making record-by-record comparisons.

4. *Record reviews* were required to determine whether printed information was factual. Supervisors were asked to check payroll listings to verify the accuracy of job titles, pay rates, and so forth; employees were instructed to check information stored on the payroll file to verify the accuracy of employee names, gender, deductions, and so forth.

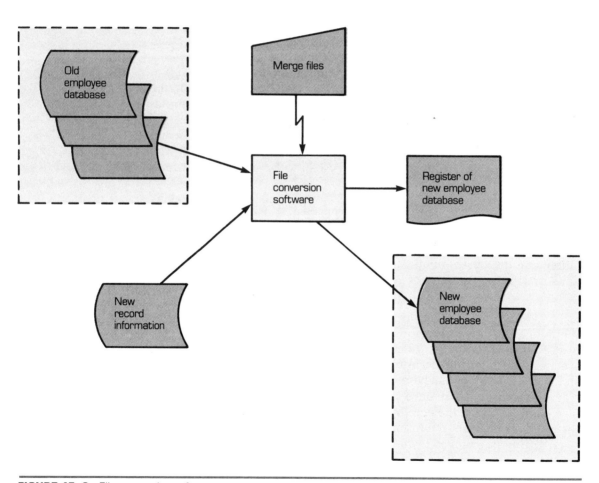

FIGURE 15–6 File conversion software

System Changeover

Following the creation of the database, system changeover is initiated. The project manager must choose among several possible methods of changeover, including:

1. direct—the "big bang method";
2. pilot—the "little bang method";
3. phased—the "modular method"; and
4. parallel—the "two systems alive method."

Let's begin with the weakest of the four, the direct method.

The *direct method* is nothing more than a complete, one-time conversion from the old to the new system. Some managers refer to this method as pulling the plug on the old system. A more apt description is that direct conversion is a "go," "no go" approach. If the software works as expected, a "go" situation is experienced. Unfortunately, if it does not work, the "no go" situation can resemble a "big bang." In practice there are times when the direct method is the only feasible alternative. The first launch of any new spacecraft illustrates direct conversion—from test phase to implementation phase. (Note that if the spacecraft fails there will be a "big bang.") Many types of equipment conversion are also handled directly. Suppose that Mansfield decides to install a larger central processing unit (CPU). The vendor advises the company that the upgrade will be made over the weekend. On Monday morning the new CPU is in place and the old CPU is gone. Employees are told: "Either the new CPU works, or we're in trouble."

The *pilot method* involves bringing up a new system in its entirety for a single user group, but not for all user groups. In this way, a single organization determines the worth of the new software. A large statewide university system used this method to install a new financial aid design. Instead of installing the new software at all universities in the system, administrators selected one as a pilot, or test, site. They reasoned that the pilot site would be able to identify the strengths and the weaknesses of the new software. In this way, the problems with the design could be identified and corrected before it was passed along to other universities. Besides, they reasoned, if the software did not work well, it would not create havoc throughout the university system. It would create only a "little bang," so to speak.

The *phased (modular) method* seeks to divide a system into pieces and install a piece at a time. This approach works especially well if the software design or the database can be segmented. Suppose a company decides to install an integrated accounts receivable system, consisting of invoicing, cash receipts, and month-end billing. Instead of bringing up the entire system at once, management can decide to bring up invoicing first, followed by cash receipts, followed by month-end billing. Only after one piece of the system is working smoothly will the next piece be installed. As another example, suppose that a company is installing a new finished goods inventory system. Instead of bringing up the system for all products at once, management can decide to limit conversion to one product line (i.e., a group of products) at a time. In this way conversion becomes much less of a rude shock and much more of a managed process.

The *parallel method* is often the preferred method along with the phased approach. It involves the concurrent operation of both the old and the new systems until the new system is judged to be effective. The value of the parallel method, besides being more conservative, is that it permits system results to be compared and corrective action to be taken if results show any deviation. **Figure 15–7** illustrates the data flows and operations important to this process. As suggested by this figure, the greatest advantage of the parallel method is that it preserves the audit trail—new and old totals can be compared to ensure that the new system leads to correct results. Unfortunately, the disadvantages of this method

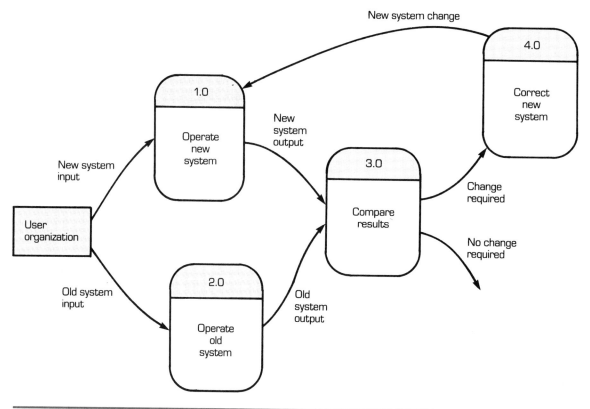

FIGURE 15–7 DFD illustrating the parallel method of conversion

are several: it is costly, places considerable pressure on employees by asking them to work two jobs, and encourages duplication of work. In addition, the results of the new system are rarely identical to the results of the old system. New codes, for instance, make cross-checking between the old and the new systems time-consuming and very tedious.

New Work Procedures

New work procedures must be installed at the same time as the new systems software. While many of these procedures are sketched out during the logical phases of analysis (see the task and technical analyses in chapter 7, for example), the final form is often delayed until system testing is near completion. Why wait this long? Since the exact steps to be taken by users are based on the last version of the software design, the systems designer must wait until final software specifications are in hand. In addition, work procedures contain references to specific sets of processing instructions and to data entry mats, forms, reports, and so forth. These references cannot be included in the procedures until the system is fully documented.

Consider the standard work procedure shown in **Figure 15–8**. Although similar in form to the task analysis worksheet shown in chapter 7, the standard procedure includes references to detailed processing instructions and samples of input mats, reports, and forms. In addition, it highlights all manual control procedures. The following control procedures are embedded in this standard procedure:

1. Compare dollar totals on payment stubs with dollar totals on personal checks.

2. Visually verify keyed data.

3. Select a random sample and compare payment totals before and after processing.

4. Count the number of checks and compare this count to the number of payments processed.

5. Forward the cash-control slip to the cash-management supervisor to permit the deposit amount to be compared with the bank statement.

6. File the cash-control slip and payment stubs to allow for an auditing of cash payment processing procedures.

Standard procedures such as the one shown on Figure 15–8 can take considerable time to write, test, and rewrite following their review. To assist in this process, the systems and programming staff is well advised to seek the services of a *technical writer*. The role of the technical writer is to translate technical matters into words and expressions that can be understood by users. Does this mean that programmers cannot be expected to write work procedures? While the answer to this question is no, a note of caution is in order. Although programmers can write work

Standard Procedure

Department: Accounts Receivable	Issued: July 15, 19XX	Number: CR101
Subject: Processing of customer payments	Frequency: Daily	Page: 1 of 1
Distribution: Accounting clerk, cash-management supervisor	Prepared by: Pamela Bennett	Supersedes: AR2-10

Abstract — Explains procedure used in making bank deposit from customer payments

Responsibility | **Action**

Accounting Clerk
1. Opens customer mail
2. Compares dollar total on payment stub with dollar total on customer check
3. Revises payment stub if customer payment does not match amount shown on stub
4. Adds check number and date of check to payment stub (see user's guide, p. xx).
5. Key-enters payment stub information (see user's guide, pp. xx–xx for data entry mats and instructions)
6. Visually verifies key-entered data
7. Prints bank deposit slip and cash-control slips (see user's guide, pp. xx–xx)
8. Selects random sample of five payment stubs and compares payment totals before and after processing
9. Counts number of checks and compares against number printed on deposit slip
10. Removes cash-control slip and forwards to cash-management supervisor
11. Files cash-control slip and payment stubs
12. Makes bank deposit

Cash-management Supervisor
1. Files cash-control slip for later comparison with bank statement

FIGURE 15–8 Standard work procedure

procedures, they tend to be better at writing and testing computer software. Technical writers, meanwhile, are better at writing work procedures.

An alternative to using technical writers is requiring users to write their own standard work procedures. *User-written work procedures* place the responsibility for conducting a formal task analysis with the individual performing the task. Does this practice lead to satisfactory results? In many instances, yes, provided a skilled systems analyst, designer, or technical writer is available to train users in writing procedures and reviewing (and editing) written results. Moreover, if the number of procedures to write is large, and staff resources are small, then asking users to prepare work procedures may be the only feasible alternative.

Completion of System Documentation

Closely tied to the writing of work procedures is the completion of system documentation. While most of the system design documentation at this point is in near-final or final form, two additional types of documentation typically need further attention. These are the operator's guide (operations documentation) and the user's guide (user documentation).

The *operator's guide* includes step-by-step instructions for setting up, operating, and distributing the results of processing. It may specify several diagnostic tests to perform if problems in processing are encountered. These instructions and tests are needed by data center operations, which is responsible for making the software operational on a day-to-day basis, and by internal auditing, which is responsible for determining whether processed totals are correct.

The *user's guide* includes step-by-step instructions for telling users how to execute the software. As with the operator's guide, "how to" materials must be prepared to tell users how to set up the software, run, audit, produce output, and exit from processing. The guide may also specify diagnostic tests to perform if problems in processing occur.

PREPARING THE OPERATOR'S GUIDE

In preparing the operator's guide, the project manager must prepare a set of procedures to show the steps people must perform in order to install and keep operational a software system. Consider the following table of contents from an operator's guide.

1. Introduction
2. System start-up
3. System failure
4. Minor maintenance
5. Operations processing functions
6. System backup
7. System troubleshooting and reporting
8. System shutdown

While a detailed explanation of these eight topics falls outside the scope of the text, a brief explanation of each helps to clarify what types of materials are to be placed in the operator's guide.

Introduction introduces the operator to the new software, the computer language used in writing the software, the computer to be used in processing the software, and the operator's portion of the design, if such a portion exists.

System start-up tells the operator how to start the system and compile computer programs. It specifies the files needed before programs are compiled. Operator instructions should also be written into the software to tell the operator that start-up was successful (or needs to be repeated).

System failure tells the operator how to restart the system in the event of system failure. Prior to restart, the operator may have to determine the reason for the failure. If so, procedures are needed to tell the operator how to track down a problem.

Minor maintenance tells the operator how to make small modifications to the software. For example, the software may need to be changed to handle a different type of computer terminal. A terminal reset procedure tells the operator how to adjust the software so that it can be accessed by a specific type of terminal.

Operation processing functions describe those parts of the systems software written specifically for the operator or administrator of the system. Suppose the designers of the system want the operator to change the user log-on message. The operator's guide would explain how the software was constructed to permit this change. In this instance, the operator is viewed as another user of the system.

System backup explains how the files required by the software design are to be backed up and when backup is required. Daily backup of all transaction files and weekly backup of all master files, for example, might be specified.

System troubleshooting and reporting describes steps the operator can take to record and, when possible, correct system problems. The operator might be asked to make several diagnostic tests, for instance, to determine whether either hardware or software are faulty. Test data are required if the operator is asked to determine the correctness of a processing run.

System shutdown tells the operator how to shut down the system following its use. The operator is instructed on how to close files, print summary reports, and backup system files as part of system shutdown.

PREPARING THE USER'S GUIDE

In preparing the user's guide, the project manager must decide which format best meets user needs, interests, aptitude, and knowledge. Dolan[1] describes five typical formats and their advantages and disadvantages:

1. Narrative or text—with this format, the user's guide reads like a book. The main advantage is that the reader is familiar with the narrative style. The main disadvantage is that the guide contains lengthy paragraphs.

2. Cookbook—in this format, the user's guide reads like a recipe contained in a cookbook; it is also similar to the imperative sentences used in writing structured English and pseudocode. Although in this form the user's guide is easy to write and interesting to read, it may be too cryptic for the reader to understand. The instruction "Move down to the third option and turn it on," for example, may be confusing to the user. Two instructions are necessary in this instance to clarify the steps important to data entry: For instance, "Space down to the third option, 'Set field length' " and "Turn option on by typing the letter X." Even with these additional instructions, a clarifying sentence is helpful, such as "The option can be turned off by typing X again."

3. Outline—with this format, the organization of the material is presented. This kind of guide is easy to prepare, but it is boring to read—and the outline format does not describe how a procedure is to be performed.

4. Playscript—here, the person responsible for an action is indicated along with the action to be performed. The main advantage is that it pinpoints responsibilities and is similar to the format used in preparing standard procedures. For example, the following playscript describes a single action to be performed by the accounting clerk:

Responsibility   *Action*
Accounting clerk   Visually reviews keyed data for error

The main disadvantage is that it is difficult to prepare if responsibilities change frequently.

5. Tutorial—with this format, a description is given of how something works, followed by an example that encourages users to try for themselves. The main advantage is that the guide explains how a system works and provides *experiential learning*. The main disadvantage is that the guide tends to be quite long. Consider the page from a user guide shown in **Figure 15–9.** As indicated, the user is asked to fill in the term line number (TLN), department number, and course number, and in so doing gains experience in entering data into processing and moving the cursor from one field to another. However, an entire page is required to show this relatively small processing step.

SELECTING THE BEST FORMAT

Given the five possible user's guide formats, the question of which format is best is perhaps best answered by the fact that most user's guides contain a variety of formats. A guide might begin with a descriptive section, followed by one or more outlines, include a tutorial, and provide one or more recipes for using software. The only clear tendency among user's guides is the inclusion of clear examples of processing. With clear examples, users can more fully understand how the system works. **Fig-**

1.1 ADD A COURSE

When you choose #1 from the Create/Update Catalog Menu, the screen shown here is displayed.

```
------------------------------------------------------------
                    *** COURSE ENTRY SCREEN ***


         TLN:▓▓▓▓       DEPARTMENT:▓▓▓▓▓▓▓        NUMBER:▓▓▓▓

       TITLE:▓▓▓▓▓▓▓▓▓▓▓▓▓▓▓▓▓▓▓             INSTRUCTOR:▓▓▓▓

        TIME:▓▓▓▓▓▓▓▓         DAY:▓▓▓▓▓▓▓         NOTES:▓▓▓▓

                       MAX SIZE:▓▓▓▓▓▓




         * * * * * * * * * * * * * * * * * * * * * * * * *
         *                                                 *
         *     Please fill in the highlighted portions     *
         *                                                 *
         * * * * * * * * * * * * * * * * * * * * * * * * *
------------------------------------------------------------
```

The cursor is located at the beginning of the first highlighted field, labeled TLN. Type in the TLN, the department, and the course number of the course you want to add. Notice that when you completely fill in the highlighted portion, the cursor automatically jumps to the beginning of the next field. Otherwise, you can move to the next field by pressing the return key.

FIGURE 15–9 Sample page from a user's guide

ure 15–10 illustrates another page from a user's guide. In this instance, a picture-by-picture description of processing was prepared. The speaker codes in this instance are samples. They resemble what actual speaker codes would look like.

User Training

User training (and operations training if required) is one of the final and most important parts of conversion. It is designed to provide users with hands-on experience with the new system. With interactive systems, users can try out software directly, observing how it responds to input entered correctly and incorrectly. Under the watchful eye of the trainer,

UPDATE SCORES

Prose Jr
Round 3

   Score Tiebreak

Section Number (Enter zero (0) if update is complete)■

Now you are ready to enter actual scores and tiebreaks. The screen above shows one example for the junior prose event, round 3. Enter the section number from your score sheet.

UPDATE SCORES

Prose Jr
Round 3, Section 4

   Score Tiebreak
1. 11111
2. 11107
3. 13114
4. 6106
5. 14101
6. 13101
7. 0

Enter scores at cursor position
Enter zero (0) if speaker code number is 0

The speaker code numbers here are samples. Those that actually appear on the screen should correspond to those on your score sheet.

FIGURE 15–10 Picture-by-picture description of processing

users gain experience in the use of a system before it becomes operational.

Successful training programs are carefully planned and organized. Let's consider the five steps important to such programs:

1. Identify the objectives of the program. The objectives of training are usually tied to the performance of users following a training period. An example of a specific training objective is as follows:

> Upon completion of training, the user should be able to
> a. Enter fifteen customer orders within ten minutes.
> b. Correct five orders in error within eight minutes.
> c. Print all system reports, using a sample test file.

2. Identify the users who require training and the trainers. Users and their supervisors are most often the people who require training, while project managers, lead analysts, and designers are typically designated as trainers. This approach works well provided the number of people to be trained is small. For a large group, however, a different method is advised. The *chief trainer method* works as follows. First, a chief trainer (let's say the project manager) trains representatives from different user groups. Second, once the representatives are trained, they become responsible for training all other members of their group.

3. Design a comprehensive, progressive training program. In the actual design of the training program, all system functions (enter a record, update a record, print/display a record, and so forth) must be covered. In addition, most training programs begin with a very simple demonstration of processing and progress, step by step, to more advanced topics. One of the most dangerous training program strategies is to begin with an advanced topic telling the group, "If you learn this part of the system, you will have no problem with the rest." This "sink or swim" strategy is often disastrous. Finding the system too complex, users back away, commenting, "We give up. This system is too difficult."

4. Select the most appropriate method of presentation. On-the-terminal training, self-paced instruction, classroom training, videotape instruction, and so on are all workable methods of presenting new system material. As with user's guide format selection, most training programs use a variety of training methods. On-the-terminal training can be supplemented with classroom training, for example. The classroom is especially useful for sharing reactions to and experiences with a new system.

5. *Determine whether users meet or exceed program objectives.* Tests conducted at the end of the training session are advised to determine whether users can perform as specified in advance. For example, will the user be able to enter fifteen orders in ten minutes?

ACCEPTANCE TESTING

Acceptance testing, the final testing of a new system, is different from either program or system testing in that it is done by user groups. In some organizations, acceptance testing is a symbolic act, in which user groups accept a system through its use more than because of its completeness and reliability. In other organizations, acceptance testing must satisfy both software and manual procedures acceptance criteria, carefully prescribed in advance and included in the technical design specification. Let's briefly consider both types of acceptance criteria.

Software Acceptance Criteria

Before a new system is accepted by user groups, predefined *software performance measures* must be satisfied. What are these measures? In practice, they vary from one project to another and from one organization to another. However, most software acceptance criteria include factors such as the following:

1. Processing (or cycle) speed—the time required to process a batch of transactions, to transform data into a more desired form, or to complete a processing cycle. The question to be raised by users is: How do actual processing speeds compare with the estimated times for different processing volumes?

2. Response time—the time required for the software to react to user instructions and queries. In this instance, users must ask: Is response time acceptable? How do peak periods affect response time?

3. Error (rerun) rate—the rate at which the software fails and must be repeated or rerun. Users must determine whether the rerun rate is acceptable (e.g., is it less than 5 percent?). Is it increasing or decreasing?

4. Completeness of system—the degree to which all aspects of the system are complete. Users must ask whether all features of the design have been fully implemented.

5. Completeness of documentation—the degree to which all system documentation is complete. Users contend, and rightfully so, that software is not complete if documentation is incomplete. They must thus ask: Have all types of system documentation been completed and are the materials clear?

Manual Procedures Acceptance Criteria

Equally important are manual procedures acceptance criteria. These criteria measure the degree to which user work procedures are acceptable in light of the new system. Factors here include:

1. Data collection—the time required by users to collect, enter, verify, and modify data keyed to processing. The central questions become:

Are data collection procedures acceptable? Do they agree with initial expectations?

2. *Output distribution*—the time required by users to verify and distribute processed results. Much like data collection, output distribution determines whether output procedures are acceptable and whether they agree with initial expectations.

3. *Data validation*—the time required by users to scan printed registers and error reports in an effort to determine the causes of errors. In this instance, the user seeks to avoid tying up valuable staff time to find errors that should have been spotted earlier.

4. *Duplication of effort*—the extent (generally measured in hours) to which data must be handled by users more than once as a result of processing errors. With reruns, for example, users may be required to repeat the same manual chores.

5. *Crisis index*—the degree to which the new system creates a crisis in the user area because of lateness in processing, errors, and so forth. In this case, the user must ask whether the new system resolves processing problems or creates more problems than it resolves.

The Mansfield, Inc., Case Study

The conversion of Mansfield's new integrated billing and receivables system was slowed by the system testing requirements set by John Seevers. John commented: "Conversion can be as smooth or as rough as we want to make it. If we release the software before all bugs are found and removed, we can expect high user and programmer frustration. We might even lose programming staff members if frustration is too great. To be on the safe side, I would rather spend extra time on system testing."

After thoroughly testing each of the twenty programs written for the new billing and receivables system, including the eighteen programs listed on Figure 7–17, the time came to consider system changeover. John described the situation.

"A combined phased and parallel method of conversion is perhaps best," he indicated to the conversion team. "The first step will be to create the database relations—the relations needed to store customer, product, and accounts receivable information. Once these relations are in place, we can move ahead and begin to create the sales summary records and invoice transaction files."

Walter remarked, "What about the sales summary file? Isn't this file created in advance, too?"

John clarified, "The sales summary keeps records of sales by type of customer and by line of products. Since our current method of processing does not generate this type of information, there is nothing to convert. All we can do in advance is create relations with account numbers. Sales and profit totals would be set to zero."

Database Conversion

Because the implementation team continued to be uncertain about the steps important to creating the new database, John decided to outline the database conversion plan.

"Several steps are required to create the database in advance of processing," he began. "First, we must design utility programs to create the customer, product, and accounts receivable database relations. Fortunately, the files from the current system can be used in building the new database. The main difference between the new system and the old is that the new database is limited to fixed-length records. Each receivable account, for example, will consist of a receivable account record, one or more invoice summary records, and one or more payment records (see Figure 11–20). Second, we need to list the contents of the old files. Third, we must list and compare the contents of the new database with the contents of the old files. Finally, we need to make corrections to the new database, as required."

System Changeover

Finding this outline format to be helpful, John continued by showing the steps important to system changeover. "Once the database is converted," he said, "we need to operate the old and new systems in parallel—at least until one entire billing cycle has been completed. The seven steps important to system changeover are:

1. Process customer invoices, payments, and accounts receivable adjustments using the old and the new systems.
2. Conduct system tests to determine whether invoices, payments, and accounts receivable adjustments are processed properly.
3. Discard the old system.
4. Purge bad debts from the accounts receivable files.
5. Conduct system tests to determine whether the purge was successful.
6. Print sales summary reports and reset monthly totals to zero.
7. Conduct system tests to determine whether the sales summary portion of the system works correctly."

Conversion System Tests

While most of John's conversion plan was clear to the group, the question of additional system testing required further explanation. Pamela asked, "John, what do you mean by the need to conduct system tests to determine whether processing was successful?"

John explained, *"Conversion system tests* are like any other system test, except they are conducted during or after conversion. Consider a black-box test for testing the completeness of the accounts receivable database, once it is created:

1. Print a register of the old accounts receivable file.
2. Print a register of the new accounts receivable database.
3. Compare the registers to verify the accuracy and completeness of processing.

4. Compare control totals to verify the accuracy and completeness of accounts receivable counts and dollar amounts.

"As another example, consider the black-box testing associated with processing customer invoices:

1. Process a batch of customer orders using the old system.
2. Process the same batch of customer orders using the new system.
3. Print invoice registers for both the old and the new system and compare results.

"Finally, consider the black-box test for determining the accuracy of the sales summary file:

1. Process a batch of customer orders by hand, determining the sales by type of customer account and by product line.
2. Process the same batch using the new system.
3. Print the sales summary register.
4. Compare the sales summary register to the totals prepared by hand."

SUMMARY

This chapter has examined the close relationship between system testing and conversion. Before a system is let go (handed over to users with the understanding that it is fully operational), a full battery of system tests is strongly advised. While most of these tests are conducted before conversion, some are also conducted during or following conversion. Users are important in this process. They are asked to make black-box tests as part of acceptance testing.

The major objective of system testing is to ensure that software is of high quality. To this end, two types of system tests—black-box and white-box tests—can be used. Black-box tests include stress tests, normal path tests, and error path tests. These tests make no attempt to determine how instructions are executed within programmed modules. White-box tests examine the execution of programmed instructions. These tests are conducted to determine whether specific patterns of processing are faulty.

A plan of system testing is advised to avoid hit-or-miss test practices. Such a plan includes a description of all tests and a test log. The log documents which tests were conducted and describes the results and actions taken as a consequence of testing.

Conversion begins during the final stages of system testing. The new database is typically created as a first step. Once the new database is fully tested, system changeover begins. Of the four methods of changeover—direct, pilot, phased, or parallel—a combined phased and parallel method is most often favored. Even then, actual conversion plans vary considerably between new systems.

New work procedures and system documentation must be completed during the system changeover period. While most of the system design is fully documented—based on the logical and the technical design specification—the operator's guide and the user's guide are most often not yet in their final forms.

The user's guide, in particular, may be difficult to prepare. Besides selecting the format for the guide—narrative, cookbook, outline, playscript, or tutorial—the writer of the guide must prepare clear examples of actual processing. With clear examples, users can better understand how the system works.

Once guides are written, user training is possible. In designing the training program, training objectives should be set in advance. Following training, testing is recommended. Testing allows the trainer to determine whether test objectives have been achieved.

Acceptance testing, the final leg in the overall conversion process, encourages users to test the newly designed and implemented system. The purpose of this testing is to determine whether systems software and manual procedures are acceptable. Provided they are, the new system is released from development. The systems development life cycle finally comes to an end.

REVIEW QUESTIONS

15–1. What is meant by the term *conversion*? Name the activities important to the conversion effort.

15–2. What is a primary objective of system testing?

15–3. In system testing, why do programmers sometimes swap code?

15–4. What is the difference between black-box and white-box testing?

15–5. In system testing, what is the significance of coverage?

15–6. What is a stress test?

15–7. What is the difference between a normal path test and an error path test?

15–8. What information is included in a test plan?

15–9. Which two activities occur simultaneously during conversion?

15–10. Why is the direct method of system changeover also known as the "big bang" method?

15–11. What is the difference between the pilot and phased system changeover?

15–12. How does the information shown on a standard work procedure differ from the information shown on a task analysis worksheet?

15–13. What is the difference between a cookbook user's guide format and a tutorial user's guide format?

15–14. Briefly describe the chief trainer method of user training.

15–15. What two types of criteria are important to acceptance testing?

EXERCISES

15–1. Suppose you are given the following pseudocode for an algorithm designed to find the value of x in a quadratic equation. The general form of the equation is

$$ax^2 + bx + c = 0$$

The general form of the equation to find the value of x is

$$x = \frac{-b \pm \sqrt{b^2 - 4ac}}{2a}$$

```
Start
Read A, B, C
IF A is not equal to 0
    THEN D = B² - 4AC
        IF D is greater than 0
            THEN Root 1 = (-B + √D)/2A
                 Root 2 = (-B - √D)/2A
            ELSE Write 'THIS EQUATION HAS
                        NO REAL ROOTS'
        ENDIF
    ELSE Write 'A IS EQUAL TO ZERO'
ENDIF
Stop
```

First, describe at least three white-box tests for this algorithm. Second, describe at least two black-box tests for this algorithm.

15–2. Examine the following pseudocode:

```
Start
Read loop value
DOWHILE loop value is not 0
    Read blood donor type and donor's name
    IF type is 0
        THEN Write donor's name
        (ELSE)
    ENDIF
    Subtract 1 from loop value
END-DOWHILE
Stop
```

First, describe at least three white-box tests for this algorithm. Second, describe at least two black-box tests for this algorithm.

15–3. Examine the following pseudocode:

```
Start
Write report heading
Read Income
DOUNTIL Income is greater than $10,000
    Interest = Income × .10
    Write Income and Interest
    Income = Income + Interest
END-DOUNTIL
Stop
```

Describe at least two white-box tests for this algorithm. Then describe at least one black-box test for this algorithm.

15–4. Define the equivalence classes for a telephone number, with the following specification:

Data item: TELEPHONE NUMBER

Specifications:
1. Number is numeric.
2. Number is positive.
3. Area code is blank or a three-digit number.
4. Prefix is a three-digit number, not beginning with a 0 or 1.
5. Suffix is a four-digit number.

15–5. Write the specification for the data term *income* that appears in exercise 15–3. Assume that the amount of income you begin with is always less than $1,000. Define the equivalence classes based on this specification.

15–6. Critique the following test plan. What are its strengths? What are its weaknesses?

The test plan consists of four phases.

In phase 1, programmers remove syntax errors and easily found bugs from their code. This informal phase ends when a program can run correctly once.

In phase 2, the test log is started. Programmers are required to document all tests of their programs. This phase ends when programmers are satisfied that their code is free of bugs.

In phase 3, code is swapped by test teams. Each test team is required to document all tests of swapped programs. Bugs found in this phase are generally corrected by the original programmer to reduce "error ripple," unless the bug is trivial. Stress testing is conducted during this phase. This phase ends when test teams are confident that the swapped code is free of bugs.

In phase 4, users test the new system while observed by programmers.

15–7. Draw a data flow diagram showing the procedure associated with a phased (modular) method of system changeover.

WORLD INTERIORS, INC.—CASE STUDY 15
CONVERTING TO THE NEW SYSTEM

Introduction

John Welby thought long and hard about how to go about converting to the new method of processing customer orders and factory invoices. He muttered to himself: "Rena is in a big hurry to pull the plug on the old system and begin with the new. Can't say I blame her. The old system seems to be more troublesome with each passing day. Even so, we must be careful. Installing a new system full of bugs would really give Rena something to worry about. Before we do anything, we'd better continue with system testing."

System Testing

John instructed Jane to work on system testing while he concentrated on drafting a conversion plan. When Jane asked for clarification, John responded, "Let's concentrate on equivalence testing of all inputs to processing. If we can block all invalid types of data from entering the system, we should be in good shape. What I want you to do is develop the equivalence classes for all key-entered data items and to test each entry in each class. Why don't you start work on customer order input data types? If I have time, I'll do some testing of factory invoice input data types."

Jane accepted her new assignment with enthusiasm. She began by noting the specifications for each data item to be key-entered from a customer order. Her careful inspection of all data items led to some interesting findings. For example, a person's first name could be hyphenated, such as the name Chu-Ling, the suffix to the postal code was optional, finish numbers were always greater than one hundred, item descriptions were never numeric (but were frequently alphanumeric), and the terms code was limited to the five letters *A, B, C, D,* and *E.*

System Changeover

While Jane was developing equivalence classes for input data items, John sketched out a conversion plan. "We will begin by creating the new database," John commented. "Following this, we will undertake either a parallel or a phased method of system changeover. Neither a direct nor a pilot method is workable. The direct method is too dangerous and a pilot would create extra work."

The creation of the database troubled John. He shared his thoughts with Jane. "There are definite problems with transforming stored data from the old system to the new. Take the current supplier product file, for example. This file contains incomplete supplier information. With the new system, however, the supplier file is designed to store supplier identification information, item information, and finish information, plus quantity sold and orders taken for each item by finish number."

"To make matters worse," added John, "the new open-order file is considerably different than the current customer order file. With the new file, we will be able to tell which orders need to be filled, which orders are problem orders, and which orders were filled within a six-month period."

When Jane asked John whether utility programs could be written to transform data from the current files to the new system, John touched on one of his key concerns. "We could do that, but what would we gain?" he questioned. "Since the current supplier product file contains so little information, we would not gain much with a transfer to a new file. In fact, it would be easier to key-enter and verify all supplier information. With the customer order file, the situation is even more hopeless. Since this file currently contains both current and obsolete customer order information—and at present we cannot separate which is which—it would be most unwise to transfer these data. All in all, I think it's best to start fresh."

CASE ASSIGNMENT 1

Develop equivalence classes for testing the following data items:

1. customer first name,
2. postal code,
3. finish number,
4. item description, and
5. terms code.

These items and others must be key-entered for each customer order. The field lengths for each item were shown earlier (see Figure 10–22).

CASE ASSIGNMENT 2

Suppose the decision is made to start fresh, as John suggests. With this direct method of conversion (data from existing files are not used in creating new files), there are some obvious problems. Explain how John might design a conversion strategy to create the new supplier file and open-order file.

CASE ASSIGNMENT 3

Help John decide on the method of conversion to use. Should a phased method, a parallel method, or some combination of the two be recommended? Design a step-by-step plan using this changeover method.

CASE ASSIGNMENT 4

Black-box system tests are required during and after system changeover. These tests include:

1. a test to determine whether new customer orders are processed correctly (see program 1.0, enter customer orders);
2. a test to determine whether factory invoice information is processed correctly (see program 5.0, enter factory invoices);
3. a test to determine whether problem orders are processed correctly (see program 7.0, enter/clear problem orders); and
4. a test to determine whether filled customer orders are successfully purged (removed) from the open-order file (see program 10.0, reset open-orders file).

Describe one black-box system test for each of these test situations.

REFERENCES

1. K. Dolan, *Business Computer Systems Design* (Santa Cruz, Calif.: Mitchell Publishing, 1984); see also D. C. Boger and N. R. Lyons, "The Organization of the Software Quality Assurance Process," *Data Base* 16, 2 (Winter, 1985): 11–15; J. D. Cooper and M. J. Fisher, eds., *Software Quality Management* (Petrocelli Books, 1979); M. Deutsch, *Software Verification and Validation: Realistic Project Approaches* (Englewood Cliffs, N.J.: Prentice-Hall, 1982); and G. Myers, *The Art of Software Testing* (New York: John Wiley and Sons, 1979).

16
Systems Maintenance

INTRODUCTION

ONCE a system has been installed and is fully operational, the responsibility for system maintenance must be assigned. Most often this assignment is given to a systems and programming group, consisting of *maintenance programmers*—individuals responsible for keeping a system operational. Typically, maintenance programmers are not involved in the systems development effort.

As might be expected, the transfer of responsibility—from a systems development team to a maintenance programming team—can be difficult, especially if the software has been written and implemented in haste. In such instances, the maintenance staff is most likely to find the new system difficult to understand and fix when the software fails. Does most software fail? Unfortunately, yes. Currently there are no program testing techniques that make software infallible.

This final chapter serves in part as a review of the previous chapters. Often, the maintenance team is asked to undertake a mini–systems development project in maintaining an operational system. When you complete this chapter, you should be able to

- understand why maintenance is needed and why even carefully designed software must be maintained;
- describe the various types of systems maintenance and indicate under which circumstances a mini–systems development effort is recommended;
- explain how DFDs, Warnier-Orr diagrams, and structure charts aid in the maintenance of a system;

- describe common types of system housekeeping tasks;
- explain different classes of corrective maintenance errors;
- explain why module repairing must be carefully administered; and
- describe several ways of improving processing efficiency.

What Is Systems Maintenance?

Before we continue, we should understand the meaning of systems maintenance. As stated in chapter 1, *systems maintenance* involves making changes to software to keep it operational; it includes maintenance projects that closely resemble mini–systems development projects.

The reasons why maintenance is required are sometimes difficult to understand. First, maintenance is required to keep software current with changing processing needs and requirements. That is, as the internal and external environment of a business change, so too must its internal systems. If a company takes on a new line of products, then sales, billing, and inventory systems must be changed to accommodate those new data and processing requirements. Are such changes major? It depends on how data processing software was designed to begin with. Second, maintenance is required to fix errors or bugs in the software. Experienced programmers realize that no software can be guaranteed to be free of bugs. A program might run successfully ninety-nine times out of one hundred, only to fail on the hundredth time. In fairness, some bugs result from attempts to extend a system beyond its design limits. For instance, an inventory system designed to handle 1,000 items might not handle 1,001 items. Error messages such as THE NUMBER OF PRODUCTS HAS EXCEEDED 1,000 or FILE SPACE EXCEEDED indicate that to remain operational, the software (or the hardware) must be changed. Third, maintenance is required to keep up with changes in computer technology. Changes in software are required as decisions are made to use different ways of inputting, storing, processing (including sorting), outputting, and controlling data. The decision to transmit data directly to a computer instead of storing keyed data on paper tape for later batch transmission, for example, leads to a maintenance project. Input routines must be rewritten to accept one transaction at a time. Fourth, maintenance is required to perfect or fine-tune coded procedures. During the design of software, some procedures are written poorly primarily because of project deadlines and pressure from users and superiors. While these modules test successfully and perform as expected, they are nontheless inefficient. The responsibility of the maintenance programmer is to spot these inefficient procedures, rewriting them as necessary. In so doing, greater processing efficiency is realized; fewer computer resources are expended on the software product.

Types of Systems Maintenance

How is systems maintenance classified? Swanson[1] suggests three main types:

1. *The adaptive* type is performed in response to changes in the business environment. It requires programmers to modify programs to incorporate new system requirements. Included in this category are housekeeping tasks that require programmers to make routine changes to software to keep it operational.

2. *The corrective* type is performed in response to failures in processing. It requires programmers to monitor and fix problems with computer programs.

3. *The perfective* type is performed in response to the need for improving or maintaining program efficiency. It requires programmers to identify the expensive parts of processing and to improve their efficiency.

Which types of maintenance are the most important and which types receive the most attention? Clearly, adaptive and perfective maintenance are the most desired; however, corrective maintenance requirements often take most of the maintenance programmer's time. Helms and Weiss[2] report that of all maintenance requests, more than half involve requests to monitor and fix problems with software. This is not an unusual finding. Most studies report that more than half of all systems and programming time is devoted to maintenance. Of this total, corrective maintenance receives the most attention. Thus, with this fact in mind, let's examine the three different types of maintenance.

ADAPTIVE MAINTENANCE

Adaptive maintenance—maintenance performed to modify programs to incorporate new system functions—should typically require the greatest amount of time from the maintenance programming staff. As internal and external business conditions change, leading to different types of data processing requirements, an organization must modify its library of operational software. Fortunately, making adaptive changes to software can be carefully managed by the systems staff. Most maintenance requests are as carefully documented as requests for service. Once they are approved, a mini–systems development project begins.

User Requests for Service

User requests for service define and help rank adaptive maintenance requirements. As shown in **Table 16–1**, the format for a user request for service comprises seven items, beginning with the identification of the system to be maintained:

TABLE 16–1 User request-for-service format

User Request for Service

1. Name of system/subsystem
2. Description of change
3. Reason for change
4. Date needed
5. Cost/benefits
6. Priority
7. Other considerations

1. The name of the system/subsystem identifies by name the area requiring change. If a specific report is to be modified, for example, the name of the report is identified.

2. The description of change clearly defines the work required. If a new output format is needed, a drawing showing row and column headings and sample data should be attached. If equations used in processing are to be modified, they should be supplied.

3. The reason for change clearly indicates why change is required at this time. For instance, a user might report that a change in federal reporting requirements mandates that a change in system output be made.

4. The date needed provides an estimate of when the change must be fully implemented.

5. Cost/benefits indicate whether the completion of the project will yield cost savings or other benefits.

6. The priority item requires users to assign a priority to all requests. One priority scheme might be as follows:
Code A Emergency condition—take immediate action.
Code B Mandatory condition—take action to complete work before date needed.
Code C Highly desired—take action as soon as possible to implement request.

7. Other considerations include any other factors that need to be taken into account. For example, a highly desired request might be combined with an emergency request to save time and expense.

The processing of user requests for service may be undertaken directly by the systems department, as in the clearing-house approach to organization discussed in chapter 1, or it may involve a committee, such as a systems maintenance user group committee. In either case, user requests for service must be assigned priorities, following which personnel can be assigned to implement the specified change requirements.

Mini–Systems Development

Maintenance programmers typically treat each adaptive maintenance request as a mini–systems development assignment. As such, the steps important to the project life cycle are followed, with some important modifications. **Table 16–2** shows a modified life cycle for adaptive maintenance projects. (You may find it useful to compare this listing of activities to the list shown for new projects in Table 1–1.) With maintenance, the intent is not to redo an entire system but rather to show how a change alters the logical and technical design of a system. Suppose that Mansfield, Inc., discovers that quarter-to-date (QTD) and year-to-date (YTD) pension dollar figures for all employees must be reported quarterly to the federal government. Chris Steward, one of Mansfield's top maintenance programmers, explained how the change was made.

"First, we defined the system modification requirements," he said. "We determined that a new QTD pension report was to be produced. The layout of the report was specified by the federal government.

"Second, we modified the logical system. We constructed a revised DFD to show new processing requirements (see **Figure 16–1**) and modified the Warnier-Orr diagram to show the new decomposed view of an employee's record (see **Figure 16–2**). As indicated by the DFD, three new procedures were required to handle this federally mandated reporting change. We had to prepare a programmed procedure to print the new QTD pension report, a manual procedure to verify the QTD pension report, and a second programmed procedure to reset the QTD fields in the employee record to zero. The Warnier-Orr diagram indicated that the change to the data structure would be minimal. Only one new field was required, namely, the QTD field.

"Third, we modified the data dictionary to show the changes to the

TABLE 16–2 Life cycle activites for user requests for service

Project Activity—Mini–Systems Development

Systems analysis	1. System modification requirements 2. Logical design change requirements 3. Data dictionary modifications
Systems design	1. Technical design change requirements 2. Data dictionary modifications 3. User guide modifications 4. Maintenance manual modifications
Systems implementation	1. Coding and debugging 2. Program and system testing 3. Implementation and user evaluation

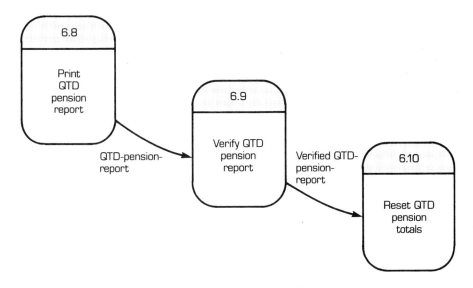

FIGURE 16–1 DFD showing the implications of producing the QTD pension report

logical design. This modification was easy. It consisted of defining the term *pension, quarter-to-date.*

"Fourth, we made technical design changes. These changes were more substantial. We determined that adding a new report would mean changing the main processing menu, designing a new report layout (following the specifications provided by the federal government), altering

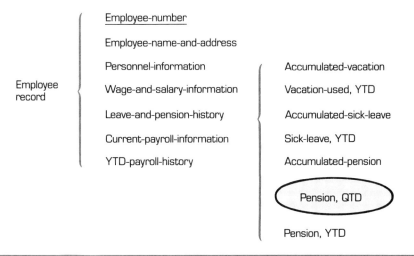

FIGURE 16–2 Revised Warnier-Orr diagram showing addition of pension, QTD field

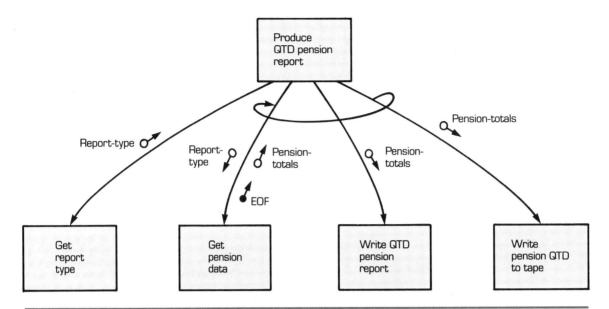

FIGURE 16–3 Structure chart describing the procedure to produce the QTD pension report

the layout of physical files, designing new structure charts, adding new module specifications, adding new processing controls, and pinpointing areas in the code where the employee record format was to be changed.

"Consider the new structure charts for this maintenance request. As shown by **Figure 16–3,** after we determined which type of pension report was required, pension data were read from the employee master file (get pension data). Once obtained, we printed the QTD pension report and placed a copy of the report on magnetic tape. However, this step did not complete all processing requirements. As shown by **Figure 16–4,** a second step required us to reset the pension, QTD field to zero. To do this, a read-write operation was required. We also printed a processing control total report to indicate the number of employee records reset and the total dollar amount reset. These totals were compared with the totals printed on the QTD pension report."

Housekeeping Tasks

Besides handling user requests for service and completing mini–systems development assignments, adaptive maintenance includes routine tasks designed to keep software operational. Known as housekeeping tasks, most of these routine maintenance tasks are handled using *utility pro-*

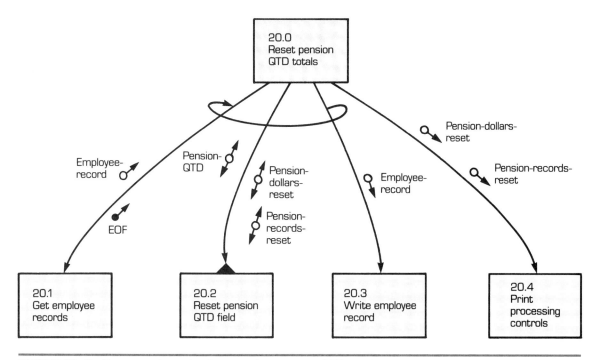

FIGURE 16–4 Structure chart showing procedure to reset QTD pension totals to zero

grams (programs not directly related to the main functions of processing). Common housekeeping utility programs include the following:

1. File copying and backup procedures—procedures that provide an exact copy of a data file or database. The copy is created and stored in a safe location to permit processing to be restored if necessary.
2. Purge procedures—procedures that remove records from a file in order to reduce its size. Inactive customers, for example, should be periodically separated from active customers.
3. File reorganization procedures—procedures that modify files to remove deleted records and records written to overflow areas. The reorganization of all master files, for instance, must be done periodically.
4. Index and directory reorganization procedures—procedures that change the contents of indexes and directories. These changes must be made to provide more uniform access to groups of items.
5. File conversion procedures—procedures for transferring data from one file storage medium to another, such as from magnetic disk storage to magnetic tape storage.

6. Reset account total procedures—procedures by which historical totals in records are shifted, allowing new totals to be stored. Fields that store such items as QTD totals must be shifted and reset to zero at the end of each three-month period.

7. Table update procedures—procedures by which values stored in tables are updated. Maintaining current tax rates, employee pay rates, subscription rates, depreciation rates, and so on reflect this type of housekeeping task.

This listing of common tasks suggests that maintenance programmers must identify the areas within a design that require change as business rates vary and as reporting periods come and go. Chris commented on Mansfield's payroll system.

"There are critical housekeeping tasks that must be performed to keep the payroll system operational," he said. "We must modify the employee pay rate table, for instance, whenever management decides to change the wage of a labor grade. Increasing the secretary III wage rate by 8.5 percent is an example of a recent maintenance requirement. We wrote a short utility program to take the old value in the rate table and to increase it by this percentage amount.

"Various account totals must be reset to zero at regular intervals," continued Chris. "Month-to-date (MTD), QTD, and YTD totals must be shifted as each occurs. If we fail to reset any of these totals back to zero when we should, still another maintenance program is required to reset and add the correct totals.

"The purging of records from the system is an especially critical step in processing," he added. "We cannot simply delete an employee's record whenever a person decides to leave the company. Instead, we must retain all employee records who have worked during a year, up until W-2 federal reporting forms are printed. The problem with this is that W-2 forms are printed on or near the fifteenth of January. This means that we have to maintain two master employee files for the first month in a new year."

CORRECTIVE MAINTENANCE

Corrective maintenance—maintenance performed in response to a software failure—should take the least amount of programmer staff time; however, it may require the greatest amount of time. This type of maintenance requires programmers to be on call to fix software when it fails. What type of failure will occur? When will a failure occur? How many hours will be required to fix the problem once it becomes known? These are the tough questions associated with corrective maintenance. With unreliable software, several types of errors, ranging from simple errors to complex ones, may be evident and may occur any time the software is executed. In such circumstances, the hours required to fix the software will also vary, ranging from several minutes to several hours.

Critical Incident Reporting

One method of tracking troublesome software is known as *critical incident reporting*. As shown in **Figure 16–5,** a critical incident report logs software failures, their probable cause, and the corrective action taken. In addition, the report shows the date and time of the incident, the party to whom the problem can be traced, and the name of the person who filed the report. The main value of critical incident reports is that they help identify four major classes of errors: program logic errors, system errors, user errors, and operations (or data center) errors.

Program logic errors stem from weaknesses in the software product itself. Inadequate data validation procedures, data type errors, boundary errors, inconsistent use of variable names, incomplete logical paths, and so forth are examples of software design problems. Even well-designed software will contain this type of error. A hyphenated last name (the designer only permitted alphabetic characters to be entered), twenty-

Critical Incident Report   Report #_____

Date of incident_____ Time_____ AM/PM
System_____ Program_____

Type of problem:
- ☐ Scheduling
- ☐ Documentation
- ☐ Job setup specifications
- ☐ System error
- ☐ Program logic
- ☐ Bad, late, missing data
- ☐ Tape/disk use
- ☐ Printing
- ☐ Punching
- ☐ Output preparation
- ☐ Delivery service

Problem traceable to:
- ☐ User department_____
- ☐ Systems programming
- ☐ Data entry
- ☐ Data center management
- ☐ Operations
- ☐ Systems programming
- ☐ Vendor_____

Problem description (if necessary):

Action taken: ☐ Repaired program ☐ Rewrote documentation ☐ Discussed with employee
☐ Initiated training   ☐ Notified_____
☐ Other:

What must be done to prevent a repetition of this incident?

Filed by_____ Date_____

FIGURE 16–5 Report for tracking problems with software

nine days in February (the designer forgot about leap years), and ten children in a family (the designer assumed that nine or fewer children would be found) all illustrate conditions that were never considered by the designer but that lead to program errors. Maintenance programmers are required to fix such errors. They must not only identify the problem but also take proper corrective action.

System errors are problems reported by the computer system and may represent hardware problems or problems with the software design. A loss of electrical power, for example, usually leads to a system error, as does a faulty track on a magnetic disk. The message OUT OF PAPER describes a simple system problem, while the message UNPRINTABLE ERROR gives the programmer reason for concern.

System problems associated directly with the software design often surface when a software design is pushed to the edge of its capabilities. The messages LINE BUFFER OVERFLOW, TOO MANY FILES, DISK FULL, FIELD OVERFLOW, or OUT OF MEMORY illustrate conditions that occur when the computer system cannot handle processing requirements. Can some of these problems be avoided? Remember that stress tests were supposed to describe overflow problems such as these. Thus, the ways of handling overflow problems should be well documented; ways of anticipating overflows should be built into the software design.

User errors can be directly traced to user groups. With this type of error, user mistakes or misunderstandings are usually the cause. Failure to properly schedule jobs, visually reviewing and accepting erroneous data, failure to purge historical files, and failure to add new data to files are typical user errors. What do these errors suggest? They indicate that the training demonstration and user guides and manuals developed for the new system may be inadequate from the user's point of view. Since users tend to be reluctant to read manuals, additional training aids, such as *interactive help screens*, may need to be included in the design. Help screens allow users to ask for help if they run into trouble. A function key stationed alongside the terminal keyboard may be used to activate the help screen. The screen itself may instruct the user to type control-Q to quit the routine being worked on, or control-B to back up and start again. In either case, the help screen offers the advantage of providing processing instructions directly, without interfering with system operation.

Operations (data center) errors can be traced to computer operators, computer schedulers, tape librarians, and other members of the operations staff. Failure to change printer ribbons, loading the incorrect tape file, improper backup of computer files, and incorrect job setup instructions all illustrate different types of operations errors. How are these types of errors controlled? Typically, by writing clear work procedures and checking to make sure they are followed. Requiring computer operators to run through a check-out procedure before starting a job often eliminates job setup errors. Following a similar check-out procedure at the end of each processing run helps ensure that files are backed up and that output is distributed as specified.

In summary, the four types of corrective maintenance errors—program logic errors, system errors, user errors, and operations errors—must be controlled if the maintenance staff is to find time to work on more purposeful types of maintenance. To this end, the critical incident report helps the maintenance programmer identify weaknesses in the processing environment as well as pinpoint the most troublesome programmed modules.

Module Repairing

Module repairing is the redesign of error-prone modules in an implemented software system. It involves changing either the computer code or the data structures called for by the code. Suppose that the section of a program intended to compute average scores leads to various errors. The problem discovered by the maintenance programmer is that scores are never verified as being correct (see **Figure 16–6a**). To handle this problem, the programmer adds a new module, entitled verify each score (see **Figure 16–6b**).

Module repairing, especially when it involves changing a system's internal structure, should be carefully administered by systems and programming managers. Imagine what would happen to the software so carefully designed for Mansfield if maintenance programmers were permitted to make changes without documentation.

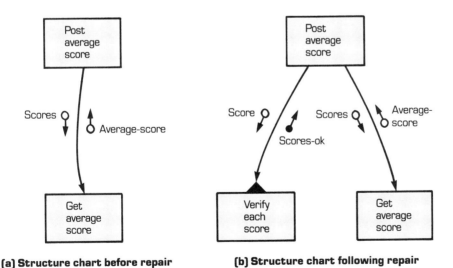

(a) Structure chart before repair   **(b) Structure chart following repair**

FIGURE 16–6 Documenting the repair of computer code

Chris commented, "If a repair leads to a change in the structure of the code, all modifications must be graphically represented using revised structure charts. Likewise, if new terms are introduced as a result of the change, these are to be added to the data dictionary. To do otherwise would mean disaster. Consider a situation in which the structure charts depict one thing and the source code describes something quite different. Before long, no one would have any confidence in the technical specification prepared for the software design. What this tells us is that maintenance without documentation is just the same as systems analysis and design without documentation. What we wind up with are systems which no one really understands, much less how to modify them."

PERFECTIVE MAINTENANCE

Perfective maintenance—maintenance performed to improve or maintain program efficiency—is second only to adaptive maintenance in importance (see Bentley[3] for a complete discussion of this subject). Its value is realized when software can be modified to require substantially reduced computer resources. Suppose that a large software program contains numerous inefficient inner loops—say 5 percent of the code contains these inner loops. Suppose next that the maintenance programmer fine-tunes these loops to remove their inefficiencies. What happens to the overall efficiency of the entire program? Run-time comparisons reveal that by improving 5 percent of the code, a 20 percent improvement in run time is realized. Are such situations unusual? Knuth[4] reports that less than 4 percent of a program accounts for 50 percent of the overall program run time. Thus, the run times of many computer programs can be greatly improved provided maintenance programmers know where to look.

Modifying Program Data Structures

Modifying program data structures is one way to improve overall program efficiency. What the maintenance programmer looks for are opportunities to reduce program run time by increasing program size or by simplifying the organization of the data structure itself. Blocking records is one example of reducing run time at the expense of increasing program size. It is typically much more efficient to read records into memory a block at a time, where they can be processed individually, than to read records into processing one by one. Consider the merging of two sets of records. One alternative would be to read a record, say from file A, and another record, say from file B; to compare record A to record B to merge the two records; and to write the pair to a third file, say file C. **Figure 16–7a** shows this relationship. A more efficient procedure (based on increasing the size of the program) is to read a block of records from

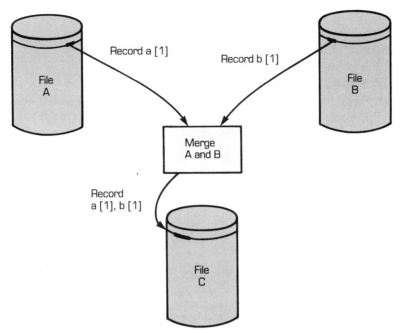

(a) Merge with unblocked records

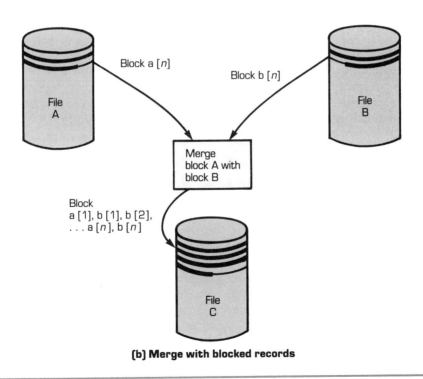

(b) Merge with blocked records

FIGURE 16–7 Increasing program size to improve efficiency: the merging of blocked records

file A, read a block of records from file B, merge the two blocks, and write the merged block to file C (see **Figure 16–7b**).

The storing of precomputed results is another way to improve efficiency at the expense of large program size. Suppose that the maximum and minimum values of an array need to be retrieved several times during a program's execution. One alternative is to search the array, finding the maximum and minimum values each time they are required. A more efficient option is to search the array just once to find the maximum and minimum values, to store them, and to use the stored values directly. The additions to the revised code might read:

```
MAX-VALUE = FUNDS [MAX,J]
MIN-VALUE = FUNDS [MIN,J]
```

where maximum value and minimum value are the values to be stored.

A third way to improve efficiency is to eliminate temporary work files. Parkin[5] believes that this is the main strategy for what he terms "hardcore tuning" of software. How does this type of improvement take place? In this instance, the maintenance programmer must determine whether intermediate steps in processing can be eliminated, as the following example illustrates.

Suppose that a bill-paying design requires a two-step procedure (see **Figure 16–8a**). The first step is to split a file of outstanding bills into two smaller files—an outstanding bills, held over file and a work file (shown as outstanding bills to be paid). The second step is to add a check payment control number and a check number to each bill; still other tasks are to sum the total amount to be paid, print the payables check register, and write all revised records to a payables check-writing file.

Suppose next that the maintenance programmer discovers that the entire file of outstanding bills to be paid can be stored internally, thus avoiding the need to create a separate work file. As a consequence, the writing to and reading from the work file are eliminated (see **Figure 16–8b**); overall processing efficiency is improved by more than 50 percent.

Simplification of data structures is another way to improve program efficiency. Suppose that a program contains several procedures that check the validity of a key value contained in an index. Suppose next that the index can be decomposed into partitions that contain identical values. One procedure would be to store each key value and to search the entire index to determine whether a key value was valid (see **Figure 16–9a** to visualize this approach). A more efficient alternative would be to *parse* the index—that is, to decompose the index into common parts—in order to save time in the search process. As shown in **Figure 16–9b,** with three-level parsing, only twenty-two digits are stored (compared with forty-eight for the nonparsed list); the time required to find a single key value is reduced considerably for large indexes.

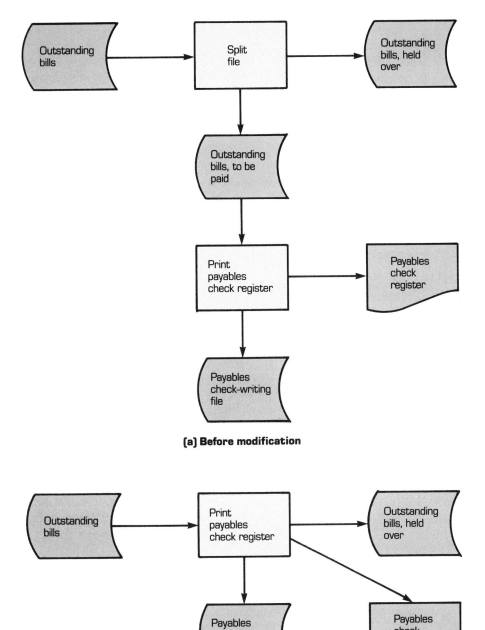

FIGURE 16–8 Eliminating work files to improve program efficiency

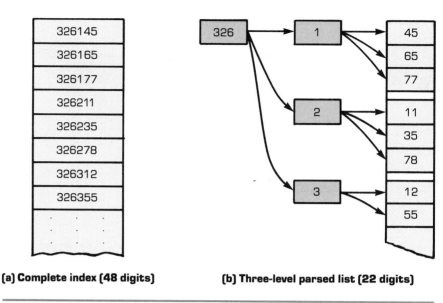

(a) Complete index (48 digits)   **(b) Three-level parsed list (22 digits)**

FIGURE 16–9 Simplifying search procedures by changing the data structure

Modifying Programmed Procedures

Modifying program procedures is another way to improve overall program efficiency, provided that the maintenance programmer can identify the most expensive parts of a system (that is, the most time-consuming parts). What are these parts? Generally, they tend to be the procedures that deal with loops, test conditions (especially when they are associated with loops), calls to other procedures, and the evaluation of algebraic expressions. While the scope of this book precludes an in-depth discussion of any of these so-called hot spots of a software system, we can nonetheless summarize each type of procedure and provide examples of how the maintenance programmer might improve these various sections of computer code.

Loop improvement consists of evaluating each coded loop in a system, attempting to minimize the work performed. Simplification of the innermost loops (in a set of nested loops) is especially important. How can loop performance be improved? Consider the example shown in **Figure 16–10**. In Figure 16–10a, the square root (SQRT) value is computed for each iteration of a fairly large loop (N = 1,000). In Figure 16–10b, the square root value is computed only once, outside the loop. In processing, the computed value is multiplied by the old value of X(I), to determine the new value of X(I). What is the savings in this case? The revised code will run fifty to seventy times faster than the original code. This is because of the high overhead cost associated with repeatedly computing the square root value.

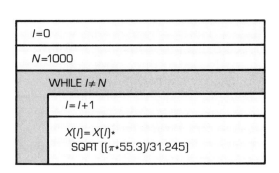 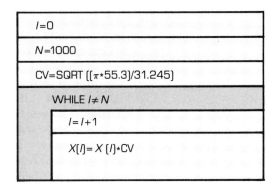

(a) Before modification **(b) After modification**

FIGURE 16–10 Simplifying program loops

Test improvement consists of testing for a conclusion once instead of each time within a loop and combining tests when possible. **Figure 16–11a** illustrates a situation in which a test is made each time a loop is executed, compared with the revised code in which the test is made only once. Suppose that the test limit is exceeded on the first iteration (even though it would not be as shown). This leads to setting the flag time after

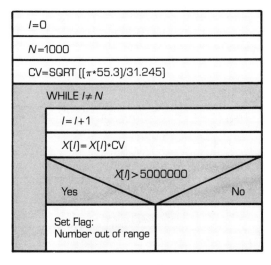 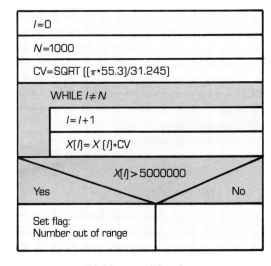

(a) Before modification **(b) After modification**

FIGURE 16–11 Test improvement

time. Compare this procedure with one in which the test is made only once (see **Figure 16–11b**). Clearly, the revised code is more efficient.

Combining tests is another way to improve program efficiency. Consider the same example, with an additional processing requirement. Assume that if $X(I) > 5000000$, we will want to discontinue processing. **Figure 16–12a** illustrates one way of coding this requirement. For each iteration, two tests are required: one to test for the end of the loop and another to test for the upper loop value. A modification to this design is to require only a single test. As shown in **Figure 16–12b,** a combined test checks for both loop end and loop value. With some compilers, such a reduction can lead to significant time savings.

External call improvement requires the maintenance programmer to determine whether a call to an external procedure (or to the user) can be simplified or eliminated. Jalics[6] found that 90 percent of the calls to a routine to calculate the Julian date returned a date that was the same as before. He suggested that the designer create two routines to avoid unnecessary calls: one to handle common occurrences and another to handle exceptions.

Interactive dialogue also provides examples of where calls to users can be simplified or avoided. Suppose the original design displays the message:

ENTER CUSTOMER ORDER DATE (MMDDYY):＿＿＿＿＿＿

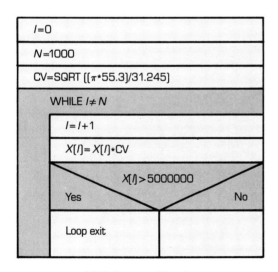

(a) **Before modification**

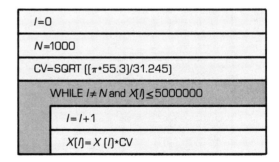

(b) **After modification**

FIGURE 16–12   Combining tests

On examination, the maintenance programmer determines that 95 percent of the entered dates are the same as the previous date. In perfecting this code, the programmer adds a new message, which reads

IS CUSTOMER ORDER DATE THE SAME (Y/N)?■

If it is the same, the enter customer order date requirement is skipped. If not (the exception), the user is required to add the date of the order.

Expression improvement is the refining of algebraic expressions and in many ways resembles loop improvement. The objective is to reduce the computer time needed to evaluate expressions. Consider the before-and-after code shown in **Figure 16–13**. In the before case, the X raised to the second power, followed by X raised to the fourth power, is found to be inefficient. The total time needed to evaluate this expression is reduced by using the precomputed value of X raised to the second power.

Another way to simplify expressions is a carry-over to the basics of algebra—namely, the factoring of expressions to reduce their computational complexity. Suppose the following code is found in a computer program:

$$Z = (X**2) - (X*Y) - (6*Y**2).$$

The maintenance programmer discovers that more efficient code can be written. The new code reads

$$Z = (X + 2*Y) * (X - 3*Y).$$

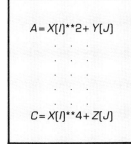
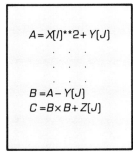

(a) **Before change**   (b) **After change**

FIGURE 16–13   Simplification of expressions

The Mansfield, Inc., Case Study

In reviewing the new billing and receivables system prepared by John Seevers and his staff, Chris Steward, the lead programmer assigned to maintain the system, was impressed. "The system looks very complete," he remarked. "You can tell that the staff attempted to do it right the first time. I especially like the way in which the design is documented. It should be a joy to maintain this system."

Request for Service

Chris got the chance to test his feelings about the new system when he received a request for service from finance and accounting. The request stated that there was a problem with the monthly remittance display (see Figure 10–21). Users wanted the customer name field deleted from the screen; they specified that the name should appear as a message within the message window. When Chris questioned why this change was desired, Tom Vlahovish from accounting remarked: "With the exception of the customer name, all the information shown on the screen is to be key-entered. The name is added by the program, provided the customer number is entered correctly. This is a problem because placement of the name field on the screen tends to confuse the operators. They see the field customer name and try to add one."

Chris responded to this request quickly and with a moderate amount of difficulty. Besides changing the terminal screen mat to remove the name, he altered the source code that controlled the printing of the name. In documenting the solution to this problem, Chris modified the monthly statement remittance display contained in the technical specification (see **Figure 16–14**). Since no new terms were added, the data dictionary did not require modification. The only other change was to add notes to the specification, stating the action taken and the effect of the action.

Module Repair

A more serious problem confronted Chris when he discovered that the no-record-on-file flag would not turn off when a customer record could not be found (see Figure 12–21). Once the flag was set, all key-entered customer identification codes were declared to be invalid.

When Chris examined the cause of the problem, he found another. The pseudocode (see Figure 14–20) and the source code written for module 2.2 did not agree with the structure chart. The pseudocode indicated that module 2.0 would check the setting of the flag; the structure chart suggested that module 2.2 would control the setting of the flag.

Chris decided that it would be best if the pseudocode and the source code were in agreement with the structure chart. Accordingly, he revised the code for module 2.2 so that it would turn off both the no-record-on-file and valid-account flags (see **Figure 16–15**). He also modified the structure chart to clearly show the display error and abort requirements (see **Figure 16–16**).

Perfecting Code

During his spare time, Chris worked on perfecting the code written for the new system. The main aspect he disliked about the structure of the code was the way in which some loops were programmed. For example, it was common to

```
Remittance number XXXXXX                    MONTHLY STATEMENT REMITTANCE

Customer number: XXXXXXXXX          Customer name:  XXXXXXXXXXXXXXX
Date of check:   XX-XX-XX           Check number:   XXXXXXXXXXXXXXX

         No.    Invoice number   Cash payment   G/L code
         1      XXXXXXXX         XXXXX.XX       XXXXXXXX
         2      XXXXXXXX         XXXXX.XX       XXXXXXXX
         3      XXXXXXXX         XXXXX.XX       XXXXXXXX
         ⋮          ⋮                 ⋮              ⋮
TOTAL PAYMENT----------------------XXXXXX.XX        Message window


         CUSTOMER NAME IS: ANTHONY POSSELLI (CORRECT? Y,N)■



PRESS CONTROL-E TO EXIT.


Notes:
Remove customer name field from display mat.
Display customer name within window and ask for visual verification.
```

FIGURE 16-14 Revised monthly statement remittance display

```
*MODULE 2.2-ADD CUSTOMER INFORMATION*
*RECEIVES CUSTOMER-ID                  *
*RETURNS CUSTOMER-SOLD-TO-NAME          *

*TURN FLAGS OFF*

NO-RECORD-ON-FILE = FALSE
VALID-ACCOUNT = TRUE

CALL 2.2.1 GET-CUSTOMER-RECORD

    *CHECK SETTING OF FIRST FLAG*
    IF NO-RECORD-ON-FILE = TRUE
        THEN CALL DISPLAY ERROR AND ABORT
        ELSE CALL VERIFY CUSTOMER RECORD

            *CHECK SETTING OF SECOND FLAG*
            IF VALID-ACCOUNT = FALSE
                THEN DISPLAY ERROR AND ABORT
            END-IF
    END-IF

*END-MODULE 2.2-ADD CUSTOMER INFORMATION*
```

FIGURE 16-15 Pseudocode for repaired module 2.2, add customer information

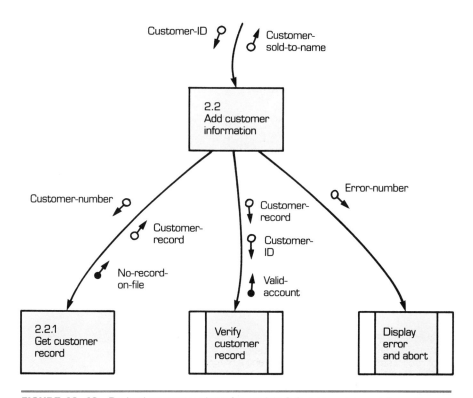

FIGURE 16–16 Revised structure chart for module 2.2, add customer information

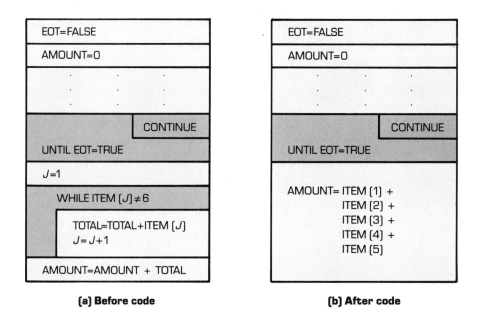

FIGURE 16–17 Before (a) and after (b) loop unrolling

find loops where the dimensions of the innermost loop were known in advance (see **Figure 16–17a**) where the value of J is shown to vary from 1 to 6.

Chris improved the efficiency of the code by using a technique called *loop unrolling*. He identified the loops where loop dimensions were known in advance and made these loops disappear (see **Figure 16–17b** to understand Chris's approach). Chris commented: "With loop unrolling, the loop overhead is totally eliminated (the second figure shows that the WHILE loop is no longer needed)."

SUMMARY

This final chapter has not only provided a description of systems maintenance but has also explained why maintenance is required for all designed software. Besides attending to failures in processing (corrective maintenance), keeping the software current (adaptive maintenance) and improving program efficiency (perfective maintenance) are key concerns of the maintenance programmer.

The value of structured analysis and design is more fully realized once a software product reaches the operations and maintenance stage. With clear specifications to work with, the maintenance programmer has a much easier time handling requests for service.

Mini–systems development often characterizes the processing of adaptive maintenance requests for service. The maintenance programmer must define system modification requirements, alter the logical system, revise the data dictionary to record modifications to the logical design, and make technical design changes.

Standard housekeeping tasks must be performed (in addition to mini–systems development projects) to keep software current. Often the maintenance programmer must design various utility programs to perform this routine maintenance.

Fixing software when it fails is better known as corrective maintenance. Unlike adaptive maintenance, here the maintenance programmer strives to eliminate all types of problems calling for corrective action. Critical incident reporting offers one way of tracking software problems and failures. Once the reasons for failure are understood, corrective maintenance is easier to manage.

Module redesign and repair are often required of error-prone programmed procedures. This kind of repair, however, must be carefully managed to preserve the integrity of the design specification. All changes to the source code must be carefully documented; any design changes must be reflected by revised structure charts, a revised data dictionary, and so forth.

Perfective maintenance seeks to improve or maintain program efficiency. The objective is to require fewer computing resources in the running of a software product. One way to improve efficiency is to improve the data structure designed for the software. Blocking records, storing precomputed results, and eliminating temporary work files are examples of ways to shorten processing run times.

Another way to improve efficiency is to modify expensive parts of a system, thereby improving their efficiency. Improving loops, test comparisons, calls to external procedures, and evaluation of algebraic expressions illustrates ways of improving the overall speed of processing.

By working more on adaptive and perfective maintenance tasks (and less on corrective maintenance tasks), the maintenance programmer strives toward the ideal: keeping operational software products current, thereby meeting or exceeding user expectations, while holding to a high level of processing efficiency.

REVIEW QUESTIONS

16–1. What is meant by systems maintenance, and why is such maintenance necessary even for carefully designed software?

16–2. What is the difference between adaptive and corrective maintenance?

16–3. Of the three types of maintenance, which two are the most desirable? Which type accounts for more than 50 percent of all requests for maintenance?

16–4. What is a user request for service?

16–5. Why are adaptive maintenance requests viewed as mini–systems development projects?

16–6. Why must maintenance programmers identify the areas within a design that require change as business rates vary and as reporting periods come and go?

16–7. How do system errors differ from program logic errors?

16–8. What does module repairing involve?

16–9. Why must module repairing be closely administered?

16–10. What is the main benefit of perfective maintenance?

16–11. Why is blocking records an example of trading space for time?

16–12. What tend to be the most expensive parts of programmed procedures (and thus draw the attention of the maintenance programmer)?

16–13. What is the objective of loop improvement?

16–14. What is meant by loop unrolling?

EXERCISES

16–1. An index found in a recently completed software design is limited to a listing of social security numbers, arranged in numeric order. The current method of processing is to search the index to find a desired number. The problem with this technique is that the entire index may need to be searched to find a number in error.

Using the data shown below, design a three-level parsed list. Compare this list with the original index. How many fewer digits are needed in the parsed list?

Suppose that a search is made for the number 481–17–1845. How many comparisons must be made with the original index to determine that the number cannot be found? How many comparisons are needed with the parsed list?

$$
\begin{array}{c}
469-38-1062 \\
469-42-1053 \\
469-42-1019 \\
471-14-1136 \\
471-83-1142 \\
481-16-1143 \\
481-16-1145
\end{array}
$$

16–2. A maintenance programmer discovered a coupling problem in the design of a program to produce grade reports. As shown on page 548, teacher ID was sent from module 2.0 to module 2.1, at which point the validity of the ID was checked by a check-digit comparison test. A segment of the pseudocode written for the module was documented as follows:

```
*Module 2.1 – Get Teacher Name          *
*Receives teacher-ID                    *
*Returns  teacher-name (if found),      *
*         invalid-teacher-ID,           *
*         no-teacher-on-file            *


INVALID-TEACHER-ID = FALSE
NO-TEACHER-ON-FILE = FALSE

CALCULATE CHECK-DIGIT REMAINDER
     IF REMAINDER NOT EQUAL TO 0
        THEN SET INVALID-TEACHER-ID TO TRUE
           .         .
           .         .
           .         .
```

Revise the structure chart on page 548 so that only a valid teacher ID is sent to the get teacher name module. In your design, create a new module 2.1, verify teacher ID. This module will receive the teacher ID and return a valid ID flag (note the difference). In addition, write the pseudocode for the new module.

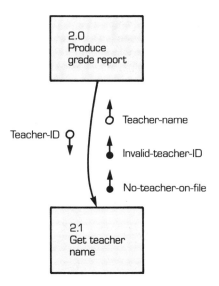

16–3. The logic used to calculate the tax totals for the PAYGOSHOP drug chain is shown below. Once the tax rates are known, the three types of tax—TAX (J)—are computed for each of PAYGOSHOP's two corporate divisions.

Rewrite this procedure (prepare pseudocode) to improve processing efficiency.

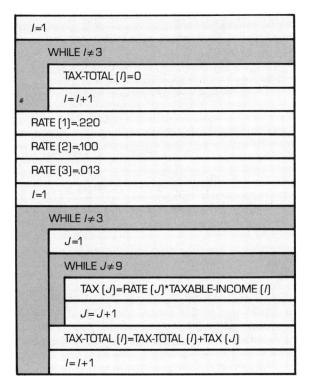

16–4. Besides designing the new structure chart showing the procedure to reset QTD pension totals to zero (see Figure 16–4), new pseudocode for all modules must also be written. Write the pseudocode for module 20.2, reset the pension, QTD field. This module receives and updates three data couples: pension QTD (a dollar value), pension dollars reset (a dollar value), and pension record reset (a numeric value).

WORLD INTERIORS, INC.—CASE STUDY 16
MAINTAINING THE NEW SYSTEM

Introduction

Once the new system was fully implemented, John Welby decided to move on to other design problems suggested by Rena and Ray. Jane stayed with the new system, maintaining it as required.

On leaving, John stated: "Jane, you and I have designed and implemented a fine piece of software. It should be relatively easy to maintain, provided users do not demand constant change. I think each new user requirement should be considered carefully before we move to alter the design."

Jane took John's comment to heart, as requests for service began coming in. "Each requirement must be justified," she reasoned. "Otherwise, the software efficiency we now experience might be lost."

User Request for Service

Soon after Jane took on the task of maintenance programming, she received a request for service. The request dealt with the way in which information was printed on the filled-order register. The problem, as described by users, could be summarized as follows.

The filled-order register was considered incomplete. As currrently designed, it was limited to information shown on the factory invoice. While this information was important, information from the customer order and customer invoice was also needed. The register needed to report

1. the total number of filled orders (i.e., the number of filled orders compared with the number of partially filled orders);
2. the sales total (i.e., the dollar amount received from the customer);
3. the profit margin (i.e., the dollar difference between the sales total and the factory cost extended), the current design being limited to a message stating that the profit margin was low; and
4. the order cycle time (i.e., the time between the placement of the order and the date of the factory invoice).

With some reluctance, Jane approved the request for service. Her misgivings were that the information desired could be obtained elsewhere. She also questioned whether such facts as order cycle time and profit margin were all that important for each invoice.

In preparation for handling this maintenance assignment, Jane copied the Warnier-Orr diagram prepared for the filled-order register (see **Figure 16–18**). She had to agree that the information contained on the register was limited to data key-entered for factory invoices.

Corrective Maintenance

Besides handling requests for service, Jane found it necessary to fix sections of code containing small errors. One of these sections was identified as module 1.2.1.3, calculate sales tax (see chapter 12 to trace the location of this module). Jane commented: "In processing, the sales-tax-error flag doesn't work correctly. The sales tax dollar amount sent by the customer might be in error; however, the module fails to pick up the error. I guess we didn't test this module very

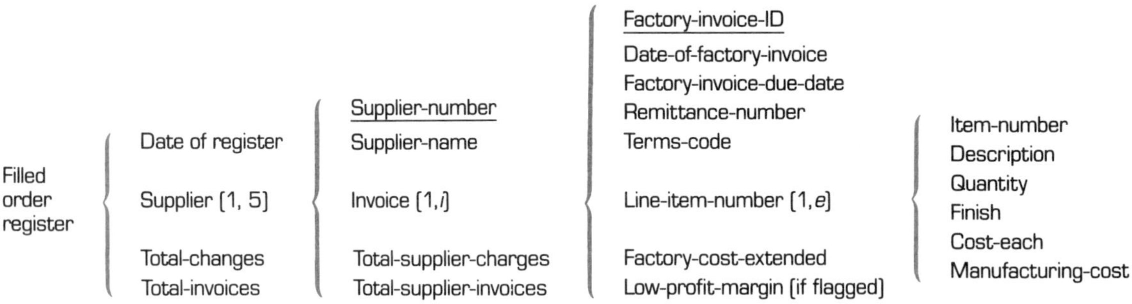

```
                                                          Factory-invoice-ID
                                                          Date-of-factory-invoice
                                                          Factory-invoice-due-date
                              Supplier-number             Remittance-number              Item-number
              Date of register Supplier-name              Terms-code                     Description
Filled                                                                                   Quantity
order         Supplier [1, 5]  Invoice [1,i]              Line-item-number [1,e]         Finish
register                                                                                 Cost-each
              Total-changes    Total-supplier-charges     Factory-cost-extended          Manufacturing-cost
              Total-invoices   Total-supplier-invoices    Low-profit-margin (if flagged)
```

FIGURE 16-18 Warnier-Orr diagram showing layout of filled orders

---

well. We may have assumed that the module was too simple to fail."

Jane studied the pseudocode written for the module, which read as follows:

```
*Module 1.2.1.3-Calculate sales
   tax*
*Receives sales-tax (paid), sales-
   total*
*Returns sales-tax-error (if tax
   is incorrect)*

RATE = .065
*TURN FLAG OFF*
SALES-TAX-FLAG = TRUE
*COMPUTE SALES TAX*
TAX = SALES-SUBTOTAL * RATE
*CHECK TAX-PAID FOR ERROR*
IF TAX = SALES-TAX
     THEN SET SALES-TAX-ERROR TO
        FALSE
END-IF
*End Module   1.2.1.3-Calculate
   sales tax*
```

Perfective Maintenance

Jane also found time to study the code written for the new design, with the objective of improving the efficiency of the design. "Some procedures need to be rewritten," she determined. "When we wrote the code we were pressed for time. While the procedures work, they are not always very efficient."

Module 1.2.1, process customer orders, was an example of a procedure that needed improvement. The pseudocode written for this procedure was documented as follows:

```
*Module 1.2.1-Process customer
   orders*

ORDER-ERROR = FALSE

*MAKE FIRST CALL*
CALL 1.2.1.1 GET SUPPLIER INFORMA-
   TION
IF NO-SUPPLIER = TRUE
     THEN SET ORDER-ERROR TO TRUE
        AND EXIT
     ELSE *MAKE SECOND CALL*
        CALL 1.2.1.2. CALCULATE-
           LINE-ITEM-CHARGES
END-IF
IF ITEM-NO-ERROR = TRUE
     THEN SET ORDER-ERROR TO TRUE
        AND EXIT
END-IF
IF FINISH-NO-ERROR = TRUE
     THEN SET ORDER-ERROR TO TRUE
        AND EXIT
END-IF
IF PRICE-EACH-ERROR = TRUE
     THEN SET ORDER-ERROR TO TRUE
        AND EXIT
     ELSE *MAKE THIRD CALL*
        CALL 1.2.1.3 CALCULATE
           SALES TAX
END-IF
```

```
IF SALES-TAX-ERROR = TRUE
    THEN SET ORDER-ERROR TO TRUE
        AND EXIT
    ELSE *MAKE FOURTH CALL*
        CALL 1.2.1.4 MATCH SALES
            AND PAYMENT TOTALS
END-IF
IF PAYMENT-ERROR = TRUE
    THEN SET ORDER-ERROR TO TRUE
END-IF

*End Module 1.2.1–Process customer
    orders*
```

CASE ASSIGNMENT 1

Prepare a revised Warnier-Orr diagram to show the new information contained in the request for service.

CASE ASSIGNMENT 2

Repair the pseudocode written for module 1.2.1.3, calculate sales tax, showing the correct way to report a sales tax error. Initially, turn the flag off, switching it on only when there is a sales tax problem.

CASE ASSIGNMENT 3

Rewrite the pseudocode written for module 1.2.1, process customer orders, to improve the efficiency of processing.

REFERENCES

1. E. W. Swanson, "The Dimension of Maintenance," *Proceedings of the 2nd International Conference on Software Engineering* (October 1976), 492–497.

2. G. Helms and I. Weiss, "Applications Software Maintenance: Can It Be Controlled?" *Data Base* (Winter 1985): 16–18.

3. J. L. Bentley, *Writing Efficient Programs* (Englewood Cliffs, N.J.: Prentice-Hall, 1982).

4. D. E. Knuth, "An Empirical Study of FORTRAN Programs," *Software—Practice and Experience 1* (April–June 1971): 105–133.

5. A. Parkin, *Systems Analysis* (Cambridge, Mass.: Winthrop Publishers, 1980).

6. P. J. Jalics, "Improving Performance the Easy Way," *Datamation 23* (April 1977): pp. 135–148.

Index

Abstract systems, 33–34
Acceptance testing, 513–514
  of manual procedures, 513–514
  of software, 477, 513
Accounts payable procedures
  audit of, 442
  data flow diagram of, 100
  perfective maintenance of, 536
  processing controls in, 259
Accounts receivable procedures.
  See Billing and accounts receivable system
Accuracy of system, defining problems in, 104
Activity centers, in Blumenthal's model of business systems, 39
Adaptive maintenance, 524–530
Addressing of records, indirect, 370
Administration of systems department, 19, 20–22
Administrative data processing, compared to management information systems, 41
Algebraic expressions, perfective maintenance of, 541
Algorithms, 7
Alternative system designs
  development of, 217–221
  evaluation of, 221–225
  project specification of, 122–123
  ranking of, 225–226
  selection of, 226
Analysis, 2–5, 9
  of alternative system designs, 221–225

causal, 109–110, 114, 118
of collected data, 3–4, 132–167
definition of, 2
design evaluation matrix in, 221
discriminative, of questionnaire responses, 148
of equipment needs, 296
of feasibility, 3, 4, 108, 118, 119–120, 204–244
functional, 113–117, 118
marginal, 110–113, 117
picture-frame, of interactive dialogue, 349–354
structured, 65–77
of system requirements, 2, 3, 101–131
testing procedures in. See Testing
time required for, 11, 12, 13, 14
traditional tools in, 86–93
weighted, 226
Analysts of systems, functions of, 2
Anthony's model of business systems, 37–38
Application software, 49, 54, 211, 212
Arithmetic/logic unit of computers, 50
Arrow symbols, in hierarchical/input-process-output charts, 299
Attributes of records, 179–180, 182, 183, 184, 185–186, 187, 364

in relational files, 374–375
Auditors, 440
Audits, 440–445
  around the computer, 440, 441–442
  audit trail in, 441, 442
  definition of, 440
  logical design specification on, 259
  physical, 440, 444–445
  software for, 443–444
  test data in, 442
  through the computer, 440, 442
  types of, 440–445
  using the computer, 440, 443–444

Backup procedures
  in file design, 381
  in housekeeping utility programs, 529
  operator's guide on, 508
Balanced decomposition of data flow diagram, 138–142
Batch processing, 88, 89, 249–250
  control procedures in, 430, 450
Billing and accounts receivable system, 27, 56
  alternative designs of, 219
  business computer in, 43–46
  causal analysis of, 124–125
  centralized and decentralized approaches to, 209–211
  context diagram of, 93–94
  conversion to new system of, 505, 514–516

553

Billing and accounts receivable system, *continued*
  data collection and analysis of, 152–154, 157–160
  data entry procedures in, 46
  data file design for, 387–390
  data flow diagrams of, 93–96
  feasibility analysis of, 234–237
  goal statement of, 207
  input and output design of, 355–357
  logical design specification of, 274–277
  maintenance of, 542–544
  organizational design strategy statement on, 208
  preliminary functional design of, 310–314
  processing controls in, 450–451
  program design for, 418–421
  programming and program testing of, 483–485
  task analysis of, 229
  technical analysis of, 230–232
  visual table of contents of, 312–313
Black-box testing, 494, 495–497
Blind spots in system, detection of, 495
Blocking of records, 365, 366, 386–387
  in perfective maintenance, 534–536
Blumenthal's model of business systems, 38–40
Boundaries of system, 34
Bubble charts, 65
Budget section
  in logical design specification, 263–264, 265–274
  in project specification, 123
  in technical design specification, 448
Budgeting system
  causal analysis of, 109–110, 114
  feasibility analysis of, 119
  functional analysis of, 114–117
  marginal analysis of, 110–111
Build versus buy decision, feasibility analysis of, 211–212

Business systems, 33–63
  administrative approaches in, 20–22
  Anthony's model of, 37–38
  Blumenthal's model of, 38–40
  business cycle model of, 41–43
  computers in, 43–49
  general models of, 37–43
  management-based, 41, 42–43
  Mansfield as example of. *See* Mansfield, Inc.
  Neumann and Hadass's model of, 40–41
  project organization in, 16–19
  system trinity in, 48–49, 62–63
  transaction-based, 41–42
  World Interiors as example of. *See* World Interiors, Inc.
Bytes of computer memory, 365–366, 367, 386

Call procedures, external, perfective maintenance of, 540–541
Calling and called modules, in structure charts, 79–80
Capacity of system
  in data entry, 344, 345
  defining problems in, 105–106
  in evaluation of alternative designs, 221–222, 225
  stress tests on, 495–496
Case statements
  in structured English, 190–191
  in structured programming, 462, 464
Case studies. *See* Mansfield, Inc.; World Interiors, Inc.
Causal analysis, 108–110, 114, 118
  compared to functional analysis, 114
  in Mansfield, Inc., 124–125
  in World Interiors, Inc., 129–131
Central processing unit, 50–51, 54
Centralization of resources, feasibility analysis on, 209–211
Change agent, 501
Change files, 381

Channels of processing, in system, 35, 36
Check-digit tests for data validation, 438–440
Clearing-house approach to systems administration, 21–22
Client, in system trinity, 48–49
COBOL programs, structure of, 474–475
Code, computer
  documentation of, in programming, 470–472
  technical design specification of, 447–448
  testing of, in programming, 472–473
  translation of design into, 285, 286
Cohesion
  in structure charts, between modules, 397, 402
  in system, testing of, 307
Collection of data, 3–4, 132–167, 334. *See also* Data collection and analysis
Combination tests for data validation, 437–438
Comment operator, in data dictionary, 174
Communication, in interactive dialogue between user and computer system, 345–355
Communications link symbol, in system flowchart, 86
Components of system, 35
Computer systems, 43–56
  applications of, 44–45, 46–47
  and audits, 440–444
  in business, 43–49
  conversion to new software in, 492, 501–512
  cost of, 273
  data file and database design for, 364–395
  data processing personnel in, 49, 55–56
  feasibility analysis on, 208–209
  goal statement on, 206–207
  hardware of, 49–54
  identification and evaluation of problems in, 130–131
  input to, 334–345. *See also* Input to system

interactive dialogue for user
  interface with, 345–355
interrelations in, 47–48
logical design specifications
  on, 251–254, 258–259
maintenance of, 522–551
operational procedures of, 49,
  54–55, 538–541
organizational design strategy
  statement for, 208
output from, 217, 323–334.
  See also Output of system
printers in, 331–332
processing controls in, design
  of, 429–445
screen design for user
  interface with, 354–355,
  360–362
software of, 6, 49, 54. See also
  Software
structure of, 46–47
system flowcharts of, 86–89,
  155–157
system trinity in, 48–49
user interface with, 345–355
Connecting symbols
  in data flow diagrams, 71
  in structure charts, 80
Consistency
  of data, testing of, 438
  of programming style, 474
Constraints on system, 37, 216,
  258, 268–269
Context diagrams
  on billing system, 93–94
  in data collection and
    analysis, 133–134, 153–154
  on output of new system,
    217–219
  in structured analysis, 67–68
Control flags, in structure charts,
  80, 406
Control record, in file-balance
  control procedures, 434–435
Control table, in file-balance
  control procedures, 433–
  434, 450–451
Control unit of computers, 51
Controls in processing, 85–86,
  429–445. See also
  Processing controls
Conversion, in business cycle,
  42
Conversion to new system, 6, 8,
  492, 501–512

acceptance testing of, 513–514
change agent in, 501
completion of system
  documentation in, 507–510
crisis index of, 514
database creation in, 501–502
definition of, 492, 501
difficulty of, 492–493, 501
direct method of, 503
in Mansfield, Inc., 514–516
new work procedures in, 505–
  507
operator's guide preparation
  in, 507–508
parallel method of, 504–505
phased or modular method of,
  503–504
pilot method of, 503
preparation of user's guide for,
  507, 508–510
system changeover in, 503–
  505
technical writer in, 505–506
user resistance to, 492–493
user training in, 510–512
in World Interiors, Inc., 520–
  521
writing of work procedures in,
  506–507
Corporate planning approach to
  systems administration, 22
Corrective maintenance, 524,
  530–534, 549–550
Correspondence of data flow,
  testing of, 306
Cost considerations
  in budget section of
    specifications. See Budget
    section
  in budgeting system. See
    Budgeting system
  computer costs in, 273
  in data processing, defining
    problems in, 104–105
  in economic feasibility
    analysis, 119
  in evaluation of alternative
    designs, 222
  labor costs in. See Labor costs
  in model of business cycle,
    41–42
  in personal interviews and
    questionnaires, 145
  preliminary estimates on, 123
  in project overhead, 273

in resource management
  environment, 22–25
in return on investment, 24,
  25, 117, 119
and trade-offs in design, 273–
  274
Cost/responsibility center, in
  systems administration, 22
Coupling of modules, in
  structure charts, 397, 399–
  402, 419–421
Coverage of software, white-box
  testing of, 495
Critical incident reporting, and
  corrective maintenance,
  531–533
Critical path
  in program testing, 477, 481–
    482
  in time schedule
    determination, 269
Cursor, in computer display,
  347, 354, 355

Data capture, in input
  procedures, 334
  minimum set of data for, 334,
    335–337
  selection of appropriate
    medium for, 341–345
Data collection and analysis, 3–
  4, 132–167
  data flow diagrams in, 133–
    142
  on forms and procedures
    used, 149–157
  in input procedures, 334
  at Mansfield, Inc., 157–160
  in new system, acceptance
    testing on, 513–514
  in personal interviews and
    questionnaires, 142–148
  system flowcharts in, 155–157
  time required for, 14
  at World Interiors, Inc., 163–
    167
Data couple symbol, in structure
  charts, 80, 290–291
Data description hierarchy, 149–
  150
Data dictionary, 73, 169–177
  contents of, 169–170
  conventions used in, 171–174
  at Mansfield, Inc., 193–194
  nested listings in, 174

Data dictionary, *continued*
  organization of, 174–176
  partitioned terms in, 169–170
  purpose of, 176–177
  redundancy problem in, 174
  self-defining terms in, 73, 169–170
Data elements, 73, 149–150, 324
  definition of, in data dictionary, 73, 169–177
Data entry. *See* Input to system
Data files. *See* Files
Data flow diagrams
  context, 67–68
  in data collection, 133–142
  data dictionary and, 73, 169–177
  decomposition of, 71–73, 78, 135–142
  functional primitives in, 71, 136–138
  level-0, 68–71
  lower-level, 71–73
  at Mansfield, Inc., 93–96
  in preliminary functional design, 287–293
  in structured analysis, 65–73, 76
  structured English describing, 76
  symbols in, 70–71
  at World Interiors, Inc., 99–100
Data flows
  duplicate, 155
  testing correspondence of, in system review, 306
  tracing of, 142
Data manager, 177
Data organization and documentation, 3, 4, 168–203
  data dictionary in, 169–177
  data store descriptions in, 177–187
  at Mansfield, Inc., 193–197
  time required for, 14
  transform descriptions in, 188–193
  at World Interiors, Inc., 202–203
Data processing. *See* Processing
Data search flag, in structure charts, 408

Data specification, 169
Data storage in computers, 48, 50, 51
  bytes of, 365–366, 367, 386
  hardware for, 50
Data stores, 70, 177–180, 364
  attributes of records in, 179–180, 182, 183, 184, 185–186, 187
  data structure diagrams of, 180–187
  definition of, 70, 177, 179
  description of, 177–180
  inverted files in, 183–186
  linked list in, 183
  logical design specifications on, 254–258
  logical external pointers in, 181–183
  logical internal pointers in, 183
  records in, 179, 364
  symbol for, 70
  Warnier-Orr diagrams on, 186–187, 324–327
Data structure, 324
  analysis of, 149–151
  diagrams of, 73–74, 180–187
  of input, 334–338
  key and nonkey fields in, 74
  logical, documentation of, 324–327, 334–337
  of output, 324–329
  of Mansfield, Inc., 194–197
  modification of, in perfective maintenance, 534–536
  physical, 327–329, 337–338
  titles and headings in, 324
Database, 382–385
  creation in system conversion, 501–502
  definition of, 180, 382
  layout in structured design, 84–85
  logical design specifications of, 254–258
  management system for, 211, 212, 382–385, 387–388
  requirements for, 5
Debugging of system
  maintenance procedures in, 523
  testing procedures in, 476–482, 492–501

Decentralization, feasibility analysis of, 209–211, 232
Decision construct
  in structured English, 76, 171, 173, 189, 190
  in structured programming, 462, 463
  white-box testing of, 495
Decision-making in business systems
  in Blumenthal's model, 38–39, 40
  in Neumann and Hadass's model, 40–41
  in system trinity, 48–49
Decomposition of data flow diagrams, 71–73, 78, 135–142
  balancing of, 138–142
DeMarco data dictionary conventions, 171, 173
Descriptive flags, in structure charts, 80, 406, 408
Design, 10, 216–226
  coding process in, 285, 286
  of computer programs, 5–6, 396–428
  of computer screen display, 354–355, 360–362
  constraints affecting, 216
  construction of dialogue tree in, 347–349
  of data files and database, 5, 364–395
  definition of, 2
  detailed, 285, 286, 303, 396
  evaluation of, 221–225
  feasibility analysis on, 216–226
  goals of, 35, 206–207, 208
  hierarchical/input-process-output charts in, 296–302, 307–308, 313–314, 319–321
  of input, 5, 334–345
  of interactive dialogue between user and computer, 345–355
  logical design specification on, 3, 4, 5, 245–283
  maintenance package in, 286, 287
  objectives of, 208, 214–216
  of output, 5, 217–226, 323–334
  preliminary, 285, 286, 287–293, 303, 306

Index **557**

process of, 5–6, 285–296
of processing controls, 5, 6, 429–445
ranking alternatives to, 225–226
requirements in, 5–6
reviews on, internal and external, 303, 304, 306
selection of, 226
strategy in, 208–214
structure charts in, 397–406
structured approach to, 77–86
structured walkthroughs in, 302–310
technical design specification on, 5, 284–285, 446–449
testing of, 285–286, 287
time required for, 11, 12, 13, 14, 15
traditional tools in, 86–93
transaction-centered and transform-centered, 247–248
visual table of contents in, 297–298, 312–313
Design evaluation matrix, 221
weighted, 226
Designer of system
functions of, 2
in system trinity, 48–49
Detailed designs, 285, 286, 303
of computer programs, 396
structured walkthrough of, 303
Development of systems, 9–25
administration of, 20–22
and mini–systems development in adaptive maintenance, 526–528
organization of functions and tasks in, 16–19
planning of, 12–15
project life cycle in, 10–19
in resource management environment, 22–25
time required for, 11–12
Dialogue, interactive, between user and computer, 345–355
dialogue tree in design of, 347–349
perfective maintenance of, 540–541
picture-frame analysis of, 349–354
Dialogue tree
construction of, 347–349

nonrestricted node on, 349
Direct labor costs, logical design specification on, 272–273
Direct organization of files, 370–371
Direct system conversion, 503
Discriminative analysis, of questionnaire response, 148
Disks, floppy and hard, 51
Display screen. *See* Screen display of computer
Display symbol, in system flowchart, 86
DO-UNTIL statements
in structured English, 191–192
in structured programming, 462, 463
white-box testing of, 495
DO-WHILE statements
in structured English, 192
white-box testing of, 495
Document(s), as data source, processing controls on, 85, 430–433, 450
Document symbol, in system flowchart, 86
Documentation, 10
of acceptance testing of system, 513
of collected and organized data, 3, 4, 168–203
of computer programs, 470–472, 476
of conversion to new system, 8, 507–510
of logical data structure of input, 334–337
of logical data structure of output, 324–327
of module repairs, 533–534
of operations, 8, 265, 507–508
of system, 8, 507–510
of system testing, 499–501
for user, 8, 265, 507, 508–510
Driver program, 292
Dummy activity, in preparation of time schedule, 268–269
Duplicate data flows, 155
Dynamic user instructions, 345

Echo checks, in input controls, 433

Edit tests, for data validation, 436–438
Education and training of users, 296
causal analysis of, 109
in conversion to new system, 8, 510–512
functional analysis of, 114
logical design specifications on, 265
project specification on, 123
technical design specifications on, 449
user's guide in, 507, 508–510
Efficiency of system
defining problems in, 106
perfective maintenance improving, 534–541, 542–544
programming for, 475–476
Egoless programming, 458–459
Employee records
check-digit test in, 440
payroll information in. *See* Payroll system
Warnier-Orr diagrams on, 186–187
End-of-stream flag, in structure charts, 408
Engineering of software, 9
English, structured, 76, 188–193
conventions in, 189–193
in data dictionary, 171
on data flow diagrams, 76
logical constructs of, 76, 171–173, 189
at Mansfield, Inc., 197
rules of, 188–189
in transform descriptions, 76, 188–193
Entity-relationship diagrams, 373–375, 377, 378, 388–389
Environment of system, 34
Equipment requirements
analysis of, 296
installation and testing of, 296
specification of, 296
and vendor selection procedures, 43, 296
Equivalence partitioning, in system testing, 497–498
Error-checking routines in testing, 7. *See also* Testing
Error path testing of software, 496–497

Executive summary
  in logical design specification, 262
  in project specification, 120–121
  in technical design specification, 447
Existing system, description of
  in logical design specification, 262
  in project specification, 121–122
Exit procedures, in data entry, 337
Expenditure system, in business cycle, 42
External call procedures, perfective maintenance of, 540–541
External reviews of system design, 303, 304, 306

Failure of system
  corrective maintenance in, 530–534
  critical incident reporting on, 531–534
  operator's guide on, 508
Feasibility analysis, 3, 4, 108, 118, 119–120, 204–244
  on build versus buy decision, 211–212
  on centralization versus decentralization decision, 209–211
  on computer-based versus manual systems, 208–209
  on job enlargement or enrichment, 233
  logical design specification on, 262
  at Mansfield, Inc., 234–237
  on new design and design alternatives, 216–226
  organizational analysis in, 232–234
  preliminary, 120
  on project, 204–205
  project specification on, 122
  on prototype versus fully functional system, 213–214
  structured approach in, 226–228
  on system, 204–205
  task analysis in, 228–229, 232, 233
  technical analysis in, 229–232
  on tool sharing, 234
  user's performance definition in, 205–216
  at World Interiors, Inc., 241–244
Fields of data
  key and nonkey, 74
  tests for validation of, 437
Files, 5, 364–395
  adaptive maintenance of, 529, 530
  backup, 381
  blocking of records in, 365, 366, 386–387
  change, 381
  conversion to new system of, 501–502
  direct organization of, 370–371
  entity-relationship diagrams on, 373–375, 377, 378, 388–389
  estimation of storage space for, 386–387
  in file-balance control procedures, 433–435, 450–451, 455
  fixed-length records in, 367
  indexed-sequential organization of, 371
  interrecord gap in, 369
  inverted, 183–186
  logical design specification on, 254–258
  of Mansfield, Inc., 387–390
  master, 380
  multidimensional records in, 367–369
  physical and logical records in, 365–371, 385–387
  relational, 371–379
  requirements for, 5, 380–381
  rules on design of, 365, 376–379
  sequential organization of, 369–370
  software for management of, 212, 345, 346
  storage device for, 365
  structure of, 369–371
  structured design of, 84–85
  summary, 380–381
  suspense, 381
  system, 380–387
  transaction, 380
  variable-length records in, 367
  of World Interiors, Inc., 394–395
Fixed-length records, 367
Flags, in structure charts, 80, 406–408, 419–421
  control, 80, 406
  descriptive, 80, 406, 408
  nag, 80
  rules on, 406–408
Flashback checks, in input controls, 433
Flow diagrams and charts
  collection and analysis of, 155–157
  data flow diagrams, 65–73
  program flowcharts, 91–93, 412–415
  symbols in, 86
  system flowcharts, 86–89, 155–157
Forms used in system
  collection and analysis of, 150–151
  duplicate, 155
  processing controls concerning, 432–433
  for record layout, 385–386
Functional analysis, 113–117, 118
  compared to causal analysis, 114
  of goals, 207
Functional organization in systems development, 18–19
Functional primitives, in data flow diagrams, 71, 136–138
Functional units, in Blumenthal's model of business systems, 38, 39, 40

Gane and Sarson data dictionary conventions, 171
Gantt chart, in preparation of time schedule, 266–269
Goals, in design of system, 35, 206–207, 208
  written statement of, 206–207

Hard copy
  of input, 341
  of output, 331

Hardware, computer, 49–54
  corrective maintenance of, 532
  logical design specification of, 264–265
  technical design specification on, 448
Hashing scheme, in direct organization of files, 370
Hat symbol, in structure charts, 405
Help screens, interactive, 532
Hierarchial approach
  to conflicting objectives, 215–216
  to database management, 383–384
  to description of data structure, 149–150, 326, 327
  to evaluation of alternative designs, 225–226
  to input-process-output charts, 296–302, 307–308, 313–314, 319–321
HIPO (hierarchical/input-process-output) charts, 296–302, 307–308, 313–314, 319–321

Identification section of computer programs, 470–472
Identification test of users, 436
Imperative statements, in structured English, 76, 188
Implementation of system, 10, 457, 458
  components of, 6–9
  modules of, 408–409, 411–416
  time required for, 14
Incremental program testing, 477, 478–481, 483–485
Indexed-sequential organization of files, 371
Indirect labor costs, logical design specification on, 273
Information systems group, 19, 26–27
Initiative approach to systems administration, 20
Input to system, 5, 34, 46, 334–345
  data capture requirements in, 334, 335–337, 341–345
  hardware for, 49–50, 51, 53

HIPO charts on, 296–302, 307–308, 313–314, 319–321
with key-to-disk procedures, 341, 342
keystroke requirements in, 334–335, 343, 344
labor costs in, 334–335, 343–344
logical data structure of, 334–337
logical design specification on, 250–254
at Mansfield, Inc., 355–357
minimum set of data of, 334, 335–337
module specifications on, 410–411
with optical character recognition, 342–345
physical data structure of, 337–338
processing control procedures in, 85, 433–435, 450, 455
selection of appropriate medium for, 341–345
shortcuts in design of, 339–341
structured design of, 83, 85
system flowchart on, 86–88
technical design specification on, 284, 285
transcription and transposition errors in, 439
validity of, 437–442
visual representation of, 338–341, 355–357
at World Interiors, Inc., 360–362
Instructions
  file processing, 345, 346
  on help screen, 532
  in interactive dialogue, 345, 346
  in modules, reasonable number of, 404–405
  in operator's guide, 265, 507–508
  program, 89–91, 474
  in user's guide, 265, 507, 508–510
Intent of modules, description of, 408–411
Interactive processing, 88, 89, 249–250
  dialogue in, 345–355

formal messages in, 345–347
perfective maintenance of, 540–541
picture-frame analysis of, 349–354
Interfaces in system
  interactive dialogue for, 345–355
  screen design for, 354–355, 360–362
  testing of, 477, 479
Internal reviews on system design, and external reviews, 303, 304, 306
Interrecord gap in files, 369
Interrelations in system, 35, 47–48
Interviews, in data collection and analysis, 143–145
  compared to questionnaires, 145, 148
  open- and closed-ended questions in, 143–144
  structured and nonstructured, 143–144
  in three sessions, 144–145
Inventory procedures, in physical audit, 444–445
Inverted files, in data stores, 183–186
Investment in system, and return on, 24, 25, 117, 119
Invoice processing
  alternative designs of, 219
  computerized, 44, 46, 47, 48, 54
  data dictionary on, 173
  forms used in, collection and analysis of, 150–151

Job enlargement or enrichment, feasibility analysis on, 233
Journaling procedure, in input control, 433

Key attributes of records, 179, 180, 183, 374–375
Key-to-disk data capture system, 341, 342
  combined with optical character recognition, 342–345
Key field in data structure diagram, and nonkey fields, 74

Keystrokes, in data entry procedures, 334–335, 343, 344
  verification of, 344

Labeling of structure charts, clarity of, 397, 403–404
Labor costs. *See also* Payroll system
  in data entry procedures, 334–335, 343–344
  direct, 272–273
  indirect, 273
  in life cycle of project, 12–15
  logical design specification on, 272–274
  project specification on, 123
  technical design specification on, 448
Language, of computer software, 6–7, 54
  in database management system, 382, 383
  query, 211
  and structure of COBOL programs, 474–475
Language, in structured analysis in design
  pseudocode, 82–83
  structured English, 76
Layouts
  of files and databases, in structured design, 84–85
  in logical design specification, 251–254
  of records, in preparation of physical file design, 385–386
  in structured design, 83
Legal feasibility analysis, 119
Level-0 data flow diagrams, 68–71, 93, 94–96, 134
Level-0 structure chart, 78
  in preliminary functional design, 287–288, 289
Library records, entity-relationship diagrams on, 373–375
Library routine of computer software, 405
Life cycle of project, 10–19
  in mini-systems development and adaptive maintenance, 526
Likert-type scale, 147

Line printers, in computer system, 331, 332
Linked list
  in data store, 183
  in database management system, 383
Log, on system testing, 499–501
Logic errors in program, corrective maintenance of, 531–532
Logical data structure
  documentation of, 324–327, 334–337
  hierarchy of set rankings in, 326, 327
  of input, 334–337
  of output, 324–327
Logical design specification, 3, 4, 5, 245–283
  compared to technical design specification, 284
  on computer program requirements, 258–259
  on data file and database requirements, 254–258
  design assumptions in, 260
  input/output requirements in, 250–254
  at Mansfield, Inc., 274–277
  organizational requirements in, 246–250
  proposed logical system description in, 246
  on schedule and budget, 263–264, 265–274
  structured, 4, 261–265
  system design requirements in, 246–260, 263
  table of contents for, 261
  trade-offs in, 273–274
  at World Interiors, Inc., 282
Logical records, 365–366, 386–387
Loops in program
  perfective maintenance of, 534, 538, 544
  white-box testing of, 495
Lower-level data flow diagrams, 71–73

Maintenance of systems, 522–551
  adaptive, 524–530
  compared to systems development, 10

  corrective, 524, 530–534, 549–550
  critical incident reporting in, 531–533
  data structure modification in, 534–536
  definition of, 523
  design of package on, 286, 287
  documentation of, 533–534
  expression improvement in, 541
  external call improvement in, 540–541
  housekeeping tasks in, 528–530
  loop improvement in, 534, 538, 544
  maintenance programmers in, 522
  at Mansfield, Inc., 542–544
  mini–systems development in, 526–528
  module repairing in, 533–534, 642
  operator's guide on, 508
  perfective, 524, 534–541, 542–544, 550–551
  procedural modifications in, 538–541
  test improvements in, 539–540
  types of, 524
  user requests for, 524–525, 542, 549
  at World Interiors, Inc., 549–551
Management information systems, compared to administrative data processing, 41
Management of system, 11, 19–27
  administrative approaches to, 20–22
  in Anthony's model, 37–38
  in Blumenthal's model, 38–40
  in business cycle model, 41, 42–43
  of data processing personnel, 55–56
  feasibility analysis on, 119
  in resource management environment, 22–25
Managers
  of data, 177

of programs, 16
in project organization, 16–19
of projects, 16
of systems, 16
Mansfield, Inc., 25–27, 56
  causal analysis of, 124–125
  computer program design for, 418–421
  data collection and analysis on, 157–160, 193–197
  data file design for, 387–390
  data flow diagrams on, 93–96
  feasibility analysis on, 234–237
  input and output design in, 355–357
  logical design specification on, 274–277
  organization of, 25–27
  preliminary functional design in, 310–314
  processing controls in, 450–451
  programming and program testing in, 483–485
  structured analysis of, 65–76
  system conversion in, 514–516
  systems maintenance in, 542–544
Manual input symbol, in system flowchart, 86
Manual procedures
  acceptance criteria on, 513–514
  compared to computerized systems, in feasibility analysis, 208–209
Marginal analysis, 110–113, 117
Master files, 380
Matrix organization, 16–18, 26–27
Memory of computers, 50, 51
  bytes of, 365–366, 367, 386
Menu instructions, 345
  file processing, 345, 346
  program processing, 89–91
Microcomputer system, hardware of, 51
Microfiche, 332, 333
Microfilm, 331, 332–334
Milestones in systems development, 11
Minimum set of input data, 334, 335–337

Mini–systems development, in adaptive maintenance, 526–528
Modeling software systems, 211, 212
Models of systems, 34–43
  in business firms, 37–43
  general, 34–37
Modular method of system conversion, 503–504
Modules in structure charts, 77, 78, 79–80, 291–292, 293, 397
  clear labeling of, 397, 403–404
  cohesive, 397, 402
  coupling of, 397, 399–402, 419–421
  error-prone, repair of, 533–534, 542
  hatted, 405
  implementation of, 408–409, 411–416
  input-output function description of, 410–411
  intent of, 408–411
  Nassi-Shneiderman charts on, 415–416
  program flowcharts on, 412–415
  pseudocode describing, 411–412
  reasonable size of, 397, 404–405
  shared use of, 397, 405–406
  span of control of, 397, 404
  specifications on, 408–416
Modulus-11 method of data validation, 438–439
Mouse, for user interface with computer, 355
Multidimensional records, 367–369

Nassi-Shneiderman charts, in program design, 415–416, 462, 467
Nested decision statements, in structured English, 190
Nested listing in data dictionary, 174
Network approach to database management, 384
Network chart in project planning, 15

Neumann and Hadass's model on business systems, 40–41
New system, organization of, 284–321
  feasibility analysis on, 232–234
  hierarchical/input-process-output charts on, 296–302
  logical design specification on, 246–250
  in Mansfield, Inc., 310–314
  packaging of, 246–250
  preliminary functional design of, 287–293
  process in design of, 285–296
  structured walkthroughs of, 302–310
  technical design specification on, 447
  transaction-centered and transform-centered, 247–248
  in World Interiors, Inc., 319–321
Node, nonrestricted, on dialogue tree, 349
Nonkey fields in data structure diagrams, and key field, 74
Nonprocedural module specifications, and procedural specifications, 409, 411
Nonrestricted node, on dialogue tree, 349
Nonstructured interviews in data collection, and structured interviews, 143–144
Normal path testing of software, 496, 497
Normalization of relations, 375

Objectives
  of system design, 208, 214–216
  of system testing, 498
Online storage symbol, in system flowchart, 86
Operating system of computer, 54
Operational procedures of computer system, 49, 54–55
  modification of, in perfective maintenance, 538–541
Operations control, in Anthony's model of business systems, 38

Operations documentation, 8, 265, 507–508
Operators of system
 corrective maintenance of errors related to, 532–533
 logical design specification on, 265
 preparation of, in system conversion, 507–508
Optical character recognition, for data entry, 341
 combined with key-to-disk system, 342–345
Optional operator, in data dictionary, 173
Organization
 of collected data, and documentation of, 3, 4, 168–203
 of data elements and data structure, in data description hierarchy, 149–150
 functional form of, 18–19
 of functions and tasks, 16–19
 of information systems group, 16–19, 26–27
 matrix form of, 16–18, 26–27
 pure form of, 16
 in systems development, 16–19
Organization, in business systems
 Anthony's model on, 37–38
 Blumenthal's model on, 38–40
 in business cycle model, 41–43
 of data processing personnel, 55–56
 of Mansfield, Inc., 25–26
 Neumann and Hadass's model on, 40–41
 placement of systems department in, 19, 20–22
 system trinity in, 48–49, 62–63
Organization charts, 16, 26, 27, 30–32
Organizational design strategy, 208
Output of system, 34–35, 323–334
 acceptance testing of, 514
 alternatives to, 217–221
 and computer output microfilm, 331, 332–334
 and computer printers, 331–332, 333
 determination of, 217
 evaluation of, 221–225
 hardware for, 50, 51, 53
 HIPO charts on, 296–302, 307–308, 313–314, 319–321
 logical data structure of, 324–327
 logical design specification on, 250–254
 at Mansfield, Inc., 355–357
 module specifications on, 410–411
 physical data structure of, 327–329
 processing control procedures in, 85–86, 435, 450, 455
 requirements for, 5; 323–324
 selection of appropriate medium for, 331–334
 short cuts in design of, 329–331
 in structured design, 83, 85
 system flowchart on, 88
 technical design specification on, 284, 285
 validity of, 437, 438, 441, 442
 visual representation of, 329–331
Overhead costs, logical design specification on, 273

Packaging of system
 logical design specification of, 246–250
 technical design specification on, 285
 types of packages in, 286, 287
Page printers, in computer system, 331–332
Parallel method of system conversion, 504–505
Parameter passing
 in coupling of modules, 399–402
 ripple effect of errors in, 402
Parsing, in perfective maintenance, 536
Partitioning
 of data into equivalence classes, in system testing, 497–498
 in decomposition of data flow diagrams, 71–73, 78, 135–142
 of modules in structure charts, 397, 398
 of terms in data dictionary, 169–170
Payback of projects
 functional analysis of, 117
 in resource management, 25
Payroll system
 adaptive maintenance of, 526–528, 530
 computer program design for, 396
 context diagram on, 67–68
 conversion to new system of, 501–502
 data dictionary of, 73, 169–170, 176
 data flow diagrams on, 133–142
 data store descriptions on, 177–179, 180
 data structure diagrams on, 73–74, 182–183, 184–186
 data validation procedures in, 437
 decomposition in, 71–73
 file-balance control procedures in, 434–435
 HIPO chart on, 297–298, 299, 300
 interactive and batch processing methods in, 89
 level-0 diagram on, 68–71
 menu-based, 345–346
 Nassi-Shneiderman chart on, 415–416
 output of, 331, 435
 preliminary functional design of, 287
 processing controls in, 85–86, 430–432, 434–435
 pseudocode describing, 82
 records in, data structure of, 150
 relational file on, design of, 378–379
 source-document control procedures in, 430–432
 structure chart on, 78–82
 structured analysis of, 65–76, 85–86
 structured English describing, 76

structured walkthrough of, 304–308
system flowchart on, 89, 156–157
top-down program design for, 416–417, 466–467
visual table of contents on, 297–298
Perfective maintenance, 524, 534–541, 542–544, 550–551
Performance of new system, definition of user's expectations of, 205–216
Phased system conversion, 503–504
Physical audits, 440, 444–445
Physical data files, design of, 365–371
preparation of, 385–387
Physical data structure
of input, 337–338
of output, 327–329
Physical records, 365–366, 386–387
Physical systems, 33–34
Picture checks for data validation, 437
Picture-frame analysis of interactive dialogue, 349–354
Pilot method of system conversion, 503
Planning process, 12–15
in Anthony's model of business systems, 37
compared to systems development, 9–10
network chart in, 15
project planning and control chart in, 13–15
project specification on, 122
in system development, 12–15
in system testing, 498–501
Plex approach to database management, 384
Pointers, in data stores, 179, 180, 182, 184, 185–186
logical external, 181–183
logical internal, 183
Preliminary design of system, 285, 286, 287–293, 303
preparation of, 287–293
structured walkthrough of, 303, 306

Pretesting of questionnaires, 147–148
Printed registers, in output controls, 435
Printers, in computer system, 331–332
compared to computer output microfilm, 333
Probability tests, for data validation, 438
Problems in system
causal analysis of, 108–110, 114, 118
classification and definition of, 102–108
determining system requirements, 101, 102–118
expected benefits of solutions for, 107
functional analysis of, 113–117, 118
identification and evaluation of, 102–108
marginal analysis of, 110–113, 117
tools used in identification of, 108–118
in World Interiors, Inc., 129–131
Procedural module specifications, and nonprocedural specifications, 409, 411
Process symbol, in system flowchart, 86
Processing
batch method, 88, 89, 249–250, 430, 450
channels in system for, 35, 36
commands on, 345
computer hardware for, 50–51, 53
constraints in, logical design specification on, 258
efficiency of, programming for, 475–476
identification and evaluation of problems in, 102–108
interactive, 88, 89, 249–250, 345–355
personnel in, 49, 55–56
and placement of data processing department in business system, 19
software for, 54, 89–91, 345, 346

stations in system for, 35–36
system flowchart on, 86–89
Processing controls, 5, 6, 54, 85–86, 429–445
audit considerations in, 440–445
batch control procedures, 430, 450
computer program controls, 85, 435–436
data validation procedures in, 435, 436–440
file-balance controls, 433–435, 450–451, 455
forms control, 432–433
input controls, 85, 433–435, 450, 455
logical design specification on, 259–260
at Mansfield, Inc., 450–451
output controls, 85–86, 435, 450, 455
source-document controls, 85, 430–433, 450
in structured design, 85–86
transaction controls, 430
types of, 430–436
on unauthorized use of system, 436
at World Interiors, Inc., 455–456
Production control report
logical data structure of, 324–327, 335–337
physical data structure of, 327–329, 338
Productivity of computer programmers, 11
Profits table module, intent of, 409–410
Program, computer, 6. See also Software
Programmers, computer
functions of, 7
in maintenance, 522
productivity of, 11
Programming, 6–7, 457–491
consistency of style in, 474
definition of, 457, 459
documentation of computer code in, 470–472
of efficient programs, 475–476
language used in, 6–7, 474–475
at Mansfield, Inc., 483–485

Programming, *continued*
  program testing in, 472–473, 476–482
  of readable programs, 460–462
  specifications on, 473–476
  standard constructs in, 462–466
  structure charts in, 467–469
  structured, 462–466
  in team approach, 458–459
  technical design specification on, 447–448
  top-down development of programs in, 466–469
  at World Interiors, Inc., 489–490
Project control chart, 13–15, 282
Project feasibility analysis, 204
  compared to system feasibility, 204–205
Project life cycle, 10–19, 526
Project specification, 118, 120–123, 204
Prompts, in interactive dialogue, 347
Proposed system, description of
  in logical design specification, 263
  in project specification, 122
Prototype system, compared to fully functional system, in feasibility analysis, 213–214
Pseudocode, 91
  compared to program flowcharts, 93
  describing module implementation, in program design, 411–412
  in structured design, 82–83
  structured walkthrough of, 303
  technical design specification on, 284
Purge procedures, in system maintenance, 529, 530

Query language, 211
Questionnaires, in data collection and analysis, 145–148
  compared to interviews, 145, 148
  postanalysis of, 148
  pretesting of, 147–148

summated scales in analysis of response to, 145–147
Range tests for data validation, 437
Ranking. *See* Hierarchical approach
Readability of computer programs, 460–462
Reasonableness of data, tests for, 437–438
Records of data, 179, 364
  appended portion of, 367
  attributes of, 179–180, 182, 183, 184, 185–186, 187, 364, 374–375
  blocking of, 365, 366, 386–387, 534–536
  conversion to new system of, 501–502
  direct organization of, 370–371
  estimation of storage space required for, 386–387
  in file-balance control procedures, 434–435
  fixed-length, 367
  indexed-sequential organization of, 371
  indirect addressing of, 370
  and interrecord gap, 369
  layout forms for, in preparation of physical file design, 385–386
  logical, 365–366, 386–387
  logical design specification on, 254–258
  mandatory portion of, 367
  multidimensional, 367–369
  perfective maintenance of, 534–536
  physical, 365–366, 386–387
  physical fields of, 367
  sequential organization of, 369–370
  structure of, 366–369
  variable-length, 367
Redundancy problem
  in data dictionary, 174
  in duplicate data flows, 155
Registers, printed, in output controls, 435
Relational files, design of, 371–379
  entity-relationship diagrams in, 373–375, 377, 378

  normalization of relations in, 375
  rules on, 376–379
Relational model of database management, 384–385
Reliability of systems, defining problem in, 103
Repetition construction
  in structured English, 76, 171, 173, 189, 191–192
  in structured programming, 462, 463
Report-writer software system, 211, 212
Requirements of system, 2, 3, 5–6, 101–131
  causal analysis of, 108–110
  as criteria in evaluation of alternative designs, 221
  feasibility analysis of, 118, 119–120
  for file processing, determination of, 380–381
  functional analysis of, 113–118
  in logical design specification, 246–260, 263
  at Mansfield, Inc., 124–125
  marginal analysis of, 110–113
  problem classification and definition determining, 101, 102–118
  in project specification, 120–123
  specifications on, 118–123
  in technical design specification, 447
  at World Interiors, Inc., 129–131
Resistance of users, in conversion to new system, 492–493
Resources
  centralization and decentralization of, 209–211
  management of, in systems development, 22–25
  marginal analysis of, 111, 113
  user requirements for, specifications on, 123
Return on investment in system, 24, 25, 117, 119
Revenue system, in business cycle, 41

Reviews of system design, internal and external, 303, 304, 306
Ripple effect in parameter passing, 402

Sales, processing of information on
 data collection and analysis of, 163–167
 data file design for, 394–395
 data flow diagram on, 99–100
 feasibility analysis on, 241–244
 HIPO charts on, 319–321
 input design for, 360–362
 logical design specification on, 282
 organization and documentation of data on, 202–203
 problems in, identification and evaluation of, 130
 processing controls in, 455–456
 program structure charts on, 425–428
 programming and program testing for, 489–490
 system conversion in, 520–521
Scales
 on alternative designs, 225–226
 Likert-type, 147
 on questionnaire response, 145–147
Schedule section
 in logical design specification, 263–264, 265–274
 in project specification, 123
 in technical design specification, 448
Scope of project, specification of, 123
Screen display of computer
 cursor in, 347, 354, 355
 design of, 354–355, 360–362
 help instructions in, 532
Self-defining terms, in data dictionary, 73, 169–170
Sequential construct
 in structured English, 76, 171, 189
 in structured programming, 462, 463
 white-box testing of, 495
Sequential files, 369–370
 indexed, 371
Serial printer, in computer system, 331, 332
Service center approach to systems administration, 33
Service level, marginal analysis of, 113
Set rankings, hierarchy of, in logical data structure, 326, 327
Shared tools, feasibility analysis of, 234
Shared use of modules, in structure charts, 397, 405–406
Shutdown of system, operator's guide on, 508
Size of structure charts, reasonable, 397, 404–405
Social feasibility analysis, 119
Soft copy
 of input, 341
 of output, 331
Software, 6, 49, 54
 alternative designs of, 219–221
 application programs, 49, 54, 211, 212
 for auditing, 443–444
 build versus buy decision on, 211–212
 clear labeling of modules in, 397, 403–404
 in COBOL language, 474–475
 cohesion of modules in, 397, 402
 control procedures in, 54, 85, 435–440
 conversion to new system of, 492, 501–512
 coupling of modules in, 397, 399–402, 419–421
 for data dictionary, 176–177
 in database management system, 382–383
 design of, 396–428
 documentation of, 470–472, 476
 engineering of, 9
 error checking of, in structured design, 85
 file conversion programs, 502
 flags in, 406–408
 identification section of, 470–472
 in input design, 339–341
 logic errors in, corrective maintenance of, 531–532
 maintenance of, 10, 522–551
 of Mansfield, Inc., 418–421, 483–485
 Nassi-Shneiderman charts on, 415–416
 operator's guide on, preparation of, 507–508
 in output design, 329–331
 partitioning of functions in, 397, 398
 processing efficiency of, 475–476
 processing programs, 54, 89–91, 345, 346
 program flowcharts on, 91–93, 412–415
 programming of, 6–7, 457–491
 pseudocode in, 411–412
 quality assurance division on, 493
 readability of, 460–462
 reasonable size of modules in, 397, 404–405
 requirements for, in design of system, 5–6
 shared use principle in, 397, 405–406
 span of control principle in, 397, 404
 stepwise refinement of, 396–397
 structure charts on, 397–408, 418–421, 425–428, 467–469
 system, 49, 54
 top-down development of, 416–417, 466–469
 user's guide on, preparation of, 507, 508–510
 utility programs, 501, 528 529
 validity of, defining problem in, 103–104
 of World Interiors, Inc., 425–428, 489–490
Software specifications, 473–476
 on design and development, 474
 on documentation, 476
 on instructions, 474

Software specifications, *continued*
  in logical design specification, 258–259, 264–265
  on modules, 408–416
  in technical design specification, 446–449
Software testing, 7, 457, 472–473, 476–482
  acceptance tests in, 477, 513
  critical path approach to, 477, 481–482
  incremental approach to, 477, 478–481, 483–485
  at Mansfield, Inc., 483–485
  program integration test in, 477
  program tests in, 476
  system integration test in, 477
  system tests in, 476–477, 492–501
  traditional approach to, 477
  unit tests in, 476, 477
  at World Interiors, Inc., 489–490
Source-document control procedures, 85, 430–433, 450
  within-document, 432, 450
Source/sink symbol, in data flow diagrams, 70–71
Span of control principle, in structure charts, 397, 404
Specification
  of data, 169
  of equipment, 296
  of logical design, 3, 4, 5, 205, 245–283
  of modules in structure charts, 408–416
  of program, 473–476
  of project, 118, 120–123, 204
  of structured design, 457
  of technical design, 5, 284–285, 446–449
  of testing, 296
  of user-interface, 296
Start-up of system, operator's guide on, 508
Static user instructions, 345, 346
Stations of processing, in system, 35–36
Steering committee approach to systems administration, 20–21

Stores of data. *See* Data stores
Strategy, in design of new system, 208
  determination of, 208–214
Stress tests on software, 495–496
Structure charts, 77–82, 397–409, 411–416
  clear labeling of, 397, 403–404
  cohesion in, 397, 402
  coupling in, 397, 399–402
  design of, principles in, 397–406
  flags in, 80, 406–408, 419–421
  hat symbol in, 405
  intent of modules in, 408–411
  level-0, 78, 287–288, 289
  partitioning of functions in, 397, 398
  in preliminary functional design, 287–293
  in program design, 397–408, 418–421, 425–428, 467–469
  reasonable size of, 397, 404–405
  review of, in structured walkthrough, 304–307
  shared use in, 397, 405–406
  span of control in, 397, 404
  specifications of modules in, 408–416
  symbols and conventions in, 79–82
  technical design specification on, 284
Structure of data. *See* Data structure
Structured analysis and design, 64–100
  data dictionary in, 73
  data flow diagram in, 65–73
  data structure diagrams in, 73–74
  decomposition in, 71–73, 78
  feasibility analysis in, 226–228
  file and database layouts in, 84–85
  input and output layouts in, 83
  logical design specification in, 4
  at Mansfield, Inc., 93–96
  processing controls in, 85–86
  productivity in, 11

  pseudocode in, 82–83
  specifications on, 4, 457
  structure charts in, 77–82
  structured English in, 76–77
  at World Interiors, Inc., 99–100
Structured interviews, compared with nonstructured interviews, 143–144
Structured programming, 462–466
Structured walkthroughs, in system design, 302–310
  advantages and disadvantages of, 310
  summary of findings in, 309
Student records
  alternative designs of, 217–218
  multidimensional structure of, 367–369
  output of, 217–218
  in relational files, 371–373, 377–378
  structure charts on, 399–402, 403, 405–406, 467–469
Summary files, 380–381
Summated scale
  on alternative designs, 225–226
  on questionnaire response, 145–147
Superordinate module, in structure charts, 292
Suspense files, 381
Symbols
  in data flow diagrams, 70–71
  flags, 80, 406–408, 419–421
  in hierarchical/input-process-output charts, 299
  in structure charts, 79–82
  in system flowchart, 86
System, definition of, 33–34
System, testing of, 476–477, 492–501
  black-box tests in, 494, 495–497
  completion criteria for, 499
  equivalence partitioning in, 497–498
  error path tests in, 496–497
  log of, 499–501
  normal path tests in, 496, 497
  objectives of, 498
  requirements in, 499

strategy of, 499
stress tests in, 495–496
team approach to, 493, 497
white-box tests in, 494, 495
written plan on, 498–501

Table of contents
  on logical design specification, 261
  visual, in design process, 297–298, 312–313
Tasks, 228
  feasibility analysis on, 228–229
Team approach
  to programming, 458–459
  to system testing, 493, 497
Technical design specification, 5, 284–285, 446–449
  compared to logical design specification, 284
Technical feasibility analysis, 119, 120, 229–232
Technical writer, in system conversion, 505–506
Test data
  in audits, 442
  software generators of, 211, 212
Testing, 6, 7
  on acceptance, 477, 513–514
  check-digit, 438–440
  of computer programs, 7, 457, 472–473, 476–482. *See also* Software testing
  for data validation, 436–440
  in design process, 285–286, 287
  edit, 436–438
  of equipment, 296
  of file conversion, 502
  improvement of, in perfective maintenance, 539–540
  of plans on logical design specification, 258, 259
  of plans on technical design specification, 447–448
  of questionnaires, prior to use, 147–148
  of specifications in logical design specification, 258, 259
  of specifications in technical design specification, 447–448
  structured walkthroughs in, 302–310
  of system. *See* System, testing of
  time required for, 11–12, 14–15
  unit tests in, 476, 477
  for user identification, 436
Throughput rate in system, 106, 107
  stress tests on, 496
Time requirements in systems development, 11–12, 12–15, 123
  detailed schedule of, 266–272
  logical design specification on, 263–264, 265–274
  project specification on, 123
  technical design specification on, 448
  and trade-offs in design, 273–274
Timeliness of system
  as criteria in evaluation of alternative designs, 221–222
  defining problem in, 105
Titles and headings, in data structure, 324
Tool sharing, feasibility analysis on, 234
Top-down program design, 416–417, 466–469
Tradeoffs in system design, 273–274
Training of users. *See* Education and training of users
Tramp data, 401–402
Transaction-centered design
  of business system, 41–42
  compared to transform-centered, 247–248
  preliminary functional design of, 292–293
  processing controls in, 259, 430
Transaction files, 380
Transcription errors in data, check-digit test for, 439
Transform-centered systems
  compared to transaction-centered, 247–248
  preliminary functional design of, 287–292
  processing controls in, 259

Transformation of data
  in business computer system, 47, 48
  diagram of, 65–67
  in input procedures, 334
Transforms, 188–193
  description of, with structured English, 76, 188–193
  symbol for, 70
Transmission of data, in computer system, 47–48, 49, 50, 51–53
Transposition errors in data, check-digit test for, 439
Treasury system, in business cycle, 42
Tree approach
  to construction of interactive dialogue, 347–349
  to database management, 383–384
Trinity system, 48–49, 62–63
Troubleshooting of system, operator's guide on, 508

Unit controls, logical design specification on, 258–259
Unit tests
  plan for, in logical design specification, 258, 259
  in program testing, 476, 477
Users, 2
  in conversion to new system, 8. *See also* Conversion to new system
  corrective maintenance of errors related to, 532
  education and training of. *See* Education and training of users
  expectations of, on performance of new system, 205–216
  identification test of, 436
  in interactive dialogue with computer system, 345–355
  requests for service, and adaptive maintenance, 524–525, 542, 549
  in structured walkthroughs, 302–310
  support requirements of, 123, 265, 296, 449
  unauthorized, 436
Utility programs
  in system conversion, 501

Utility programs, *continued*
  in system maintenance, 528–529

Validation of data
  in audits, 440–445
  in new system, acceptance testing on, 514
  problems in software design affecting, 103–104
  processing control procedures in, 435, 436–440
Validation flag, in structure charts, 408
Variable-length records, 367
Vendor selection procedures, for equipment requirements, 43, 296
Visual representation
  of input, 338–341, 355–357
  of output, 329–331
Visual table of contents, in design process, 297–298, 312–313

Walkthroughs, structured, in system design, 302–310
  advantages and disadvantages of, 310
  summary of findings in, 309
Warnier-Orr diagrams, 186–187
  in adaptive maintenance, 526
  on input, 336–337, 338
  at Mansfield, Inc., 355
  on output, 324–327
  at World Interiors, Inc., 360–362
White-box testing, 494, 495
Work procedures
  acceptance testing of, 513–514
  collection and analysis of written statements on, 151–155
  feasibility analysis on job enlargement or enrichment, 233
  new, in system conversion, 505–507
World Interiors, Inc.
  causal analysis of, 129–131
  computer program design for, 425–428
  conversion to new system in, 520–521
  data collection and analysis on, 163–167
  data file design for, 394–395
  data flow diagrams on, 99–100
  feasibility analysis on, 241–244
  HIPO charts on, 319–321
  input design and Warnier-Orr diagrams on, 360–362
  logical design specification on, 282
  maintenance of systems in, 549–551
  organization chart on, 30–32
  organization and documentation of data, 202–203
  processing controls in, 455–456
  programming and program testing in, 489–490
  system trinity in, 62–63